Optical Waves and Laser Beams in the Irregular Atmosphere

T0187437

Optical Waves and Laser Beams in the Irregular Atmosphere

Edited by
N. Blaunstein and N. Kopeika

CRC Press
Taylor & Francis Group
Boca Raton London New York

CRC Press is an imprint of the
Taylor & Francis Group, an **informa** business

CRC Press
Taylor & Francis Group
6000 Broken Sound Parkway NW, Suite 300
Boca Raton, FL 33487-2742

First issued in paperback 2020

© 2018 by Taylor & Francis Group, LLC
CRC Press is an imprint of Taylor & Francis Group, an Informa business

No claim to original U.S. Government works

ISBN 13: 978-0-367-57295-2 (pbk)
ISBN 13: 978-1-138-10520-1 (hbk)

Library of Congress Cataloging-in-Publication Data

Names: Blaunstein, Nathan, editor. | Kopeika, Norman S., editor.
Title: Optical waves and laser beams in the irregular atmosphere / Nathan Blaunstein & Natan Kopeika, editors.
Description: Boca Raton : CRC Press, 2018.
Identifiers: LCCN 2017020720 | ISBN 9781138105201 (hardback : alk. paper)
Subjects: LCSH: Meteorological optics. | Physical optics. | Atmosphere--Laser observations. | Atmospheric physics. | Turbulence.
Classification: LCC QC975.2 .O67 2018 | DDC 551.56/5--dc23
LC record available at https://lccn.loc.gov/2017020720

Visit the Taylor & Francis Web site at
http://www.taylorandfrancis.com

and the CRC Press Web site at
http://www.crcpress.com

Contents

Preface

This book is intended for any scientist, practical engineer, and designer who is concerned with the operation and service of optical wireless (atmospheric) communication links, laser beam systems, and LIDAR applications. It can be useful for graduate and postgraduate students who study such topics in courses related to the operation and service of atmospheric communication links, lasers, and LIDAR applications in atmospheric optical communications.

It briefly examines the fundamentals of the atmosphere, its structure, and the content, including the effects of atmospheric turbulence and different kinds of hydrometeors, such as aerosols, rain, clouds, and snow, on optical wave propagation in the atmospheric links. The main task of this book is to explain these effects deeply for each specific situation occurring in the irregular atmosphere and for specific natural phenomena accompanying the corresponding features (e.g., turbulences and hydrometeors) that fully affect the propagation of optical rays and laser beams through the atmosphere. The book also emphasizes how to use LIDAR to investigate atmospheric phenomena and predict primary parameters of the irregular turbulent atmosphere, and to suggest what kinds of optical devices can be operated in different atmospheric situations to minimize the deleterious effects of natural atmospheric phenomena. Finally, such investigations allow prediction of imaging of various phenomena via solution of inverse problems related to irregular atmospheric communication channel.

Generally speaking, the book introduces the reader of any scientific level to the main relevant topics of optical wave propagation in the irregular turbulent atmosphere and their relations to laser beam and LIDAR applications, both for communications and for optical location.

The organization of the book is as follows. Chapter 1 illuminates the fundamental aspects of the atmosphere as an inhomogeneous gaseous structure and briefly describes the main parameters of the atmosphere. The content of the atmosphere is presented briefly to reflect on the problems. Particularly, the structures of aerosols and their dimensions, concentration, and the spatial distribution of aerosol sizes, their spectral extinction, and altitude localization are briefly presented in the chapter. Then, the existence of various water and ice particles in the inhomogeneous atmosphere, called hydrometeors, such as rain and clouds, their spatial and altitudinal distribution, and size distribution and their effects on optical wave propagation are briefly discussed in this chapter. Finally, the atmospheric turbulent structures caused by the temperature and humidity fluctuations combined with turbulent mixing by wind and convection-induced random changes in the air density of the atmosphere, as an irregular gaseous medium, is briefly discussed in this chapter.

In Chapter 2, a general presentation of optical waves, as a part of a wide electromagnetic waveband spectrum, is done. Fundamentals of optical wave propagation based on deterministic approach of Maxwell equation in the time-harmonic and phasors forms and the propagation of optical waves in free space and through the intersection between two material media are briefly presented based on the deterministic presentation of wave equations. Then, the subject of optical wave propagation, through the irregular atmosphere as a random medium is discussed briefly based on the methods of statistical physics by introducing the main random equations and functions that describe stochastic processes in the random medium. Finally, we describe the main random equations, based on Feynman's diagram theory, via the expansion of Green's functions.

In Chapter 3, the results of long-term experimental studies of properties of atmospheric turbulence in the anisotropic boundary layer are stated. It is shown that the correct assignment of turbulent characteristics of the atmosphere is an important premise for the exact forecast of the results of distribution of optical radiation in the atmosphere. It is established that the similarity theory of turbulent flows can be extended to any anisotropic boundary layer. With the use of semiempirical hypotheses of the turbulence theory, it is shown theoretically and experimentally that any anisotropic boundary layer can be considered locally weakly anisotropic, in which the weakly anisotropic similarity theory of Monin–Obukhov is applied locally (in some vicinity of each point in the layer). It is established that at the known characteristic scales of temperature and velocity, the anisotropic boundary layer can be replaced with the isotropic layer. It provides an opportunity to use the optical models of turbulence developed for the isotropic boundary layer.

Processes of origination and disintegration of the Benard's cell in air are experimentally studied. It is established that disintegration of the Benard's cell is realized according to Feigenbaum's scenario. It is shown in the chapter that the turbulence resulting from the disintegration of Benard's cell meets all the criteria characterizing the appearance of chaos in typical dynamic systems. We define the coherent structure as the compact formation including the long-living space hydrodynamic vortex cell (resulting from the long action of thermodynamic gradients) and products of its discrete coherent cascade disintegration.

It is shown in this chapter that the known processes of transition of laminar flows in turbulent (Rayleigh–Benard convection, the flow fluid around obstacles, etc.) can be considered as coherent structures (or the sums of such structures). In this case, the real atmospheric turbulence can be considered as an incoherent mix of different coherent structures with incommensurable frequencies of the main energy vortices.

Chapter 4 describes the most typical situations occurring in the irregular turbulent atmosphere, where nonlinear optical effects during laser radiation propagation should be taken into account. The stimulated Raman scattering (SRS) effect in airborne liquid droplets, as typical hydrometeors, is considered, which is connected with the excitation and amplification of Stokes waves in the whispering gallery zones in cavity droplets. Equations for the SRS excitation threshold in microparticles are given. The most attention is paid to the pump depletion case. The threshold conditions are found for the incident light intensity resulted in steady-state generation of Stokes radiation of a preset strength. The computational framework of light field intensity for thermal nonlinear effects, where the medium is heated due to the light energy dissipation or kinetically cooled, is presented here. The corresponding calculations are carried out on the basis of original algorithms based on the radiation transfer equation. The results presented in the chapter, which are based on this equation, allow to calculate the wide-aperture laser beams, which can be strongly distorted in the region of geometric shadow of a beam. The calculations of high-power laser radiation propagation along the far stratospheric paths under the SRS effect are also described in this chapter. Additionally, the diffraction effect on the nonlinear distortions of wide-aperture laser beams is discussed. Special attention is paid to the pump depletion effect under light field transformation into the Stokes components.

In Chapter 5, a brief historic insight into the development of femtosecond LIDAR technology and femtosecond lasers, as well as perspectives on the application of white-light LIDAR and fiber optic devices, is provided. In addition, this chapter describes the main mechanisms of generation of supercontinuum radiation in optic fiber and in atmospheric air. Most attention in this chapter is paid to the LIDAR's equation for the problems of atmospheric sensing by ultrashort (femtosecond) pulses with allowance made for the formation of conical emission and the use of supercontinuum radiation as a main effect for the study of gas–aerosol composition of the atmosphere. Here, we provide the basic equations and describe the Monte Carlo simulation of transport of conical broadband emission radiation in the atmosphere. We prove the proposed approach by introducing some numerical and field experiments aimed at the evaluation of possibilities of applying white-light LIDAR for the determination of the gas–aerosol composition of the atmosphere as well as on the development of the solution of inverse problems. Particularly, the possibilities of water vapor sensing with application of LIDAR based on femtosecond sources of radiation is analyzed.

Finally, we analyze the possibility to utilize the white-light LIDAR for aerosol sensing and present an iterative method for the solution of LIDAR's equation to determine the optical interaction coefficients, as well as the microphysical characteristics of clouds based on methods of neural networks and genetic algorithms. The conducted LIDAR experiment on the study of microphysical characteristics of artificial aerosol at short paths is described finally, and possibilities of reconstructing the microphysical characteristics of thin mists with the use of white-light LIDAR are evaluated and examined.

Chapter 6 introduces the development of modern theory of vision in scattering media. It reflects the mechanisms of imaging and special features of influencing scattering media on the quality of image of objects, for example, problems of correction of hyperspectral satellite images for the distorting effect of the atmosphere. It is shown that despite the fact that until recently there are number of general theoretical problems demanding unambiguous interpretation, the present framework is devoted exactly to this subject. The main problems are formulated here via the possibility to determine the optical transfer function (pulse response, point spread function, or influence function) of a scattering medium or an imaging channel in the scattering medium, or via the possibility to determine the general transfer properties inherent in a concrete scattering medium (by analogy to its optical characteristics) that can subsequently be used for imaging of objects with arbitrary optical properties and any characteristics. In this chapter, we illuminate these problems and present the corresponding models giving comprehensive answers to the imaginary problems that could take place in the irregular atmosphere. The chapter presents an additional problem that arises due to the fact that the pulsed response or the point spread function (PSF) used in vision theory is not an image of a point object, as in the theory of optical systems (though in the literature devoted to vision theory they are sometimes identified). Having solved this problem, we simultaneously answer the question on how many optical transfer functions or PSFs are required to be determined in order to restore or filter out wide-angle images of objects observed through the scattering media. The answers to these questions and problems of constructing the PSF connected with them for spherical models of the system atmosphere—Earth's surface—are the main topics of this chapter.

We can emphasize that this book is a synthesis of different backgrounds in order to present a broad and unified approach to the propagation of optical waves and laser beams in various scenarios occurring in the irregular ionosphere, consisting of turbulences, aerosols, and hydrometeors, with applications to LIDAR utilization in direct problems (communication) and in inverse problems (imaging of objects and location), based on novel technologies and numerical frameworks that have been recently proved experimentally.

Editors

Nathan Blaunstein earned his MS degree in radio physics and electronics from Tomsk University, Tomsk, former Soviet Union, in 1972, and PhD and DS degrees in radio physics and electronics from the Institute of Geomagnetism, Ionosphere, and Radiowave Propagation (IZMIR), Academy of Science USSR, Moscow, Russia, in 1985 and 1991, respectively. From 1979 to 1984, he was an engineer and a lecturer, and then, from 1984 to 1992, a senior scientist, an associate professor, and a professor at Moldavian University, Beltsy, Moldova, former USSR. From 1993 he was a researcher of the Department of Electrical and Computer Engineering and a visiting professor in the Wireless Cellular Communication Program at the Ben-Gurion University of the Negev, Beer-Sheva, Israel. Since April 2001, he has been an associate professor, and in 2005 a full professor in the Department of Communication Systems Engineering. Dr. Blaunstein has published 10 books, two special chapters in handbooks on applied engineering and applied electrodynamics, six manuals, and over 190 articles in radio and optical physics, communication, and geophysics. His research interests include problems of radio and optical wave propagation, diffraction, reflection, and scattering in various media (subsoil medium, terrestrial environments, troposphere, and ionosphere) for the purposes of optical communication and radio and optical location, aircraft, mobile-satellite, and terrestrial wireless communications and networking.

Natan Kopeika was born in Baltimore in 1944. Raised in Philadelphia, he received BS, MS, and Ph.D. degrees in electrical engineering from the University of Pennsylvania in 1966, 1968, and 1972, respectively. He and his family moved to Israel, and he began his career at Ben-Gurion University of the Negev in 1973. He chaired the Department of Electrical and Computer Engineering for two terms (1989–1993) and was named Reuven and Francis Feinberg Professor of Electro-optics in 1994. He was the first head of the Department of Electro-optical Engineering (2000–2005), which grants graduate degrees in electro-optical engineering. He has published more than 190 papers in international reviewed journals and well over 150 papers at various conferences. Recent research involves development of a novel inexpensive focal plane array camera for terahertz imaging. He is a fellow of SPIE and co-recipient of JJ Thomson Award by IEE (1998). Other areas of research include interactions of electromagnetic waves with plasmas, the optogalvanic effect, environmental effects on optoelectronic devices, imaging system theory, propagation of light through the atmosphere, imaging through the atmosphere, image processing and restoration from blur, imaging in the presence of motion and vibration, lidar, target acquisition, and image quality in general. He is the author of the textbook *A System Engineering Approach to Imaging*, published by SPIE Press (first printing 1998, second printing 2000), and is a topic editor for atmospheric optics in the *Encyclopedia of Optical Engineering*.

Contributors

Vladimir V. Belov is head of laboratory of the V. E. Zuev Institute of Atmospheric Optics Russian Academy of Sciences, Siberian Branch, Tomsk, Russia and a professor at the Tomsk State University, Tomsk, Russia. He holds a doctor of science. He has published more than 250 articles in journals and proceedings of the conferences, five collective monographs, and four author certifications.

His research interests include image and signal transfer through scattering and media, influence of the atmosphere and the Earth's underlying surface (ocean or dry land) on the intensity of upward light fluxes, influence of the processes of scattering, absorption, reflection, and luminance of the atmosphere on the accuracy of temperature estimation in passive sensing of the underlying surface.

The development of hardware and software for removal (decrease) of the distorting effect of the atmosphere on the characteristics of optical signals and images and optical communication on scattered laser radiation in atmospheric and water medium also form his areas of interest.

Gennady G. Matvienko is director of the V. E. Zuev Institute of Atmospheric Optics Russian Academy of Sciences, Siberian Branch, Tomsk, Russia. He is also professor at Tomsk State University, Tomsk, Russia. He has published more than 395 articles in journals and proceedings of the conferences, 21 collective monographs, and 10 author certifications.

His research interests involve the main scientific trend of G. G. Matvienko's activity connected with the development of methods and hardware of laser sounding of aerosol atmosphere and ecological monitoring of the environment. Special attention is given to the development of lidar methods of dynamic processes in the atmosphere and optophysical characteristics of atmospheric aerosol formations.

Victor V. Nosov graduated from Tomsk State University, Russia, and his specialties are radiophysics and optics. Since 1972, he has worked at the V. E. Zuev Institute of Atmospheric Optics SB RAS. He obtained PhD in 1978 and doctor of science in 2010. He is the author of more than 420 publications, including eight books and more than 100 papers in peer-reviewed Russian and international journals. His research interests include radiophysics, optics of the atmosphere and ocean, theory of turbulence, and mathematical physics.

Alexander Y. Sukhanov is a PhD candidate of technical sciences and a senior researcher at the V. E. Zuev Institute of Atmospheric Optics Russian Academy of Sciences, Siberian Branch, Tomsk, Russia. He is an assistant professor at Tomsk State University, Tomsk, Russia.

He has published more than 70 articles in journals and proceedings of the conferences along with four collective monographs. The main direction of his scientific activity is the direct and inverse problems of lidar sounding based on bionic methods, including restoration of gas component concentrations, microphysical aerosol characteristics from lidar signals, classification, and detection of bio-aerosols from the spectra of laser-induced fluorescence.

Mikhail V. Tarasenkov has a PhD in physico-mathematical sciences and is a senior researcher at the V. E. Zuev Institute of Atmospheric Optics Russian Academy of Sciences, Siberian Branch, Tomsk, Russia.

He published more than 40 articles in journals and proceedings of the conferences and five author certifications. His research interests include image and signal transfer through scattering and absorbing media, Monte Carlo method in atmospheric optics, atmospheric correction of satellite image for UV, visible and IR wavelengths, and optical communication on scattered laser radiation in atmospheric and water medium.

Alexander Zemlyanov is head of division of the V. E. Zuev Institute of Atmospheric Optics, Russian Academy of Science, Siberian Branch, Tomsk, Russia. He has a doctor of science and also is a professor. He has published more than 300 articles in scientific journals and 12 collective monographs.

His research interest includes problems of propagation of intense laser radiation in UV, visible, infrared spectral ranges in atmosphere, and study of the nonlinear interaction of ultrashort laser pulses with matter.

Symbols and Abbreviations

T	absolute temperature (in Kelvin)
P	pressure (in millibars, Pascals or in mmHg)
ρ	density (in kg m^{-3})
$N(z)$	number of aerosols as function of altitude
$n(r)$	distribution of aerosols' density as function of their dimensions
$\Gamma(\cdot)$	gamma function
$\alpha(\lambda)$	parameter of attenuation of optical wave as function of wavelength λ
α	wave power attenuation factor
$N(D)$	1-D distribution of aerosols over drops diameter
$C(D)$	cross-section of ray power attenuation
R_{peak}	peak rainfall rate
γ_c	ray attenuation factor in cloud
$\text{Re} = V \cdot l/\nu$	Reynolds number
V	velocity of turbulent flow (in m/s)
l	current size of turbulence (in m)
ν	kinematic viscosity (in m^2/s)
L_0	outer scale of turbulent structure
κ_0	wavenumber of outer-scale turbulent structure
l_0	inner scale of turbulent structure
κ_m	wavenumber of outer-scale turbulent structure
ε	average dissipation rate of the turbulent kinetic energy
$D_V(r)$	structure function of velocity field spatial variations
$D_n(r)$	structure function of refractive index spatial variations
C_n^2	refractive index structure parameter
C_V^2	velocity structure parameter
C_T^2	temperature structure parameter
δn	refractive index fluctuations
$\Phi_n(\kappa)$	power spectrum of refractive index fluctuations

$\Phi_V^{1\text{-}D}(\kappa)$, $\Phi_V^{3\text{-}D}(\kappa)$	one-dimensional and three-dimensional power spectrum of turbulent velocity variations, respectively
A	arbitrary vectors of electromagnetic field
E	vector of electric field component of the electromagnetic wave
H	vector of magnetic field component of the electromagnetic wave
D	vector of induction of electric field component of the electromagnetic wave
B	vector of induction of magnetic field component of the electromagnetic wave
$\mathbf{E}(z,t)$	2-D distribution of the vector of electrical component of the electromagnetic wave
$\mathbf{H}(z,t)$	2-D distribution of the vector of magnetic component of the electromagnetic wave
$\tilde{\mathbf{E}}(z)$	phasor of the electrical component of the electromagnetic wave
$\tilde{\mathbf{H}}(z)$	phasor of the magnetic component of the electromagnetic wave
j	vector of electric current density
J	vector of the full current in medium/circuit
ρ	charge density in medium
∇	operator Nabla of arbitrary scalar field
$\Delta = \nabla^2$	Laplacian of the vector or scalar field
$div = \nabla\cdot$	flow of arbitrary vector or a scalar product ("Nabla dot") of the field
$grad\Phi = \nabla\Phi$	gradient of arbitrary scalar field or effect of Nabla operator on the scalar field
$curl \equiv rot = \nabla\times$	rotor of arbitrary vector field or the vector product of the operator Nabla
Q	full charge in circuit, material or in medium
$z = a + ib$	complex number, a—its real part, b—its imaginary part
$i = \sqrt{-1}$	ort of imaginary part of the complex number
$\varepsilon = \varepsilon' + i\varepsilon''$	complex permittivity of arbitrary medium
$\varepsilon_r = \varepsilon_r' + i\varepsilon_r''' = \varepsilon_{Re}' + i\varepsilon_{Im}'''$	relative permittivity of arbitrary medium; $\varepsilon_{Re}' = \varepsilon'/\varepsilon_0$- its real part, $\varepsilon_{Im}''' = \varepsilon''/\varepsilon_0$—its imaginary part
$\varepsilon_0 = \dfrac{1}{36\pi}10^{-9}(F/m)$	permittivity of free space
$n = n' + in''$	complex refractive index of arbitrary medium
$n' = \sqrt{\varepsilon_r'}$	real part of refractive index
$n'' = \sqrt{\varepsilon_r''}$	imaginary part of refractive index

σ	conductivity of arbitrary medium or material
$\mu = \mu' + i\mu''$	complex permeability of arbitrary medium
$\mu_r = \mu'_r + i\mu'''_r = \mu'_{Re} + i\mu'''_{Im}$	relative permittivity of arbitrary medium
$\mu'_{Re} = \mu'/\mu_0$	its real part, $\mu'''_{Im} = \mu''/\mu_0$—its imaginary part
$\mu_0 = 4\pi \cdot 10^{-9} (H/m)$	dielectric parameter of free space
η	wave impedance in arbitrary medium
$\eta_0 = 120\pi \text{ Ohm} = 377 \text{ Ohm}$	wave impedance of free space
$c = \dfrac{1}{\sqrt{\varepsilon_0\mu_0}}$	velocity of light in free space
V	volume of arbitrary surface
$R_{V,H}$	coefficient of reflection from boundary of two media of vertical and horizontal polarization, respectively
$T_{V,H}$	coefficient of refraction (transfer of the wave into the medium) of vertical and horizontal polarization, respectively
$\gamma = \alpha + i\beta$	parameter of propagation in arbitrary material medium
α	parameter of wave attenuation in arbitrary medium
β	parameter of phase velocity deviation in arbitrary medium
λ	wavelength in arbitrary medium
$\Psi(r,t)$	scalar wave presentation in the space and time domains
$\mu(r)$	random function of refractive index spatial fluctuations
$C(r_1,r_2)$	covariance function of refractive index spatial variations
FT	Fourier transform
$\mu(\kappa,\omega)$	Fourier transforming function of the refractive index variations
L_0	nonrandom operator
$G^{(0)} = L_0^{-1}$	unperturbed propagator (unperturbed Green's function, $G^{(0)}(r,r')$)
L_1	random operator
$G = (L_0 + \varepsilon L_1)^{-1}$	perturbed propagator (perturbed Green's function)
$G \otimes G^*$	perturbed double propagator
$\Gamma(\cdot)$	covariance function of a multivariate Gaussian distribution
$G_{ij}^{(0)}(k)$	Fourier transform of unperturbed propagator
$K_\perp$	renormalized wave number of transverse oscillations of optical wave field
$\bar{v}_i$	average field of velocity

$\overline{T}$	average field of temperature
$\langle v_i' \cdot v_j' \rangle$	second moment of velocity deviations
$\langle v_j' \cdot T' \rangle$	second moment of temperature deviations
K_{ij}	coefficients of turbulent viscosity
$K_{Tij} - D_{uuu}(r)$	coefficients of turbulent temperature conductivity
Rf	dynamic Richardson number
$D_{rr}(r)$	structure function of fluctuations of the longitudinal velocity
$D_T(r)$	structure function of fluctuations of the temperature
C	Kolmogorov constant
C_θ	Obukhov constant
S	asymmetry of probability distribution of the longitudinal velocity difference
$D_{uTT}(r)$	third spatial moment of the longitudinal difference of wind velocities
$D'_{uTT}(\tau)$	third temporal moment of the longitudinal difference of wind velocities
$D_T(r)$	structure function in coherent turbulence
$R_F = (\lambda x)^{1/2}$	radius of the first Fresnel zone; x—distance; λ—wavelength
$\sigma_c^2 = <\rho_c^2>$	variance of displacements of the energy centroid of laser beam
$\sigma_t^2 = <\rho_t^2>$	variance of displacements of the image of optical sources
σ_α^2	variance of angular displacements (jitter) of images
F_t	focal length of the receiving telescope
S_M	meteorological visibility range
BAO	Baikal Astrophysical Observatory of the Russian Academy of Sciences (Russia)
BAT	Big Alt-azimuthal Telescope (Russia)
BPA	back propagation algorithm
FFT	fast Fourier transform
GB	Gaussian beam
IAO	V. E. Zuev Institute of Atmospheric Optics, Russian Academy of Sciences, Siberian Branch
IAO SB RAS	V.E. Zuev Institute of Atmospheric Optics SB RAS (Russia)
ISTP SB RAS	Institute of Solar-Terrestrial Physics SB RAS (Russia)
LSVT	Large Solar Vacuum Telescope (Russia)
NN	neural network
NSE	nonlinear Schrodinger equation

OIL	oxygen–iodine laser
PMT	photo multiply tube
PSF	point spread function
RB	ring beam
RSRS	rotational sublevels of atmospheric gases
S/N	signal-to-noise ratio
SAO	Special Astrophysical Observatory of the Russian Academy of Sciences
SB RAS	The Siberian Branch of the Russian Academy of Sciences
SBS	stimulated Brillouin scattering
SCR	supercontinuum radiation
SGB	super-Gaussian beam
SRS	stimulated Raman scattering
SSO	Sayan Sun Observatory of the Russian Academy of Sciences (Russia)
SSM	strongly scattering medium
TE	transverse electric
TIR	total inner reflection
TM	transverse magnetic
USSR	Union of Soviet Socialist Republics (former)
UV	ultra-violet frequency band

Chapter 1

Atmosphere Fundamentals

Nathan Blaunstein and Natan Kopeika

Contents

1

1.1 Structure and Main Parameters of the Atmosphere

The atmosphere is a gaseous envelope that surrounds the Earth from the ground surface up to several hundred kilometers. The atmosphere consists of different kinds of gaseous, liquid, and crystal structures, including gas molecules and atoms, aerosols, cloud, fog, rain, hail, dew, rime, glaze, and snow [1–16]. Except for the first two, the other are usually called in the literature as hydrometeors [9–15]. Furthermore, due to irregular and sporadic air streams and motions, that is, irregular wind motions, the chaotic structures defined as atmospheric turbulence are also present in the atmosphere [17–21].

Based mostly on temperature variations, the Earth's atmosphere is divided into four primary layers [1,8]: (1) the troposphere that surrounds the Earth from the ground surface up to 10–12 km, with the tropopause region as the isothermal layer above troposphere up to ∼20 km, which spreads upward to the stratosphere; (2) the stratosphere with the stratopause from 20 km up to ∼50 km altitude; (3) the mesosphere with the mesopause up to ∼90 km; and (4) the thermosphere (up to ∼600 km), which contains the multilayered plasma structure usually called ionosphere (70–400 km).

In our further discussions, we focus on the effects of the troposphere on optical wave propagation starting with a definition of the troposphere as a natural layered air medium consisting of different gaseous, liquid, and crystal structures.

The physical properties of the atmosphere are characterized by main parameters such as temperature, T (in Kelvin), pressure, P (in millibars, Pascals or in mmHg), and density, ρ (in kg m^{-3}). All these parameters significantly change with altitude, seasonal and latitudinal variability, and strongly depend on environmental conditions occurring in the troposphere [22].

Over 98% of the troposphere is comprised of the elements nitrogen and oxygen. The number density of nitrogen molecules, $\rho_N(h)$, at a height h can be found in Reference 22. The temperature $T(h)$ and pressure $P(h)$, where the height is measured in meters, for the first 11 km of the troposphere can be determined from the following expressions [6,8]:

$$T(h) = 288.15 - 65.45 \times 10^{-4} h \qquad (1.1)$$

$$P(h) = 1.013 \times 10^5 \cdot \left[\frac{288.15}{T(h)}\right]^{5.22} \qquad (1.2)$$

The temperature and pressure for h changing from 11 to 20 km in the atmosphere can be determined from

$$T(h) = 216.65 \qquad (1.3)$$

$$P(h) = 2.269 \times 10^4 \cdot \exp\left[-\frac{0.034164(h - 11000)}{216.65}\right] \qquad (1.4)$$

The number density of molecules can be found from [6]:

$$\rho(h) = \left(\frac{28.964 \text{ kg/kmol}}{8314 \text{ J/kmol-K}}\right) \cdot \frac{P(h)}{T(h)} = 0.003484 \cdot \frac{P(h)}{T(h)} \text{ (kg/m}^3) \qquad (1.5)$$

1.2 The Content of the Atmosphere

1.2.1 Aerosols

Aerosols are liquid or solid particles generally uniformly distributed in the atmosphere [6,13,24–43]. Aerosol particles play an important role in the precipitation process, providing the nuclei upon which condensation and freezing take place. The particles participate in chemical processes and influence the electrical properties of the atmosphere.

1.2.1.1 Aerosol Dimensions

Actual aerosol particles range in diameter from a few nanometers to about a few micrometers. When smaller particles are in suspension, the system begins to acquire the properties of a real aerosol structure. For larger particles, the settling rate is usually so rapid that they may not properly be called real aerosols. Nevertheless, the term is commonly employed, especially in the case of fog or cloud droplets and dust particles, which can have diameters of over 100 μm. In general, aerosols composed of particles larger than about 50 μm are unstable unless the air turbulence is extreme, as in a severe thunderstorm (see details in References 5–7).

From all classifications of atmospheric aerosols, the most commonly used one is according to size. General classification suggests three modes of aerosols [13]: (1) a nuclei mode, which is generated by spontaneous nucleation of the gaseous material for particles less than 0.1 μm in diameter, (2) the accumulation mode for particles between 0.1 and 1 μm in diameter, mainly resulting from coagulation and cloud processes, and (3) the coarse mode for particles larger than 1.0 μm in diameter originating from the Earth's surface, land, and ocean. The classification is quite similar to Junge's designation [24], referred to as Aitken, large, and giant particles. The particles vary not only in chemical composition and size but also in shape (spheres, ellipsoids, rods, etc.).

Aerosol concentrations and properties depend on the intensity of the sources, on the atmospheric processes that affect them, and on the particle transport from one region to another. The size distribution of the atmospheric aerosol is one of its core physical parameters. It determines the various properties such as mass and number density, or optical scattering, as a function of particle radius. For atmospheric aerosols, this size range covers more than five orders of magnitude, from about 10 nm to several hundred micrometers. This particle size range is very effective for scattering of radiation at ultraviolet (UV), visible, and infrared (IR) wavelengths. The aerosol size distribution varies according to location, altitude, and time. In a first attempt to identify geographically distinct atmospheric aerosols, Junge classified aerosols depending on their location in space and sources into background, maritime, remote continental, and rural [24]. This classification later was expanded and quantified [25–30].

From the optical standpoint, to underline scattering properties of the atmosphere the term haze aerosol was introduced [31,32]. Hazes are polydisperse aerosols in which the size range of particles extends from about 0.01 to 10 μm. Haze is a condition wherein the scattering property of the atmosphere is greater than that attributable to the gas molecules but is less than that of fog. Haze can include all types of aerosols. Cosmic dust, volcanic ash, foliage exudations, combustion products, bits of sea salt are all found to varying degrees in haze.

1.2.1.2 Aerosol Altitude Localization

Because aerosols exhibit considerable variation in location, height, time, and constitution, different concepts exist for describing aerosol loading in the atmosphere. Models for the vertical variability of atmospheric aerosols are generally broken into a number of distinct layers. In each of these layers, a dominant physical mechanism determines the type, number density, and size distribution of particles. Generally accepted layer models consist of the following [33,34]: a boundary layer that includes aerosol mixing goes from 0 to 2–2.5 km elevation, a free tropospheric region from 2.5 to 7–8 km, a stratospheric layer from 8 to 30 km, and layers above 30 km composed mainly of particles that are extraterrestrial in origin such as meteoric dust [21].

The average thickness of the aerosol-mixing region is approximately 2–2.5 km. Within this region, one would expect the aerosol concentration to be influenced strongly by conditions at ground level. Consequently, aerosols in this region display the highest variability with meteorological conditions, climate, etc. [34–43].

In Reference 16, various studies have been compiled, and a global climatology of aerosol optical properties for Arctic, Antarctic, desert, continental, urban, and maritime regions developed. Based on these aerosol distributions, they have modeled aerosol optical properties on a global scale.

1.2.1.3 Aerosol Concentration

In the free tropospheric layer that extends from 2–2.5 to 7–8 km, an exponential decay of aerosol number density is observed. The total number of their density varies as [12]:

$$N(z) = N(0)\exp\left(-\frac{z}{z_s}\right) \tag{1.6}$$

where the scale height z_s ranges from 1 to 1.4 km. Another proposed form of $N(z)$ takes into account an inversion layer (z_1) and mixing turbulent layer (z_2) as a boundary layer, and is given by the following expression [35]:

$$N(z) = \begin{cases} N(z_0) - \left[\dfrac{N(z_0) - N(z_1)}{z_1}\right] \cdot z, & z_0 \leq z \leq z_1 \\ N(z_1) = \text{const}, & z_1 < z \leq z_2 \\ N(z_2)\exp\left[\dfrac{z_2 - z}{8}\right], & z > z_2 \end{cases} \tag{1.7}$$

where the heights z_1 and z_2 for a semiarid zone are $z_1 = 0.3$ km and $z_2 = 4$ km for summer, and $z_1 = 0.4$ km and $z_2 = 1.5$ km for winter.

The few data available show that in the atmospheric boundary layer (0–2 km) and in the lower stratosphere (9–14 km altitude), different layers (known as mixing layers) with constant and increased aerosol concentration can exist [36,37]. These layers can be caused by temperature inversions at ground level and by tropopause effects where the temperature gradient changes sign. Moreover, in the stratospheric region, there is some latitude dependence of aerosol layer height (Junge layer). The concentration maximum of stratospheric aerosols near the equator is located at 22–26 km elevation altitude, but at about 17–18 km height in the polar region.

1.2.1.4 Aerosol Size Distribution

Atmospheric aerosols are polydisperse [6]. Although their sizes can vary in the 0.01–100 μm range of diameters, the number of small particles is limited by the coagulation process and the number of large particles by gravitational sedimentation. Between these limits the number of particles varies with size.

The manner in which the particle population is spread over the range of sizes is defined by the size distribution function. The size distribution of the atmospheric aerosol is one of its core physical parameters. It determines how the various properties like mass and number density, or optical scattering, are distributed as a function of particle radius. Particle size distributions are necessary inputs for models used to predict the attenuation and scatter of radiation between the transmitter and receiver in different applications (optical communication, satellite image restoration, weapons-based electro-optical systems, etc.).

The number $n(r)$ of particles per unit interval of radius and per unit volume is given by

$$n(r) = \frac{dN(r)}{dr} \tag{1.8}$$

The differential quantity $dN(r)$ expresses the number of particles having a radius between r and $r + dr$, per unit volume, according to the distribution function $n(r)$.

Because of the many orders of magnitude present in atmospheric aerosol concentrations and radii, a logarithmic size distribution function is often used:

$$n(r) = \frac{dN(r)}{d\log(r)} \tag{1.9}$$

The much used distribution function is the power law first presented by Junge [24,40]. Junge's model is

$$n(r) = \frac{dN(r)}{d\log r} = C \cdot r^{-v}, \quad r \geq r_{min} \tag{1.10}$$

Or, in a nonlogarithmic form,

$$n(r) = \frac{dN(r)}{dr} = 0.434Cr^{-v}, \tag{1.11}$$

where C is a normalizing constant to adjust the total number of particles per unit volume, and v is the shaping parameter. Most measured size distributions can best be fit by values of v in the range $3 \leq v \leq 5$ for hazy and clear atmospheric conditions, and for aerosols whose radii lie between 0.1 and 10 μm [41]. According to the power-law size distribution, the number of particles decreases monotonically with an increase in radius. In practice, there is an accumulation in the small particle range. Actual particle size distributions may differ considerably from a strict power-law form.

The modified power-law distribution was given by McClatchey et al. [41] and was modified in Reference 42 by use of the gamma probability density function (PDF) to

$$n(r) = \frac{dN(r)}{dr} = ar^{\alpha} \exp(-br^{\beta}), \tag{1.12}$$

where a is the total number density, and α, β, and b are shaping parameters.

The total particle concentration given by the integral over all particle radii according to Reference 39 is, for this distribution:

$$N = a \cdot \beta^{-1} \cdot b^{-(\alpha+1)/\beta} \Gamma\left(\frac{\alpha+1}{\beta}\right) \tag{1.13}$$

The mode radius for this distribution is given by [6,39]

$$r_m^\beta = \frac{\alpha}{b \cdot \beta} \tag{1.14}$$

The value of the distribution at the mode radius is [6,39]

$$n(r_m) = a \cdot r_m^\alpha \cdot \exp(-\alpha/\beta) \tag{1.15}$$

Because it has four adjustable constants, (1.15) can be fitted to various aerosol models. The gamma PDF is usually employed to model haze, fog, and cloud particle size distributions.

The tropospheric aerosols above the boundary layer are assumed to have the same composition, but their size distribution is modified by eliminating the large particle component because of the higher elevation.

To utilize the extensive measurements, a series of aerosol models for different environmental conditions and seasons were constructed [34]. The models have been divided into four altitude regimes, as described above. Many experiments were carried out over the last three decades to test and modify the present aerosol models. The new data available show that the distribution varies dramatically with altitude, often within meters. Large variations exist in the data from different locations [16].

1.2.1.5 Aerosol Spectral Extinction

It is often found that within the optical subrange ($\sim$0.16–1.2 μm), the light scattering coefficient, $\alpha(\lambda)$, and aerosol size distribution in form (1.15), obey the following power-law relationship [12]:

$$\alpha(\lambda) = C \cdot \lambda^{-b} \tag{1.16}$$

where b is referred to as the exponent power and equals $b = v - 3$ [12]. Thus, if α_λ depends strongly on wavelength (large b), then the size distribution function (1.15) decreases with particle size. Measurements have shown that values of b tend to be higher for continental aerosols than for clean marine aerosols [6,12].

Multiple aerosol scattering induces loss of optical wave after it has traversed the scatterers. In satellite imagery, the aerosol blur is considered to be the primary source of atmospheric blur (the turbulence blur usually can be neglected, see the next chapter), and is commonly called the adjacency effect [21]. In optical communication, the dense aerosol/dust layers and clouds as part of the communication channel can cause signal power attenuation, as well as temporal and spatial signal fluctuations (i.e., fading [6,8]).

1.2.2 Hydrometeors

Hydrometeors are any water or ice particles that have formed in the atmosphere or at the Earth's surface as a result of condensation or sublimation. Water or ice particles blown from the ground into the atmosphere are also classed as hydrometeors. Some well-known hydrometeors are rain, fog, snow, clouds, hail, and dew, glaze, blowing snow, and blowing spray. Scattering by hydrometeors has an important effect on signal propagation.

1.2.2.1 Fog

Fog is a cloud of small water droplets near ground level and sufficiently dense to reduce horizontal visibility to less than 1,000 m. Fog is formed by the condensation of water vapor on condensation nuclei that are always present in natural air. This can result as soon as the relative humidity of the air exceeds saturation by a fraction of 1%. In highly polluted air, the nuclei may grow sufficiently to cause fog at humidities of 95% or less. Three processes can increase the relative humidity of the air:

1. Cooling of the air by adiabatic expansion
2. Mixing two humid airstreams having different temperatures
3. Direct cooling of the air by radiation (namely, cosmic ray radiation)

According to physical processes of fog creation, there are different kinds of fogs usually observed: advection, radiation, inversion, and frontal. We do not discuss their creation, but refer the reader to literature [5–7].

When the air becomes nearly saturated with water vapor (relative humidity RH $\rightarrow$ 100%), fog can form assuming sufficient condensation nuclei are present. The air can become saturated in two ways, either by mixing of air masses with different temperatures and/or humidities (advection fogs), or by the air cooling until the air temperature approaches the dew point temperature (radiation fogs).

Fog models, which describe the range of different types of fog, have been widely presented based on measured size distributions [33]. The modified gamma size distribution (1.9) was used to fit the data. The models represent heavy and moderate fog conditions. The developing fog can be characterized by droplet concentrations of 100–200 particles per cm^3 in the 1–10 µm radius range with mean radius of 2–4 µm. As the fog thickens, the droplet concentration decreases to less than 10 particles per cm^3 and the mean radius increases from 6 to 12 µm. Droplets less than 3 µm in radius are observed in fully developed fog. It is usually assumed that the refractive index of the fog corresponds to that of pure water. Natural fogs and low-level clouds are composed of spherical water droplets, the refractive properties of which have been fairly well documented in the spectral region of interest.

1.2.2.2 Rain

Rain is precipitation of liquid water drops with diameters greater than 0.5 mm. When the drops are smaller, the precipitation is usually called drizzle. Concentration of raindrops typically spreads from 100 to 1000 m^{-3}. Raindrops seldom have diameters larger than 4 mm because, as they increase in size, they break up. The concentration generally decreases as diameters increase. Meteorologists classify rain according to its rate of fall. The hourly rates relate to light, moderate, and heavy rain, which correspond to dimensions less than 2.5 mm, between 2.8 and 7.6 mm, and more than 7.6 mm, respectively. Less than 250 mm and more than 1500 mm per year represent approximate extremes of rainfall for all of the continents (see details in References 1, 6–10).

Thus, if such parameters of rain, as the density and size of the drops are constant, then, according to References 6–10, the signal power P_r at the optical receiver decreases exponentially with optical ray path r through the rain, with the parameter of power attenuation in e^{-1} times, α, that is,

$$P_r = P_r(0)\exp\{-\alpha r\} \tag{1.17}$$

Expressing (1.17) on a logarithmic scale gives

$$L = 10\log\frac{P_t}{P_r} = 4.343\alpha r \tag{1.18}$$

Another way to estimate the total loss via the specific attenuation in decibels per meter was shown by Saunders in Reference 10. He defined this factor as

$$\gamma = \frac{L}{r} = 4.343\alpha \tag{1.19}$$

where now the power attenuation factor α can be expressed through the integral effects of the one-dimensional (1D) diameter D of the drops, defined by $N(D)$, and the effective cross section of the frequency-dependent signal power attenuation by rain drops, $C(D)$ [dB/m], as

$$\alpha = \int\limits_{D=0}^{\infty} N(D) \cdot C(D) dD \tag{1.20}$$

As mentioned in References 8–10, in real tropospheric conditions the drop size distribution $N(D)$ is not constant and can be accounted for in the range dependence of the specific attenuation. The range dependence $\gamma = \gamma(r)$ can be integrated over the whole optical ray path length r_R to find the total path loss

$$L = \int\limits_{0}^{r_R} \gamma(r) dr \tag{1.21}$$

To resolve Equation 1.20, a special mathematical procedure was proposed in Reference 10 that accounts for the drop size distribution. This procedure yields an expression for $N(D)$ as

$$N(D) = N_0 \exp\left\{-\frac{D}{D_m}\right\} \tag{1.22}$$

where $N_0 = 8 \times 10^3$ m^{-2} mm^{-1} is a constant parameter [10], and D_m is a parameter that depends on the rainfall rate R measured above the ground surface in millimeters per hour, as

$$D_m = 0.122 \cdot R^{0.21} \text{ mm} \tag{1.23}$$

As for the attenuation cross section $C(D)$ from Equation 1.20, it can be found using the Rayleigh approximation that is valid for lower frequencies, when the average drop size is small compared to the wavelength. In this case only absorption inside the drop occurs and the Rayleigh approximation is valid giving a very simple expression for $C(D)$, that is,

$$C(D) \propto \frac{D^3}{\lambda} \tag{1.24}$$

Attenuation caused by rain increases more slowly with frequencies approaching a constant value known as the optical limit. Near this limit, scattering forms a significant part of attenuation, which can be described using the Mie scattering theory discussed earlier.

In practical situations, an empirical model is usually used, where $\gamma(r)$ is assumed to depend only on rainfall rate R and wave frequency. Then according to References 1, 9, 10, we can obtain

$$\gamma(f,R) = a(f)R^{b(f)} \qquad (1.25)$$

where γ has units dB/km and $a(f)$ and $b(f)$ depend on frequency.

The cell diameter appears to have an exponential probability distribution of the form [8–10]

$$P(D) = \exp(-D/D_0) \qquad (1.26)$$

where D_0 is the mean diameter of the cell and is a function of the peak rainfall rate R_{peak}. For Europe and the United States, the mean diameter D_0 decreases slightly with increasing R_{peak} when $R_{peak} > 100$ mm/h. This relationship appears to obey a power law

$$D_0 = a\,R_{peak}^{-b}, \quad R_{peak} > 10\,\text{mm/h} \qquad (1.27)$$

Values for the coefficient a ranging from 2 to 4, and the coefficient b from 0.08 to 0.25, have been reported.

1.2.2.3 Clouds

Clouds have dimensions, shape, structure, and texture, which are influenced by the kind of air movements that result in their formation and growth, and by the properties of the cloud particles. In settled weather, clouds are well scattered and small and their horizontal and vertical dimensions are only a kilometer or two. In disturbed weather, they cover a large part of the sky, and individual clouds may tower as high as 10 kilometers or more. Growing clouds are sustained by upward air currents, which may vary in strength from a few cm/s to several m/s. Considerable growth of the cloud droplets, with falling speeds of only about one cm/s, leads to their fall through the cloud and reaching the ground as drizzle or rain. Four principal classes are recognized when clouds are classified according to the kind of air motions that produce them:

1. Layer clouds formed by the widespread regular ascent of air
2. Layer clouds formed by widespread irregular stirring or turbulence
3. Cumuliform clouds formed by penetrative convection
4. Orographic clouds formed by ascent of air over hills and mountains

We do not discuss how such kinds of clouds are created, because it is part of meteorology. Interested readers can find information in References 1–8.

There have been several proposed alternative mathematical formulations for the probability distribution of sky cover, as an observer's view of the sky dome. Each of them uses the variable x ranging from zero (for clear conditions) to 1.0 (for overcast). There are four cloud cover models: (1) the first cloud cover model; (2) the second cloud cover model; (3) the third cloud cover model; and (4) the ceiling cloud model. We do not enter into discussions on each of these models because this is out of scope of the present monograph, and we refer the reader to the excellent works [4,6–8].

We only will note that in all cloud models a distinction between cloud cover and sky cover was not explained, and therefore should be briefly explained here. Sky cover is an observer's view of cover of the sky dome, whereas cloud cover can be used to describe areas that are smaller or larger than the floor space of the sky dome. It follows from numerous observations that in clouds and fog the drops are always smaller than 0.1 mm, and the theory for the small size scatters is applicable [7,20–24]. This gives for the attenuation coefficient

$$\gamma_c \approx 0.438\, c(t) q / \lambda^2, \text{dB/km} \tag{1.28}$$

where λ is the wavelength measured in centimeters, and q is the water content measured in grams per cubic meter. For visibility of 600, 120, and 30 m, the water content in fog or cloud is 0.032, 0.32, and 2.3 g/m^3, respectively. The calculations show that the attenuation, in a moderately strong fog or cloud, does not exceed the attenuation due to rain with a rainfall rate of 6 mm/h.

Due to lack of data, a semiheuristic approach is presented here. Specifically, we assume that the thickness of the cloud layer is $w_c = 1$ km, and the lower boundary of the layer is located at a height of $h_c = 2$ km. The water content of clouds has a yearly measured percentage of [7]

$$P(q > x) = p_c \exp\left(-0.56\sqrt{x} - 4.8x\right), \% \tag{1.29}$$

where p_c is the probability of cloudy weather (%).

We should notice that although the attenuation in clouds is less than in rain, the percentage of clouds can be much more essential than that of the rain events. Thus, the additional path loss due to clouds can be estimated as 2 and 5 dB for 350 km path and $h_2 = 6$ km, and for the time availability of 95% and 99%, respectively.

1.2.2.4 Snow

Snow is the solid form of water that crystallizes in the atmosphere and, falling to the Earth, covers permanently or temporarily about 23% of the Earth's surface. Snow falls at sea level poleward of latitude 35°N and 35°S, though on the west coast of continents it generally falls only at higher latitudes. Close to the equator, snow-fall occurs exclusively in mountain regions, at elevations of 4900 m or higher. The size and shape of the crystals depend mainly on the temperature and the amount of water vapor available as they develop. In colder and drier air, the particles remain smaller and compact. Frozen precipitation has been classified into seven forms of snow crystals and three types of particles: graupel, that is, granular snow pellets, also called soft hail, sleet, that is, partly frozen ice pellets, and hail, for example, hard spheres of ice (see details in References 1–4, 6–9).

1.2.3 Atmospheric Turbulence

The temperature and humidity fluctuations combined with turbulent mixing by wind and convection induce random changes in the air density and in the field of atmospheric refractive index in the form of optical turbules (or eddies), which is one of the most significant parameters for optical wave propagation [35–47]. Random space–time redistribution of the refractive index causes a variety of effects on an optical wave related to its temporal irradiance fluctuations (scintillations) and phase fluctuations. A statistical approach is usually used to describe both atmospheric turbulence and its various effects on optical wave propagation in the nonlinear irregular atmosphere.

Atmospheric turbulence is a chaotic phenomenon created by random temperature, wind magnitude variation, and direction variation in the propagation medium. This chaotic behavior results in index-of-refraction fluctuations. The turbulence spectrum is divided into three regions by two scale sizes [6,8,17–20]:

- ■ The outer scale (or macro size) of turbulence: L_0
- ■ The inner scale (or micro size) of turbulence: l_0

These values vary according to atmosphere conditions, distance from the ground, and other factors. The inner scale l_0 is assumed to lie in the range of 1–30 mm and near the ground, it is typically observed to be around 3–10 mm, but generally increases to several centimeters with increasing altitude h. A vertical profile for the inner scale has still not been established. The outer scale L_0 near the ground is usually taken to be roughly kh, where k is a constant on the order of unity. Thus, L_0 is usually either equal to the height from the ground (when the turbulent cell is close to the ground) or in the range of 10–100 m or more. Vertical profile models for the outer scale have been developed based on measurements, but different models predict very different results.

1.2.3.1 Turbulence Energy Cascade Theory

To investigate physical properties of turbulent liquids, in the earliest study of turbulent flow, Reynolds used the theory "of similarity" to define a nondimensional quantity $Re = V \cdot l/\nu$, called the Reynolds number [6,8,17–20], where V and l are the characteristic velocity (in m/s) and size (in m) of the flow respectively, and v is the kinematic viscosity (in m²/s). The transition from laminar to turbulent motion takes place at a critical Reynolds number, above which the motion is considered to be turbulent. The kinematic viscosity v of air is roughly 10^{-5} m² s⁻¹ [6,8]. Air motion is considered highly turbulent in the boundary layer and troposphere, where the Reynolds numbers Re $\sim 10^5$ [6,8,14]. Richardson [8] first developed a theory of the turbulent energy redistribution in the atmosphere, which he called the energy cascade theory. It was noticed that smaller-scale motions originated as a result of the instability of larger ones. A cascade process, shown in Figure 1.1 taken from Reference 8, in which eddies of the largest size are broken into smaller and smaller ones, continues down to scales in which the dissipation mechanism turns the kinetic energy of motion into heat.

Let us denote by l the current size of turbulence eddies, by L_0 and l_0 their outer and inner scales, and by $\kappa_0 = (2\pi/L_0)$, $\kappa = (2\pi/l)$ and $\kappa_m = (2\pi/l_0)$ the spatial wave numbers of these kinds of eddies, respectively. Using this notation, one can divide turbulences in the three regions:

$$\text{Input range:} \quad L_0 \leq l, \quad \kappa < \kappa_0 \equiv \frac{2\pi}{L_0}$$

$$\text{Inertial range:} \quad l_0 < l < L_0, \quad \frac{2\pi}{L_0} \equiv \kappa_0 < \kappa < \kappa_m \equiv \frac{2\pi}{l_0} \quad (1.30)$$

$$\text{Dissipation range:} \quad l \leq l_0, \quad \frac{2\pi}{l_0} \equiv \kappa_m \leq \kappa$$

These three regions induce strong, moderate, and weak spatial and temporal variations, respectively, of signal amplitude and phase, referred to in the literature as *scintillations* [6,8,20].

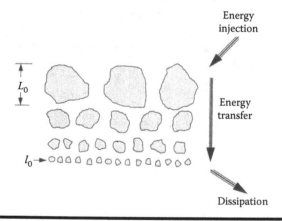

Figure 1.1   Richardson's cascade theory of turbulence.

Kolmogorov [44] introduced a hypothesis stating that during the cascade process the direct influence of larger eddies is lost and smaller eddies tend to have independent properties, universal for all types of turbulent flows. Following Kolmogorov, the energy cascade process consists of an energy input region, inertial subrange, and energy dissipation region, as sketched in Figure 1.1.

At large characteristic scales or eddies, a portion of kinetic energy in the atmosphere is converted into turbulent energy. When the characteristic scale reaches a specified outer scale size, L_0, the energy begins a cascade that forms a continuum of eddy size for energy transfer from a macroscale L_0 to a microscale l_0. The scale sizes l bounded above by L_0 and below by l_0 form the inertial subrange.

Kolmogorov proposed that in the inertial subrange, where $L_0 > l > l_0$, turbulent motions are both homogeneous and isotropic and energy may be transferred from eddy to eddy without loss, that is, the amount of energy that is being injected into the largest structure must be equal to the energy that is dissipated as heat [44].

The term homogeneous is analogous to stationary process [6,8,20] and implies that the statistical characteristics of the turbulent flow are independent of position within the flow field. The term isotropic requires that the second and higher order statistical moments depend only on the distance between any two points in the field.

The inertial subrange is dominated by inertial forces and the average properties of the turbulent flow are determined only by the average dissipation rate ε (in units m^2/s^3) of the turbulent kinetic energy. When the size of a decaying eddy reaches l_0, the energy is dissipated as heat through viscosity processes. It was also hypothesized that the motion associated with the small-scale structure l_0 is uniquely determined by the kinematic viscosity v and ε, where $l_0 \sim \eta = (v^3/\varepsilon)^{1/4}$ is the Kolmogorov microscale [19,44–47]. The Kolmogorov microscale defines the eddy size dissipating the kinetic energy. The turbulent process, shown schematically in Figure 1.1 according to the simple theory of Richardson, was then summarized by Kolmogorov and Obukhov (called in the literature the Kolmogorov–Obukhov turbulent cascade process [5]) as follows: the average dissipation rate ε of the turbulent kinetic energy will be distributed over the spatial wavelength κ-range as [6]:

In the input range ($\kappa_0 \sim 1/L_0$): $\varepsilon \sim \kappa_0^{-5/3}$

In the inertial range ($\kappa \sim 1/l$): $\varepsilon \sim \kappa^{-5/3}$ (1.31)

In the dissipation range ($\kappa_m \sim 1/l_0$): $\varepsilon \sim \kappa_m^{-5/3}$

where, as above, L_0, l, and l_0 are the initial (outer), current, and inner turbulent eddy sizes.

In general, turbulent flow in the atmosphere is neither homogeneous nor isotropic. However, it can be considered locally homogeneous and isotropic in small subregions of the atmosphere.

1.2.3.2 Characteristic Functions and Parameters of Atmospheric Turbulence

By using dimensional analysis, Kolmogorov showed that the structure function of velocity field between the two observation points, $\mathbf{r}_1$ and $\mathbf{r}_2 = \mathbf{r}_1 + \mathbf{r}$ in the inertial subrange satisfies the universal 2/3 power law as [17–20,44]

$$D_V(r) = \left\langle \left(V(\mathbf{r}_2) - V(\mathbf{r}_1) \right)^2 \right\rangle \equiv C_V^2 r^{2/3}, \quad l_0 < r < L_0 \tag{1.32}$$

where the angle brackets denote average (time or ensemble, according to ergodic theorem [8]), $V(\mathbf{r}_1)$ is the turbulent component of velocity vector at point $\mathbf{r}_1$, $V(\mathbf{r}_2)$ is the same measure but at point $\mathbf{r}_2$, $r = |\mathbf{r}|$ is the distance between the two observation points, $\mathbf{r}_1$ and $\mathbf{r}_2$, and C_V^2 is the velocity structure constant, defined in References 1–5 as a measure of the total amount of energy in the turbulence. The structure constant is related to the average energy dissipation rate ε by $C_V^2 = 2\varepsilon^{2/3}$ [17–20].

The velocity field inner scale l_0 is on the order of the Kolmogorov microscale η, and is given by $l_0 = 12.8\eta$ [8,21]. Note that l_0 increases with kinematic viscosity, which increases with altitude. The inverse dependence of inner scale on the average rate of dissipation ε shows that strong turbulence has smaller inner scales and weak turbulence has larger inner scales. The inner scale is typically on the order of 1–10 mm near the ground and can be on the order of centimeters or more in the troposphere and stratosphere [16–19].

The outer scale L_0 is proportional to $\varepsilon^{1/2}$, and it increases and decreases directly with the strength of turbulence [44–47]. The outer scale L_0 can be scaled with height h above ground in the surface layer up to ~100 m according to $L_0 = 0.4h$ [5].

The behavior of the velocity field structure function at small-scale sizes ($r < l_0$) varies with the square of separation distance as follows [8,17–21]:

$$D_V(r) = C_V^2 l_0^{-4/3} r^2, \quad 0 < r < l_0 \tag{1.33}$$

Because the random field of velocity fluctuations is basically nonisotropic for scale sizes larger than the outer scale L_0, no general description of the structure function can be predicted for $r > L_0$.

The validity of the 2/3 power law for the structure function has been established over a wide range of experiments [16–19]. The inertial subrange is the region of greatest interest where the 2/3-power law contains essentially all information on turbulence of practical importance.

As mentioned above, the outer scale L_0 denotes the scale size, below which turbulence properties are independent of the flow and generally are nonisotropic. The source of energy at large scales is either wind shear or convection. Scale sizes smaller than the inner scale belong to the viscous dissipation range. In this regime, the turbulent eddies disappear and the energy is dissipated as heat.

The above is clearly seen from Figure 1.1 where the whole turbulent process is shown schematically, with an energy input region, inertial subrange, and energy dissipation region.

As shown in Reference 5, the physical parameter that describes the optical effects of atmospheric turbulence is random index-of-refraction fluctuations, called also in the literature as the optical turbulence [1–5]. For optical wave propagation, refractive index fluctuations are caused primarily by fluctuations in temperature (in the cases when variations in humidity and pressure can be neglected). The refractive-index structure parameter, as the measure of the "strength" or "power" of the turbulent structure, is considered the most critical parameter along the optical ray path in characterizing the effects of atmospheric turbulence.

Usually, to obtain relationships between the structure parameter C_n^2 and the atmospheric refractive index fluctuations δn, it is usually assumed [8,17–20] that the atmospheric refractive index fluctuations are stationary, homogeneous, and isotropic. At the same time, the refractive index n is a complicated function of various meteorological parameters. For example, for marine atmospheres the value of the refractive index n can be presented as [8,21]:

$$n \approx 1 + \frac{77 \cdot P}{T}\left[1 + \frac{7.53 \cdot 10^3}{\lambda^2} - 7733\frac{Q}{T}\right] \tag{1.34}$$

where P is the air pressure (here in [mb]), T is the air temperature (in [K]), Q is specific humidity (in [g/(mg)/m^3]), and λ is the wavelength. A simple approximation of relationship between the refractive index fluctuations, and the structure parameter C_n^2 is given by [8,17–20]

$$C_n^2 \approx \frac{\left\langle (\delta n)^2 \right\rangle}{r^{2/3}} \tag{1.35}$$

where r represents the distance between the source and the detector or transmitter–receiver range, and C_n^2 is the refractive index structure constant (in units m$^{-2/3}$).

The statistical description of the random field of turbulence-induced fluctuations in the atmospheric refractive index is similar to that for the related field of velocity fluctuations. Obukhov [45,47] related the velocity structure function to the structure function for potential temperature, following the concept of a conservative passive additive scalar and, finally, to the structure function for variations in the index of refraction, which satisfies the relationship

$$D_n(r) = \left\langle \left(n(\mathbf{r}_2) - n(\mathbf{r}_1)\right)^2 \right\rangle \equiv C_n^2 r^{2/3}, \quad l_0 < r < L_0 \tag{1.36}$$

where, as above, $r = |\mathbf{r}_2 - \mathbf{r}_1|$ is the distance between the two observation points $\mathbf{r}_1$ and $\mathbf{r}_2$, and $D_n(r)$ is defined as a measure of magnitude of fluctuations in the index of refraction [44–47]. It characterizes the strength of the refractive turbulence. It is the critical parameter for describing optical turbulence and it is often used synonymously with optical turbulence.

Generally speaking, the refractive index structure function describes the behavior of correlations of passive scalar field fluctuations between two given points separated by a distance r. The quantity L_0 is a measure of the largest distance over which fluctuations in the index-of-refraction are correlated, whereas l_0 is a measure of the smallest correlation distances. As described above, the correlation distances L_0 and l_0 are usually referred to as the outer and inner scale size of the turbulent eddies, respectively. The inner scale of refractive index fluctuations is related to the Kolmogorov microscale L_0 via $l_0 = 7.4 \cdot \eta = 7.4 \cdot \nu^{3/4}\varepsilon^{-1/4}$ [17–20]. In the dissipative interval, where the constraint $r < l_0$ occurs, the refractive index structure function corresponds to behavior in form (1.33). Values of refractive index C_n^2 near ground typically range from about 10^{-16} m$^{-2/3}$ and less for "weak turbulence" and up to 10^{-12} m$^{-2/3}$ or more when the turbulence is "strong," with changes of magnitude in about only a minute.

Over short time intervals at a fixed propagation distance and constant height above a uniform surface, it may often be assumed that C_n^2 is essentially constant. However, for vertical and slant-path propagation C_n^2 varies as a function of height above ground.

Moreover, the structure parameter C_n^2 also varies according to variations of meteorological parameters. Thus, for marine atmosphere [8,21]

$$C_n^2 \approx \left(79 \cdot 10^{-6} \frac{P}{T^2}\right)^2 \left(C_T^2 + 0.113 C_{TQ} + 0.003 C_Q^2\right) \tag{1.37}$$

where C_T^2 and C_Q^2 are the air temperature and water vapor structure coefficients, respectively; C_{TQ} is the combined temperature–water vapor structure coefficient or covariance. The covariance is usually positive, but can be negative and thereby decrease C_n^2 [21]. The C_T^2 term, which is the mean-square statistical average of the difference in temperature ΔT between two points along the optical ray path separated by a distance r, can be given by

$$C_T^2 = \left\langle (\Delta T)^2 \right\rangle r^{2/3} \tag{1.38}$$

The values of C_n^2 are closely related to the temperature structure parameter by the equation [17–21]

$$C_n^2 \cong \left(\frac{79}{10^6 \cdot T^2}\right)^2 C_T^2 \tag{1.39}$$

This relation is valid for propagation of optical waves in visible or near-infrared ranges over the land. From relations (1.36) and (1.39), it follows that structure functions of random fields of refractive index and temperature fluctuations in the over-the-land atmosphere differ only by constant coefficient.

1.2.3.3 Spectral Characteristics of Atmospheric Turbulence

The wavenumber power spectrum of refractive index fluctuations in the atmosphere has important consequences on a number of problems involved in the propagation and scattering of optical waves. On the basis of the above 2/3 power-law expression, it can be deduced, and the associated power spectral density for refractive-index fluctuations can be described, by the following expression [8,46]:

$$\Phi_n(\kappa) = 0.033 C_n^2 \kappa^{-11/3}, \quad \frac{2\pi}{L_0} < \kappa < \frac{2\pi}{l_0} \tag{1.40}$$

This is the well-known Kolmogorov spectrum, which was calculated in normalized form, $\bar{\Phi}(\kappa) = \Phi(\kappa)/0.033 C_n^2$ in Reference 9 for inertial and dissipation ranges. Here κ is the spatial frequency or wavenumber, $\kappa = (2\pi/l)$ (in units rad · m^{-1}). We present in Figure 1.2 the corresponding computations following Reference 8.

As shown by Tatarskii [19,45], the Kolmogorov spectrum is theoretically valid only in the inertial subrange. The use of this spectrum is justified only within that subrange or over all wave numbers $\kappa = 2\pi/l$, if the outer scale is assumed to be infinite and the inner scale negligibly small. We note that the three-dimensional (3D) classical power spectrum (1.40) with $\kappa^{-11/3}$ power law is related to 1D spectrum with $\kappa^{-5/3}$ power law due to the following relation [8,17–20].

$$\Phi_V^{3-D}(\kappa) = -\frac{1}{2\pi\kappa} \frac{d\Phi_V^{1-D}(\kappa)}{d\kappa} \tag{1.41}$$

where Φ_V^{1-D} and $\Phi_V^{3-D}(\kappa)$ are the 1D and 3D power spectrum, respectively.

The spectrum measures the distribution of the variance of a variable over scale sizes or periods. If the variable is a velocity component, the spectrum also describes the distribution of kinetic energy over spatial periods.

Using the relation between the structure function and the covariance, and the Wiener–Khinchin theorem, the relation between the structure function and the power spectrum is given by References 8, 17–20.

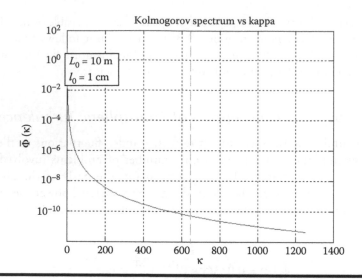

Figure 1.2 Kolmogorov normalized spectrum, shown for the inertial and dissipation ranges; the dashed vertical line indicates $\kappa = 2\pi/l_0$.

$$D_V(r) = 8\pi \int\limits_0^\infty \kappa^2 \Phi_V(\kappa) \left[1 - \frac{\sin(r\kappa)}{r\kappa} \right] d\kappa \tag{1.42}$$

We should notice that in practice of atmospheric optical communication, the velocity power spectrum is not of interest, when the optical properties of turbulence need to be characterized. In this case, Obukhov has adopted a Kolmogorov's 3D law to the 1D law by use of the concept of "passive scalar" [45–47], which states that fluctuations in passive scalar quantities (e.g., temperature field, refractive index, etc.) associated with turbulent structure function, has a similar form, as (1.36), under assumption that the turbulent field is locally homogeneous and isotropic. Therefore, he introduced the 1D spectral power function, $\Phi_S(\kappa)$, which describes the power spectrum of a given passive scalar S [45]

$$\Phi_S(\kappa) \approx C_S^2 \cdot \kappa^{5/3} \tag{1.43}$$

where $C_S^2 \cong \text{Const} \cdot \varepsilon_T \cdot \varepsilon^{-1/3}$ is the structure constant derived by Obukhov [15]. Here ε_T is the heat flux intensity over the spectrum (thermal dissipation rate).

Finally, based on Obukhov's law (1.43) and on the 2/3 power law (1.36) for the structure function, the associated 3D spectrum for the refractive index fluctuations over the inertial subrange has the same form, as that obtained by Kolmogorov and expressed by (1.40). Therefore, the Equation 1.40 is known in the literature as the Obukhov–Kolmogorov power spectrum [44–47]. The model is valid only for the inertial subrange, although it is often extended over all wave numbers by assuming the inner scale is zero and the outer scale is infinite. It is usually assumed that the inertial subrange determines the optical properties of the turbulent atmosphere.

Other spectrum models have been proposed for making calculations when inner scale and/or outer scale effects cannot be ignored. The following isotropic forms of spectra, which take into account deviations from a power law in the region of turbulence outer and inner scale, can be used in calculations [8,17–21].

Thus, for mathematical convenience, by von Kármán was assumed that the turbulence spectrum is statistically homogeneous and isotropic over all wave numbers. A spectral model that he used in this case is that combines the three regions defined by (1.1), called in literature, as the von Kármán spectrum [8,20]:

$$\Phi_{nk}(\kappa) = 0.033 C_n^2 \left(\kappa^2 + \kappa_0^2\right)^{-11/6} \exp\left(-\kappa^2/\kappa_m^2\right), \quad 0 \leq \kappa < \infty \qquad (1.44)$$

Here $\kappa_m = (5.92/l_0)$ and $\kappa_0 = (C_0/L_0)$, where $1 \leq C_0 \leq 8\pi$ is the scaling constant for the outer scale parameter [8,20]. Note that even though the last equation describes the entire spectrum, its value in the input range must be considered only approximate, because it is generally anisotropic and depends on how the energy is introduced into the turbulence. This model, unlike the models shown above, does not have a singularity at $\kappa = 0$. The von Kármán spectrum (1.44) was computed in Reference 8, and we present it here in Figure 1.3.

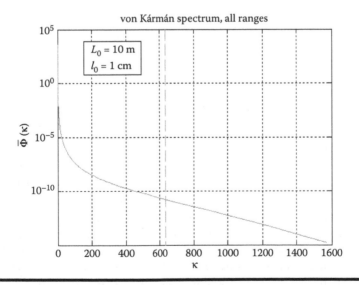

Figure 1.3 von Kármán normalized spectrum, shown for all ranges; the dashed vertical line indicates $\kappa = 2\pi/l_0$.

As shown in Reference 8, although the Tatraskii spectrum "explodes" near the origin, the von Kármán spectrum inclination is suppressed in this region. For other regions, the two spectra are almost identical. Therefore, we can emphasize that the von Kármán spectrum is almost identical to the Tatarskii spectrum except for a difference in small values of wave number.

However, we should notice that the last two spectra have the correct behavior only in the inertial range: that is, the mathematical form that permits the use of these models outside the inertial range is based on mathematical convenience, and not because of any physical meaning. The Tatarskii spectrum has been shown to be inaccurate in about 50% for predicting the irradiance variance for the strong-turbulence regime in optical propagation experiments and in more than 40% for weak turbulences [19,20,47].

Hill (see details in References 8, 20) developed a numerical spectral model with a high wave-number rise that accurately fits the experimental data. He has performed a hydrodynamic analysis that led to a numerical spectral model with the small rise (or bump) at high wave numbers near $K_m \sim 1/l_0$ that fit the experimental data of temperature spectrum [8,20]. The bump in temperature spectrum also induces a corresponding spectral bump in the spectrum of the refractive index fluctuations that can have important consequences on a number of statistical quantities important in problems involving optical wave propagation.

However, because Hill's model is described in terms of a second-order differential equation that must be solved numerically, the corresponding spectrum cannot be used in analytic developments. An analytic approximation to the Hill spectrum, which offers the same tractability as the von Kármán model (1.44), was developed by Andrews and colleagues [8,20].

This approximation, commonly called the modified atmospheric spectrum (or just modified spectrum), is given by References 22–25, and it is valid for wave numbers in the range of $0 \leq \kappa < \infty$.

$$\Phi_n(\kappa) = 0.033C_n^2 \left[1 + 1.802(\kappa/\kappa_l) - 0.254(\kappa/\kappa_l)^{7/6}\right] \frac{\exp(-\kappa^2/\kappa_l^2)}{\left(\kappa^2 + C_0^2/L_0^2\right)^{11/6}} \qquad (1.45)$$

where $\kappa_l = 3.3/l_0$. We compare the modified model (1.45) with that proposed by von Kármán and present this comparison in Figure 1.4, following computations made in Reference 8. Numerical comparison of results based on the above equation and the Hill spectrum reveal differences no larger than 6%, but generally within 1%–2% of each other. Note that these forms of the turbulence spectrum are used for computational reasons only and are not based on physical models. The modified model, which is obviously based on Hill's numerical spectral model, is closed to von Kármán model for $\kappa \leq 600$ m^{-1} and what is important, it provides good agreement with experimental results [8,20].

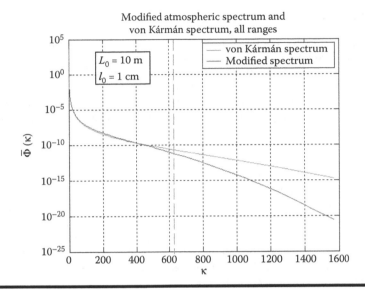

Figure 1.4 Comparison between the modified and von Kármán normalized spectra; the dashed vertical line indicates $\kappa = 2\pi/l_0$.

We also should notice that expression (1.45) is similar to the von Kármán functional form (1.44), except for the terms within the brackets that characterize the high wave number spectral bump.

We can summarize that in the simple Kolmogorov model of the turbulence, the atmosphere is usually described as a single turbulent layer in which the variations of the refractive index with temperature and pressure induce both phase and amplitude fluctuations of the propagating wavefront. In addition, it is usually assumed that the time scale of temporal changes in the atmospheric layer of wave propagation is much smaller than the time it takes the wind to blow the turbulence over the receiver aperture (Taylor's hypothesis of frozen turbulence, see Reference 8). The spatial and temporal properties of this single layer are thus linked by the wind speed V of the layer as $\omega = \kappa \cdot V$, where ω is the frequency in rad/s.

1.2.3.4 Non-Kolmogorov Turbulence

For elevations around the boundary layer (2–3 km) and above, the exponent 11/3 in Equation 1.40, and C_n^2 change as functions as elevation. This does not conform to the Kolmogorov empirical assumptions and is called non-Kolmogorov turbulence [48–51]. It affects wireless communication reliability and image quality. Some special cases of how the non-Kolmogorov processes can occur around the bounded irregular turbulent atmospheric layers are also discussed here in Chapter 3.

References

1. Pruppacher, H. R., and Pitter, R. L., A semi-empirical determination of the shape of cloud and rain drops, *J. Atmos. Sci.*, vol. 28, pp. 86–94, 1971.
2. Slingo, A., A GSM parametrization for the shortwave radiative properties of water clouds, *J. Atmos. Sci.*, vol. 46, pp. 1419–1427, 1989.
3. Chou, M. D., Parametrizations for cloud overlapping and shortwave single scattering properties for use in general circulation and cloud ensemble models, *J. Climate*, vol. 11, pp. 202–214, 1998.
4. Attenuation due to clouds and fog, *International Telecommunication Union, ITU-R Recommendation* P. 840-2, Geneva, 1997.
5. Liou, K. N., *Radiation and Cloud Processes in the Atmosphere*. Oxford: Oxford University Press, 1992.
6. Blaunstein, N., Arnon, Sh., Zilberman, A., and Kopeika, N., *Applied Aspects of Optical Communication and LIDAR*. New York: CRC Press; Taylor & Francis Group, 2010.
7. Bean, B. R., and Dutton, E. J., *Radio Meteorology*. New York: Dover, 1966.
8. Blaunstein, N., and Christodoulou, Ch., *Radio Propagation and Adaptive Antennas for Wireless Communication Networks: Terrestrial, Atmospheric and Ionospheric*. New York: Wiley InterScience, 2014.
9. Specific attenuation model for rain for use in prediction methods, *International Telecommunication Union, ITU-R Recommendation*, P. 838, Geneva, 1992.
10. Saunders, S. R., *Antennas and Propagation for Wireless Communication Systems*. New York: John Wiley & Sons, 1999.
11. Twomey, S., *Atmospheric Aerosols*. Amsterdam: Elsevier, 1977.
12. McCartney, E. J. *Optics of the Atmosphere: Scattering by Molecules and Particles*. New York: John Wiley & Sons, 1976.
13. Whitby, K. Y., The physical characteristics of sulfur aerosols, *Atmos. Environ.*, vol. 12, pp. 135–159, 1978.
14. Friedlander, S. K., *Smoke, Dust and Haze*. New York: John Wiley & Sons, 1977.
15. Seinfeld, J. H., *Atmospheric Chemistry and Physics of Air Pollution*, New York: John Wiley & Sons, 1986.
16. d'Almeida, G. A., Koepke, P., and Shettle, E. P., *Atmospheric Aerosols, Global Climatology and Radiative Characteristics*. Hampton: Deepak Publishing, 1991.
17. Ishimaru, A., *Wave Propagation and Scattering in Random Media*. New York: Academic Press, 1978.
18. Rytov, S. M., Kravtsov, Yu. A., and Tatarskii, V. I., *Principles of Statistical Radiophysics*. Berlin: Springer, 1988.
19. Tatarskii, V. I., *Wave Propagation in a Turbulent Medium*. New York: McGraw-Hill, 1961.
20. Andrews, L. C., and Phillips, R. L., *Laser Beam Propagation through Random Media*, 2nd ed. Bellingham, WA: SPIE Press, 2005.
21. Kopeika, N. S., *A System Engineering Approach to Imaging*. Bellingham, WA: SPIE Press, 1998.
22. US Standard Atmosphere. 1976, US GPO, Washington, D.C., 1976.
23. Kovalev, V. A., and Eichinger, W. E., *Elastic Lidar: Theory, Practice, and Analysis Methods*, Hoboken, NJ: John Wiley & Sons, Inc., 2004.
24. Junge, C. E., *Air Chemistry and Radioactivity*. New York: Academic Press, 1963.

25. Jaenicke, R., Aerosol physics and chemistry, in *Physical Chemical Properties of the Air, Geophysics and Space Research, 4(b)*, Fisher, G. Ed. Berlin: Springer-Verlag, 1988.

26. d'Almeida, G. A. On the variability of desert aerosol radiative characteristics, *J. Geophys. Res.*, vol. 93, pp. 3017–3026, 1987.

27. Shettle, E. P., Optical and radiative properties of a desert aerosol model, in *Proc. Symposium on Radiation in the Atmosphere*, Fiocco, G. Ed. A. Deepak Publishing, pp. 74–77, 1984.

28. Remer L. A., and Kaufman, Y. J., Dynamic aerosol model: Urban/Industrial aerosol, *J. Geophys. Res.*, vol. 103, pp. 13859–13871, 1998.

29. Crutzen, P. J., and Andreae, M. O., Biomass burning in the tropics: Impact on atmospheric chemistry and biogeochemical cycles, *Science*, vol. 250, pp. 1669–1678, 1990.

30. Rosen, J. M., and Hofmann, D. J., Optical modeling of stratospheric aerosols: Present status, *Appl. Opt.*, vol. 25(3), pp. 410–419, 1986.

31. Butcher, S. S., and Charlson, R. J., *Introduction to Air Chemistry*. New York: Academic Press, 1972.

32. Cadle, R. D., *Particles in the Atmosphere and Space*. New York: Van Nostrand Reinhold, 1966.

33. Shettle, E. P., and Fenn, R. W., Models for the aerosols of the lower atmosphere and the effects of humidity variations on their optical properties, AFGL-TR-79-0214, 1979.

34. Herman, B., LaRocca, A. J., and Turner, R. E., Atmospheric scattering, in *Infrared Handbook*, Wolfe, W. L., and Zissis, G. J. Eds. St. Petersburg, Russia: Environmental Research Institute of Michigan, 1989.

35. Zuev, V. E., and Krekov, G. M., *Optical Models of the Atmosphere*. Leningrad: Gidrometeoizdat, 1986.

36. Hobbs, P. V., Bowdle, D. A., and Radke, L. F., Particles in the lower troposphere over the high plains of the United States. I: Size distributions, elemental compositions and morphologies, *J. Climate Appl. Meteor.*, vol. 24, pp. 1344–1349, 1985.

37. Kent, G. S., Wang, P.-H., McCormick, M. P., and Skeens, K. M., Multiyear stratospheric aerosol and gas experiment II: Measurements of upper tropospheric aerosol characteristics, *J. Geophys. Res.*, vol. 98, pp. 20725–20735, 1995.

38. Berk, A., Bernstein, L. S., and Robertson, D. C., *MODTRAN: A moderate resolution model for LOWTRAN 7*, Air Force Geophysics Laboratory Technical Report GL TR-89-0122. Hanscom AFB, MA, 1989.

39. Jursa, A. S. Ed. *Handbook of Geophysics and the Space Environment*. Boston, MA: Air Force Geophysics Laboratory, 1985.

40. Junge, C. E. Ed. Atmospheric chemistry, in *Advances in Geophysics*, vol. 4. New York: Academic Press, 1958.

41. McClatchey, R. A., Fenn, R. W., Selby, J. E. A., Volz, F. E., and Garing, J. S., *Optical Properties of the Atmosphere*. L.G. Hanscom Field, Bedford, MA: Air Force Cambridge Research Lab., AFCRL-72-0497, 1972.

42. Deirmenjian, D., *Electromagnetic Scattering on Spherical Polydispersions*. New York: Elsevier, 1969.

43. Richardson, L. F., *Weather Prediction by Numerical Process*. Cambridge, UK: Cambridge University Press, 1922.

44. Kolmogorov, A. N., The local structure of turbulence in incompressible viscous fluids for very large Reynolds numbers, in *Turbulence, Classic Papers on Statistical Theory*,

Friedlander, S. K., and Topper, L. Eds. New York: Wiley-Interscience, 1961, pp. 151–155.

45. Tatarskii, V. I., (Translator), *The Effects of the Turbulent Atmosphere on Wave Propagation*, Trans. Jerusalem: For NOAA by the Israel Program for Scientific Translations, 1971.

46. Kraichman, R. H., On Kolmogorov's inertial-range theories, *J. Fluid Mech.*, vol. 62, pp. 305–330, 1974.

47. Obukhov, A. M., Temperature field structure in a turbulent flow, *Izv. Acad. Nauk SSSR Ser. Geog. Geofiz.*, vol. 13, pp. 58–69, 1949.

48. Golbraikh, E., and Kopeika, N. S., Behavior of the structural function of the refraction coefficient in different turbulent fields, *Applied Optics*, vol. 43, pp. 6151–6156, 2004.

49. Zilberman, A., Golbraikh, E., Kopeika, N. S., Virtser, A., Kuperschmidt, I., and Shtemler, Y., Lidar study of aerosol turbulence characteristics in the troposphere: Kolmogorov and non-Kolmogorov turbulence, *Atmospheric Research*, vol. 88, pp. 66–77, 2008.

50. Zilberman, A., Golbraikh, E., and Kopeika, N. S., Propagation of EM waves in Kolmogorov and non-Kolmogorov atmospheric turbulence: Three layer altitude model, *Applied Optics*, vol. 47, pp. 6385–6391, 2008.

51. Zilberman, A., Golbraikh, E., and Kopeika, N. S., Some limitations on optical communication reliability through Kolmogorov and non-Kolmogorov turbulence, *Optical Communications*, vol. 283, pp. 1229–1235, 2010.

Chapter 2

Basic Aspects of Optical Wave Propagation

Nathan Blaunstein and Natan Kopeika

Contents

2.1 Identity of Optical and Radio Waves

Electromagnetics plays a key role in modern radio and optical communication systems, including optical telecommunication systems and LIDAR [1–16]. Therefore, optical communication emphasizes the electromagnetic phenomena described mathematically by Maxwell's unified theory [1–3,10–14], which we will consider in detail shortly. We should stress here that optical communications started to be investigated just after the invention of the laser in 1960. Atmospheric propagation of optical waves was investigated parallel to improvements in optical lasers and understanding of the problems of optical wave generation in both time and space domains. Thus, solving problems with weather, line-of-site (LOS) clearance, and beam broadening and bending in the real atmosphere contributed toward removal of free-space optical communication as a major aspect in wireless communication.

Since optical waves have the same nature as electromagnetic waves (see Figure 1.1 in Chapter 1), we will start with a physical explanation of electromagnetic waves based on Maxwell's unified theory [1,2,10–13], which postulates that an electromagnetic field could be represented as a wave. The coupled wave components, the electric and magnetic fields, are depicted in Figure 2.1, from which it

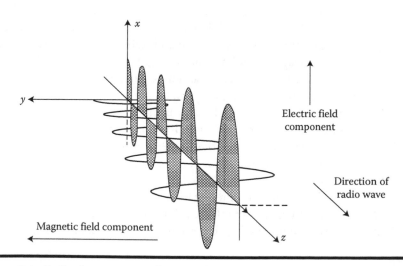

Figure 2.1 Optical wave as an electromagnetic wave with its electrical and magnetic components, wavefront, and direction of propagation.

follows that the electromagnetic (EM) wave travels in a direction perpendicular to both EM field components. In Figure 2.1, this direction is denoted as the *z*-axis in the Cartesian coordinate system by the wave vector **k**. In their orthogonal space-planes, the magnetic and electric oscillatory components repeat their waveform after a distance of one wavelength along the *y*-axis and *x*-axis, respectively (see Figure 2.1).

Both components of the EM wave are in phase in the time domain, but not in the space domain [1,2,10–13]. Moreover, the magnetic component value of the EM field is closely related to the electric component value, from which one can obtain the radiated power of the EM wave propagating along the *z*-axis (see Figure 2.1).

At the same time, using the Huygen's principle, well known in electrodynamics [10–13], one can show that an optical wave is the electromagnetic wave propagating only straightforward from the source as a ray with minimum loss of energy and with minimum propagation time (according to Fermat's principle [2,7,14]) in free space, as a unbounded homogeneous medium without obstacles and discontinuities.

Thus, if we present the Huygen's concept, as it is shown from Figure 2.2, the ray from each point propagates in all forward directions to form many elementary spherical wavefronts, which Huygens called wavelets.

The envelope of these wavelets forms the new wave fronts. In other words, each point on a wave front acts as a source of secondary elementary spherical waves,

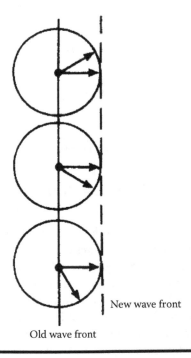

New wave front

Old wave front

Figure 2.2 Huygens principle for a proof of straight propagation of waves as rays.

described by Green's function (see References 10–13). These waves combine to produce a new wavefront in the direction of straight propagation. As we will show below, each wave front can be represented by the plane that is normal to the wave vector **k** (e.g., wave energy transfer). Moreover, propagating forward along straight lines normal to their wave front, waves propagate as light rays in optics, spending minimum energy for passing from the source to any detector, that is, the maximum energy of the ray is observed in a straight direction normal to the wave front (as seen from Figure 2.2). Kirchhoff first mathematically showed this principle based on Maxwell's general unified theory. Let us now assess all propagation phenomena theoretically using Maxwell's unified theory.

2.2 Optical Waves Propagation: Deterministic Approach

2.2.1 Main Equations

The theoretical analysis of optical wave propagation, as a part of the whole electromagnetic spectrum [1–6] (see also Figure 1.1 in Chapter 1), is based on Maxwell's equations [10–16]. In vector notation, their representations in the uniform macroscopic form are [1–6]:

$$\nabla \times \mathbf{E}(\mathbf{r},t) = -\frac{\partial}{\partial t}\mathbf{B}(\mathbf{r},t), \tag{2.1}$$

$$\nabla \times \mathbf{H}(\mathbf{r},t) = \frac{\partial}{\partial t}\mathbf{D}(\mathbf{r},t) + \mathbf{j}(\mathbf{r},t), \tag{2.2}$$

$$\nabla \cdot \mathbf{B}(\mathbf{r},t) = 0, \tag{2.3}$$

$$\nabla \cdot \mathbf{D}(\mathbf{r},t) = \rho(\mathbf{r},t). \tag{2.4}$$

Here $\mathbf{E}(\mathbf{r},t)$ is the electric field strength vector in volts per meter (V/m); $\mathbf{H}(\mathbf{r},t)$ is the magnetic field strength vector in amperes per meter (A/m); $\mathbf{D}(\mathbf{r},t)$ is the electric flux induced in the medium by the electric field in coulombs/m^3 (this is why, in the literature sometimes, it is called an "induction" of an electric field); $\mathbf{B}(\mathbf{r},t)$ is the magnetic flux induced by the magnetic field in weber/m^2 (it is also called "induction" of a magnetic field); $\mathbf{j}(\mathbf{r},t)$ is the vector of electric current density in amperes/m^2; $\rho(\mathbf{r},t)$ is the charge density in coulombs/m^2. The curl operator, $\nabla\times$, is a measure of field rotation, and the divergence operator, $\nabla\cdot$, is a measure of the total flux radiated from a point.

It should be noted that for a time-varying EM wave field, Equations 2.3 and 2.4 can be derived from Equations 2.1 and 2.2, respectively. In fact, taking the divergence of (2.1) (by use of the divergence operator $\nabla\cdot$), one can immediately obtain (2.3). Similarly, taking the divergence of Equation 2.2 and using the well-known continuity equation [1–3,10–16]

$$\nabla \cdot \mathbf{j}(\mathbf{r},t) + \frac{\partial \rho(\mathbf{r},t)}{\partial t} = 0 \tag{2.5}$$

one can arrive at Equation 2.4. Hence, only Equations 2.1 and 2.2 are independent.

Equation 2.1 is the well-known Faraday law and indicates that a time-varying magnetic flux generates an electric field with rotation; Equation 2.2 without the term $\partial \mathbf{D}/\partial t$ (displacement current term [10–13]) limits us to the well-known Ampere's law and indicates that a current or a time-varying electric flux (displacement current [10–13]) generates a magnetic field with rotation.

Because one now has only two independent Equations 2.1 and 2.2, which describe the four unknown vectors $\mathbf{E}$, $\mathbf{D}$, $\mathbf{H}$, $\mathbf{B}$, three more equations relating these vectors are needed. To do this, we introduce relations between $\mathbf{E}$ and $\mathbf{D}$, $\mathbf{H}$ and $\mathbf{B}$, $\mathbf{j}$ and $\mathbf{E}$, which are known in electrodynamics. In fact, for isotropic media, which are usually considered in problems of land radio propagation, the electric and magnetic fluxes are related to the electric and magnetic fields, and the electric current is related to the electric field, via the constitutive relations [10–13]:

$$\mathbf{B} = \mu(\mathbf{r})\mathbf{H} \tag{2.6}$$

$$\mathbf{D} = \varepsilon(\mathbf{r})\mathbf{E} \tag{2.7}$$

$$\mathbf{j} = \sigma(\mathbf{r})\mathbf{E} \tag{2.8}$$

It is very important to note that relations 2.6 are valid only for propagation processes in linear isotropic media, which are characterized by the three scalar functions of any point $\mathbf{r}$ in the medium: permittivity $\varepsilon(\mathbf{r})$, permeability $\mu(\mathbf{r})$, and conductivity $\sigma(\mathbf{r})$. In relations 2.6 through 2.8, it was assumed that the medium is inhomogeneous. In a homogeneous medium, the functions $\varepsilon(\mathbf{r})$, $\mu(\mathbf{r})$, and $\sigma(\mathbf{r})$ transform to simple scalar values ε, μ, and σ.

In free space, these functions are simply the constants, that is, $\varepsilon = \varepsilon_0 = 8.854 \times 10^{-12}$ Farad/meter, $\mu = \mu_0 = 4\pi \times 10^{-7}$ Henry/meter, and $c = \left(1/\sqrt{\varepsilon_0\mu_0}\right)$ is the velocity of light, which has been measured very accurately.

The system (2.1)–(2.4) can be further simplified if we assume that the fields are time harmonic. If the field time-dependence is not harmonic, then using the fact that Equation 2.1 are linear, we may treat these fields as sums of harmonic components and consider each component separately. In this case, the time harmonic field is a complex vector and can be expressed via its real part as [10–13]:

$$\mathbf{A}(\mathbf{r},t) = \text{Re}[\mathbf{A}(\mathbf{r})e^{-i\omega t}], \tag{2.9}$$

where $i = \sqrt{-1}$, ω is the angular frequency in radians per second, $\omega = 2\pi f$, f is the radiated frequency in Hz $= \text{s}^{-1}$, and $\mathbf{A}(\mathbf{r},t)$ is the complex vector ($\mathbf{E}$, $\mathbf{D}$, $\mathbf{H}$, $\mathbf{B}$, or $\mathbf{j}$). The time dependence $\sim e^{-i\omega t}$ is commonly used in the literature of electrodynamics and wave propagation. If $\sim e^{i\omega t}$ is used, then one must substitute $-i$ for i and i for $-i$, in all equivalent formulations of Maxwell's equations. In Equation 2.9, $e^{-i\omega t}$ presents the harmonic time dependence of any complex vector $\mathbf{A}(\mathbf{r},t)$, which satisfies the relationship:

$$\frac{\partial}{\partial t}\mathbf{A}(\mathbf{r},t) = \text{Re}[-i\omega A(r)e^{-i\omega t}] \tag{2.10}$$

Using this transformation, one can easily obtain from the system (2.1)

$$\nabla \times \mathbf{E}(\mathbf{r}) = i\omega \mathbf{B}(\mathbf{r}) \tag{2.11}$$

$$\nabla \times \mathbf{H}(\mathbf{r}) = -i\omega \mathbf{D}(\mathbf{r}) + \mathbf{j}(\mathbf{r}) \tag{2.12}$$

$$\nabla \cdot \mathbf{B}(\mathbf{r}) = 0 \tag{2.13}$$

$$\nabla \cdot \mathbf{D}(\mathbf{r}) = \rho(\mathbf{r}) \tag{2.14}$$

It can be observed that system (2.11)–(2.14) was obtained from system (2.1)–(2.4) by replacing $\partial/\partial t$ with $-i\omega$. Alternatively, the same transformation can be obtained by the use of the Fourier transform of system (2.1)–(2.4) with respect to time [1,2,10–15]. In Equations 2.11 through 2.14, all vectors and functions are actually the Fourier transforms with respect to the time domain, and the fields $\mathbf{E}$, $\mathbf{D}$, $\mathbf{H}$, and $\mathbf{B}$ are functions of frequency as well. We call them phasors of time domain vector solutions. They are also known as frequency domain solutions of the EM field according to system (2.11)–(2.14). Conversely the solutions of system (2.1) are the time-domain solutions of the EM field. It is more convenient to work with system (2.11)–(2.14) instead of system (2.1)–(2.4) because of the absence of the time dependence and time derivatives in it.

2.2.2 Propagation of Optical Waves in Free Space

Mathematically, optical wave propagation phenomena can be described by use of both the scalar and vector wave equation presentations. Because most problems of optical wave propagation in wireless communication links are considered in unbounded, homogeneous, source-free isotropic media, we can, with a great accuracy, consider $\varepsilon(\mathbf{r}) \equiv \varepsilon$, $\mu(\mathbf{r}) \equiv \mu$, $\sigma(\mathbf{r}) \equiv \sigma$, and finally obtain from general wave equations:

$$\nabla \times \nabla \times \mathbf{E}(\mathbf{r}) - \omega^2 \varepsilon \mu \mathbf{E}(\mathbf{r}) = 0$$
$$\nabla \times \nabla \times \mathbf{H}(\mathbf{r}) - \omega^2 \varepsilon \mu \mathbf{H}(\mathbf{r}) = 0 \tag{2.15}$$

Because both equations are symmetric, one can use one of them, namely that for **E**, and by introducing the vector relation $\nabla \times \nabla \times \mathbf{E} = \nabla(\nabla \cdot \mathbf{E}) - \nabla^2 \mathbf{E}$ and taking into account that $\nabla \cdot \mathbf{E} = 0$, we finally obtain

$$\nabla^2 \mathbf{E}(\mathbf{r}) + k^2 \mathbf{E}(\mathbf{r}) = 0 \tag{2.16}$$

where $k^2 = \omega^2 \varepsilon \mu$.

In special cases of a homogeneous, source-free, isotropic medium, the three-dimensional (3D) wave equation reduces to a set of scalar wave equation. This is because in Cartesian coordinates, $\mathbf{E}(\mathbf{r}) = E_x \mathbf{x}_0 + E_y \mathbf{y}_0 + E_z \mathbf{z}_0$, where $\mathbf{x}_0$, $\mathbf{y}_0$, $\mathbf{z}_0$ are unit vectors in the directions of the x, y, z coordinates, respectively. Hence, Equation 2.16 consists of three scalar equations such as

$$\nabla^2 \Psi(\mathbf{r}) + k^2 \Psi(\mathbf{r}) = 0, \tag{2.17}$$

where $\Psi(\mathbf{r})$ can be either E_x, E_y, or E_z. This equation fully describes propagation of optical waves in free space.

2.2.3 Propagation of Optical Waves through the Boundary of Two Media

2.2.3.1 Boundary Conditions

The simplest case of wave propagation over the intersection between two media is that where the intersection surface can be assumed as flat, and the second medium perfectly conductive.

If so, for a perfectly conductive flat surface, the total electric field vector is equal to zero, that is, $\mathbf{E} = 0$ [1–3,10–13]. In this case, the tangential component of electric field vanishes at the perfectly conductive flat surface, that is,

$$E_\tau = 0 \tag{2.18}$$

Consequently, it follows from Maxwell's equation $\nabla \times \mathbf{E}(\mathbf{r}) = i\omega \mathbf{H}(\mathbf{r})$ (see References 14, 46) for the case of $\mu = 1$ and $\mathbf{B} \equiv \mathbf{H}$, at such a flat perfectly conductive surface the normal component of the magnetic field also vanishes, that is,

$$H_n = 0 \tag{2.19}$$

As also follows from Maxwell's equations (2.1)–(2.4), the tangential component of magnetic field does not vanish because of the electric surface current. At the same time, the normal component of electric field does not vanish because of electrical charge density at the perfectly conducting surface. Hence, by introducing the Cartesian coordinate system, one can present the boundary conditions (2.18) and (2.19) at the flat perfectly conductive surface as follows:

$$E_x(x, y, z = 0) = E_y(x, y, z = 0) = H_z(x, y, z = 0) = 0 \qquad (2.20)$$

2.2.4 Main Formulations of Reflection and Refraction Coefficients

As was shown above, the influence of a flat material surface on optical wave propagation leads to phenomena such as reflection. Because all kinds of waves can be represented by means of the concept of the plane waves [1–3,10–13], let us obtain the main reflection and refraction formulas for a plane wave incident on a plane surface between two media, as shown in Figure 2.3. The media have different dielectric properties, which are described above and below the boundary plane $z = 0$ by the permittivity and permeability ε_1, μ_1 and ε_2, μ_2, respectively for each medium.

Without reducing the general problem, let us consider an optical wave with wave vector $\mathbf{k}_i$ and frequency $\omega = 2\pi f$ incident from a medium described by parameter n_1. The reflected and refracted waves are described by wave vectors $\mathbf{k}_1$ and $\mathbf{k}_2$, respectively. Vector $\mathbf{n}$ is a unit normal vector directed from medium with (n_2) into medium (n_1).

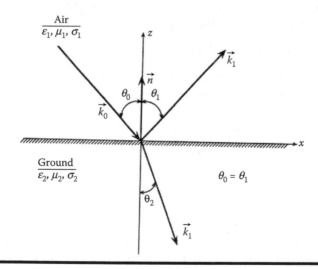

Figure 2.3 Reflection and refraction of optical wave at the boundary of two media.

We should notice that in optics, usually the designer of optical systems deal with nonmagnetized materials, so that the normalized dimensionless permeability of the two media should equal unity, that is, $\tilde{\mu}_1 = \mu_1/\mu_0 = 1$ and $\tilde{\mu}_2 = \mu_2/\mu_0 = 1$. Moreover, instead of being used in electrodynamics and radio physics, the normalized dimensionless permittivity for each medium, $\tilde{\varepsilon}_1 = \varepsilon_1/\varepsilon_0$, $\tilde{\varepsilon}_2 = \varepsilon_2/\varepsilon_0$, are expressed via the corresponding refractive index n_1 and n_2, that is, $\tilde{\varepsilon}_1 = n_1^2$ and $\tilde{\varepsilon}_2 = n_2^2$. According to the relations between electrical and magnetic components, which follow from Maxwell's equations (see system (2.1)–(2.4)), we can easily obtain expressions for the coefficients of reflection and refraction. A physical meaning of the reflection coefficient is as follows: it defines the ratio of the reflected electric field component of the optical wave to its incident electric field component. The same physical meaning applies to the refractive coefficient: it defines the ratio of the refractive electric field component to the incident electric field component of the optical wave. Before presenting these formulas, let us show two important laws that are being used in classical optics. As follows from Maxwell's equations, and the boundary conditions and geometry presented in Figure 2.3, the values of the wave vectors are related by the following expressions:

$$|\mathbf{k}| = |\mathbf{k}_r| \equiv k = \frac{\omega}{c} n_1, \quad |\mathbf{k}_t| \equiv k_t = \frac{\omega}{c} n_2 \tag{2.21}$$

From the boundary conditions described earlier by Equations 2.18 through 2.20, one can easily obtain the condition of the equality of phase for each wave at the plane $z = 0$:

$$(\mathbf{k} \cdot \mathbf{x})_{z=0} = (\mathbf{k}_r \cdot \mathbf{x})_{z=0} = (\mathbf{k}_t \cdot \mathbf{x})_{z=0} \tag{2.22}$$

which is independent of the nature of the boundary condition. Equation 2.22 describes the condition that all three wave vectors must lie in the same plane. From this equation, it also follows that

$$k \sin\theta_i = k_r \sin\theta_r = k_t \sin\theta_t \tag{2.23}$$

which is the analogue of the second Snell law:

$$n_1 \sin\theta_r = n_2 \sin\theta_t \tag{2.24}$$

Moreover, because $|\mathbf{k}_i| = |\mathbf{k}_r|$, we find $\theta_i = \theta_r$; the angle of incidence equals the angle of reflection. This is the first Snell law.

In the literature, which describes wave propagation aspects, optical waves are usually called waves with vertical and horizontal polarization, depending on the orientation of the electric field component regarding the plane of propagation, perpendicular or parallel, respectively.

Without entering into straight computations, we obtain the expressions for complex coefficients of reflection (R) and refraction (T) for waves with vertical (denoted by index V) and horizontal (denoted by index H) polarization, respectively.

For vertical polarization:

$$R_V = \frac{n_1 \cos\theta_i - \sqrt{n_2^2 - n_1^2 \sin^2\theta_i}}{n_1 \cos\theta_i + \sqrt{n_2^2 - n_1^2 \sin^2\theta_i}} \tag{2.25}$$

$$T_V = \frac{2n_1 \cos\theta_i}{n_1 \cos\theta_i + \sqrt{n_2^2 - n_1^2 \sin^2\theta_i}} \tag{2.26}$$

For horizontal polarization:

$$R_H = \frac{-n_2^2 \cos\theta_i + n_1\sqrt{n_2^2 - n_1^2 \sin^2\theta_i}}{n_2^2 \cos\theta_i + n_1\sqrt{n_2^2 - n_1^2 \sin^2\theta_i}} \tag{2.27}$$

$$T_H = \frac{2n_1 n_2 \cos\theta_i}{n_2^2 \cos\theta_i - n_1\sqrt{n_2^2 - n_1^2 \sin^2\theta_i}} \tag{2.28}$$

In the case of vertical polarization, there is a special angle of incidence, called the Brewster angle, for which there is no reflected wave. For simplicity, we will assume that the condition $\mu_1 = \mu_2$ is valid. Then from Equation 2.24, it follows that the reflection coefficients zero when the angle of incidence is equal to Brewster's angle

$$\theta_i \equiv \theta_{Br} = \tan^{-1}\left(\frac{n_2}{n_1}\right) \tag{2.29}$$

Another interesting phenomenon that follows from these formulas is called total reflection. It takes place when the condition of $n_2 \gg n_1$ is valid. In this case from Snell's law (2.24) it follows that, if $n_2 \gg n_1$, then $\theta_r \gg \theta_i$. Consequently, when $\theta_r \gg \theta_i$ the reflection angle $\theta_r = (\pi/2)$, where

$$\theta_c = \sin^{-1}\left(\frac{n_2}{n_1}\right) \tag{2.30}$$

For waves incident at the surface under the critical angle $\theta_i \equiv \theta_c$ there is no refracted wave within the second medium; the refracted wave is propagated along the boundary between the first and second media and there is no energy flow from one medium to the other.

Therefore, this phenomenon is called total internal reflection (TIR) in the literature, and the smallest incident angle θ_i for which we get TIR is called the critical angle $\theta_i \equiv \theta_c$ defined by expression (2.28).

2.3 Optical Wave Propagation in Random Media: Statistical Approach

The problem of optical wave propagation through an irregular atmosphere, consisting a lot of inhomogeneous structures (see Chapter 1) could be understood by using the statistical description of the wave field (vector and/or scalar) and quantum theory [17–37]. Because the problems of random equations are not tractable with standard mathematical tool, we should use here some special methods such as Feynman's diagram method [17–20], the method of renormalization [21,23,33,34,38], and so forth.

2.3.1 Main Wave Equations and Random Characteristics

A random medium is a medium whose parameters, such as pressure, density, temperature, etc. are random functions of position and time (all definitions, for example, can be found in References 31, 32, 39–41). This means that we are not describing the exact values of these parameters, but only the probability to find them between a given range of values at given intervals in the space and time domains. A random medium can also be thought of as a collection of inhomogeneous media, each of which may be either continuous (turbulent medium [32,33,35–37,41]) or discrete (medium with random inclusions [34,40,42,43]). In the following, we introduce the main equations that describe stochastic processes in a random medium.

2.3.2 Wave Equations

The propagation phenomena of optical waves in random medium are described by a linear differential equation with random coefficients, instead of deterministic one (see Equation 2.17), used for description of deterministic models. Thus, a scalar wave Equation 2.17 can be presented in the following form:

$$\Delta\Psi(r,t) - \frac{n^2(r,t)}{c^2}\frac{\partial^2\Psi(r,t)}{\partial t^2} = 0 \qquad (2.31)$$

where $\Psi(r,t)$ is the wave field amplitude in the space and time domains, $n(r, t)$ is the refractive index, which is a random function of space (r) and time (t), and c is the wave velocity in free space.

A compact scalar wave equation with a source $g(r)$ (described, let's say, with a Green "point source" function) can be presented as follows:

$$\Delta\Psi(r) - k^2 n^2(r)\Psi(r) = g(r) \tag{2.32}$$

Here, Equation 2.32 is presented assuming a harmonic time dependence $\sim\exp\{ickt\} \equiv \exp\{i\omega t\}$ and a time-independent refractive index n, where $k = (2\pi/\lambda)$ is the wave number and λ is the wavelength in medium under consideration. The source term $g(r)$ is assumed to be given and not randomized (e.g., deterministic). In such an assumption, a usually used electromagnetic vector wave equation can be presented in the following form:

$$\Delta\mathbf{E}(r,t) - \nabla(\nabla \cdot \mathbf{E}(r,t)) - \frac{n^2(r,t)}{c^2}\frac{\partial^2\mathbf{E}(r,t)}{\partial t^2} = 0, \tag{2.33}$$

where $\mathbf{E}(r,t)$ is the vector presentation of the electromagnetic field.

We shall always treat the refractive index as a time-independent random function, which is equivalent to the assumption that the characteristic time of index fluctuations is much longer than the period of the propagating wave. The medium in such conditions will be taken statistically as homogeneous. This assumption excludes any medium where the atmospheric turbulence is concentrated in a small volume of space.

To conclude this subsection, let us show that the scalar wave equation 2.31 and the reduced scalar wave Equation 2.32 may be treated simultaneously. Equation 2.31 corresponds to an initial value problem well-known as the Cauchy problem, and must be given as $\Psi(r,0)$ and $\partial\Psi(r,0)/\partial t$ in order to find $\Psi(r,t)$ [32,39,44]. Equation 2.32 corresponds to a radiation problem [34]. Let us introduce the Laplace transform of the wave function $\Psi(r,t)$

$$\Psi(r,q) = \int_0^\infty \Psi(r,t) \cdot \exp\{i \cdot q \cdot t\} \cdot dt, \quad \mathrm{Im}(q) > 0. \tag{2.34}$$

It satisfies the following equation, which is the Laplace transform of (2.32):

$$\Delta\Psi(r,q) - \frac{q^2}{c^2} \cdot n^2(r) \cdot \Psi(r,q) = \frac{n^2(r)}{c^2}\left[i \cdot q \cdot \Psi(r,t=0) - \frac{\partial\Psi(r,t=0)}{\partial t}\right] \tag{2.35}$$

Equations 2.35 and 2.32 are identical if we take

$$q = c \cdot k, \quad \Psi(r, t = 0) = 0, \quad \frac{\partial \Psi(r, t = 0)}{\partial t} = -\frac{c^2}{n^2(r)} g(r) \qquad (2.36)$$

We shall always choose these initial conditions for Equation 2.31 and treat it and Equation 2.32 simultaneously, interchanging q and $c \cdot k$, whenever necessary.

2.3.3 Random Functions and Their Moments

A detailed treatment of stochastic processes may be found in References 34, 40–43. By introducing any finite number of points $r_1, r_2, \ldots, r_n$, we assume that the mean value of random functions $\mu(r_1) \cdot \mu(r_2) \ldots \mu(r_n)$ always exists. A moment of order n for a random function is given by

$$\langle \mu(r_1) \cdot \mu(r_2) \cdots \mu(r_n) \rangle = \int_\Omega \mu(r_1) \cdot \mu(r_2) \cdots \mu(r_n) \cdot P(d\omega) \qquad (2.37)$$

A random function is often characterized by the infinite set of all its moments [34,40–43]. The random function $\mu(r)$ is centered if

$$\langle \mu(r) \rangle = 0 \qquad (2.38)$$

This function is stationary if the joint distribution of any finite number of random variables $\mu(r_1) \cdot \mu(r_2) \ldots \mu(r_n)$ is invariant with respect to any simultaneous transfer of its arguments. For a space-dependent random function, it would perhaps be better to call it a stationary homogeneous random function (e.g., homogeneous turbulence [32,34,39–44]). If the random function is also real valued, the second order moment

$$C(r_1, r_2) = \langle \mu(r_1) \cdot \mu(r_2) \rangle \qquad (2.39)$$

is called the covariance function.

If the random function $\mu(r)$ is stationary, the covariance function $C(r)$ is only a function of $r = (r_1 - r_2)$, that is,

$$C(r_1, r_2) = C(r_1 - r_2) \qquad (2.40)$$

The function $C(r)$ has a Fourier transform, which is a positive measure $C(k)$, called the spectral measure of the stationary random function, or spectral density function if it reduces to an ordinary function of wave number k.

Gaussian Random Function: A random function $\mu(r)$ is called Gaussian if the joint distribution of any finite number of random variables $\mu(r_1) \cdot \mu(r_2) \dots \mu(r_n)$ is Gaussian [34,40–43]. This function is of great theoretical interest and has many practical applications, especially because of the following property:

Any scalar linear function of a Gaussian random function is a Gaussian random variable.

Furthermore, we note the important property of the moments of a centered Gaussian random function [33,40,42,43]

$$\left\langle \mu(r_1) \cdot \mu(r_2) \cdots \mu(r_{2n+1}) \right\rangle = 0 \tag{2.41}$$

$$\left\langle \mu(r_1) \cdot \mu(r_2) \cdots \mu(r_{2n}) \right\rangle = \underbrace{\sum \left\langle \mu(r_i) \cdot \mu(r_j) \right\rangle \left\langle \mu(r_k) \cdot \mu(r_m) \right\rangle \cdots}_{p\ factors} \tag{2.42}$$

Here the summation extends over all $(2 \cdot n)!/(2^n \cdot n!)$ partitions of $r_1 \dots r_{2n}$ into pairs. Particularly, for $n = 2$ one can easily obtain from (2.42)

$$\left\langle \mu(r_1) \cdot \mu(r_2) \cdot \mu(r_3) \cdot \mu(r_4) \right\rangle = \left\langle \mu(r_1) \cdot \mu(r_2) \right\rangle \cdot \left\langle \mu(r_3) \cdot \mu(r_4) \right\rangle$$
$$+ \left\langle \mu(r_1) \cdot \mu(r_3) \right\rangle \cdot \left\langle \mu(r_2) \cdot \mu(r_4) \right\rangle + \left\langle \mu(r_1) \cdot \mu(r_4) \right\rangle \cdot \left\langle \mu(r_2) \cdot \mu(r_3) \right\rangle \tag{2.43}$$

2.3.4 Fourier Transform of Stationary Random Functions

Let us consider random valued measures as the Fourier transform (FT) of stationary random functions. A stationary random function on real line, $\mu(\kappa,\omega)$, with continuous covariance function has a spectral representation of

$$\mu(\kappa, \omega) = \int\limits_{-\infty}^{\infty} dZ(k,\omega) \exp\{ik\kappa\} \tag{2.44}$$

Here $Z(k,\omega)$ is a random function with orthogonal increments. This means that when the parameter values satisfy the following conditions [32–34,38,39,44]:

$$k_1 < k_2 \le k_3 < k_4$$
$$\left\langle [Z(k_2) - Z(k_1)][Z(k_4) - Z(k_3)] \right\rangle = 0 \tag{2.45}$$

the integral in Equation 2.44 is a Stieltjes integral [17–23]. With this definition, the Fourier transform of a stationary random function does not appear as another random function but as some derivative of a random function with orthogonal increments. The integral presentation of Equation 2.44 can be generalized for the case of a 3D random function.

2.3.5 The Cluster Expansion of the Centered Random Function and Its Fourier Transform

If the random function $\mu(r)$ is centered its covariance is also its two-point correlation function, but this is not true for higher moments. As was shown in References 17–23, the n-point correlation functions are not simultaneously correlated. We introduce therefore the correlation functions $h(r_1, r_2)$, $h(r_1, r_2, r_3)$, ... , $h(r_1, r_2, ... , r_p)$ through the following cluster expansions [32–34,37,45,46]:

$$\left\langle \mu(r_1)\mu(r_2) \right\rangle = h(r_1, r_2) \tag{2.46}$$

$$\left\langle \mu(r_1)\mu(r_2)\mu(r_3) \right\rangle = h(r_1, r_2, r_3) \tag{2.47}$$

$$\left\langle \mu(r_1)\mu(r_2)\mu(r_3)\mu(r_4) \right\rangle = h(r_1, r_2)h(r_3, r_4) + h(r_1, r_3)h(r_2, r_4)$$
$$+ h(r_1, r_4)h(r_2, r_3) + h(r_1, r_2, r_3, r_4) \tag{2.48}$$

$$\left\langle \mu(r_1) \cdot \mu(r_2) \cdots \mu(r_p) \right\rangle = \sum h(r_{i_1}, ... , r_{i_k}) \, h(r_{j_1}, ... , r_{j_m}) \, h(r_{l_1}, ... , r_{l_n}) ...... \tag{2.49}$$

where the summation is extended over all parameters of the set 1, 2, K, p into clusters of at least two points according to Equation 2.46. From system (2.46)–(2.49), it follows that for a centered Gaussian random function, all correlation functions except the second order one vanish.

A graphic representation in terms of Mayer (called then the Feynman) diagrams described in References 17–23, 33 were helpful to describe the matter. Thus, the correlation function $h(r_1, r_2, ... , r_p)$ was represented by a set of p points connected by p lines:

$$h(r_1, r_2) = \quad \begin{matrix} \bullet \\ | \\ \bullet \end{matrix} \begin{matrix} 1 \\ \\ 2 \end{matrix} \qquad h(r_1, r_2, r_3) = \quad \triangle \quad ,... \tag{2.50}$$

The cluster expansion is then written graphically. For example,

$$\left\langle \mu(r_1)\mu(r_2)\mu(r_3)\mu(r_4) \right\rangle = \quad \begin{matrix} | & | \end{matrix} \quad + \quad \begin{matrix} \\ \end{matrix} \quad + \quad \times \quad + \quad \square \tag{2.51}$$

This definition of correlation functions ensures that they vanish if the points r_1, $r_2, \ldots, r_p$ are not inside a common sphere of radius ℓ (see References 17–23, 33). We also need the FT of the correlation function as mentioned below:

$$b(k_1, k_2, \ldots, k_p) = \frac{1}{(2\pi)^{3p}} \int b(r_1, r_2, \ldots, r_p) \exp\{-i(k_1 r_1 + \cdots + k_p r_p)\} d^3 r_1 \cdots d^3 r_p$$

(2.52)

If the random function $\mu(r)$ is stationary, this is not a function but a measure concentrated in the hyperplane $k_1 + k_2 + \cdots + k_p = 0$ [32, 33, 38]. Hence, we can write

$$b(k_1, k_2, \ldots, k_p) = g(k_1, k_2, \ldots, k_p) \cdot \delta(k_1 + k_2 + \cdots + k_p)$$

(2.53)

and call the ordinary functions $g(k_1, k_2, \ldots, k_p)$ or simply the correlation functions in k-space. Using these functions, we can write the cluster expansion of the moments in k-space as

$$\langle \mu(k_1) \mu(k_2) \rangle = g(k_1, k_2) \cdot \delta(k_1 + k_2)$$

(2.54)

$$\langle \mu(k_1) \mu(k_2) \mu(k_3) \rangle = g(k_1, k_2, k_3) \cdot \delta(k_1 + k_2 + k_3)$$

(2.55)

$$\langle \mu(k_1) \mu(k_2) \mu(k_3) \mu(k_4) \rangle = g(k_1, k_2) \cdot g(k_3, k_4) \cdot \delta(k_1 + k_2) \cdot \delta(k_3 + k_4) + \cdots$$

(2.56)

The moment $\langle \mu(k_1) \mu(k_2) \ldots \mu(k_p) \rangle$ is thus not only concentrated in the hyperplane $k_1 + k_2 + \cdots + k_p = 0$, but it appears as sums of products of terms that are concentrated in a hyper plane of lower dimensions [35–37, 45, 46].

2.3.6 Random Equations

A random equation such as

$$\Delta \Psi(r) - k^2 n^2(r) \Psi(r) = g(r),$$

(2.57)

describes linear waves and does not constitute a linear problem because the mean solutions do not satisfy the mean equation. This is because

$$\langle n^2(r) \Psi(r) \rangle \neq \langle n^2(r) \rangle \langle \Psi(r) \rangle$$

(2.58)

In other words, the wave function and the refractive index are not statistically independent. If we try to evaluate $\langle n^2(r)\Psi(r)\rangle$, we must multiply Equation 2.57 by $n^2(r)$ and average afterward; this will yield the form $\sim \langle n^2(r_1)n^2(r)\Psi(r)\rangle$, and so on.

Keller [30] has obtained an equation for a function generating the entire set of moments. This equation helps with new approximation procedures, but does not solve the problem. The fact that even the lowest order moment of the wave function $\langle \Psi(r)\rangle$ depends upon the infinite set of moments of the refractive index seems to make the problem hopelessly difficult. However, it happens that in certain limiting cases, one may obtain solutions that do not depend upon the entire set of moments of the refractive index.

The perturbation method, described in References 29, 30, gives Bouret's equation, which depends only on the mean value and the covariance of the refractive index. It is only valid for wavelengths that are long compared to the range of index correlations. Conversely, for the random Taylor expansion (see References 30–32, 45), we need only the probability distribution of the refractive index and some of its derivatives at one fixed point. It is valid for wavelengths that are very short compared to the range of index correlations.

Another case of great interest is when $n^2(r)$ a Gaussian random function is. In this case, as was shown in Reference 45, it is then possible to get an exact solution of Equation 2.57 through functional integration, which gives all the moments of the wave functions in terms of mean value and the covariance of $n^2(r)$ (see References 45, 46, Chapter 3). Unfortunately, this method cannot be generalized to other equations such as the electromagnetic vector wave equation (2.33). Finally, it must be noted that no rigorous mathematical treatment of Equation 2.57 has been presented until now. This is mainly because we are not able to solve linear partial differential equations with non-constant (e.g., variable) coefficients [45].

All the aspects mentioned above can be usefully used for the description of both scalar and vector stochastic equations for different kinds of random media—quasi-homogeneous isotropic and anisotropic and irregular turbulent. We briefly present below some of these methods for the scalar and electromagnetic stochastic wave equations, based on different approximations. For a detailed description of the problem, we direct the reader to References 45, 46.

2.3.7 Scalar Wave Equation Presentation

Here, to make things simple, we shall only consider the scalar wave equation

$$\Delta\Psi(r,t) - \frac{n^2(r)}{c^2}\frac{\partial^2 \Psi(r,t)}{\partial t^2} = 0, \tag{2.59}$$

together with the initial conditions

$$\Psi(r,0) = 0, \quad \frac{\partial \Psi}{\partial t} = -\frac{c^2}{n^2(r)} j(r) \tag{2.60}$$

In Section 2.3.2, it was shown that this problem is equivalent to the random variable problem described by formula (2.57). We shall assume that the refractive index $n(r)$ is a stationary random function of position and is time independent. The assumption of strict stationarity (i.e., not only for the two first moments) is essential. We separate now the constant mean value of $n^2(r)$ and its random part.

$$n^2(r) = \langle n^2(r) \rangle [1 + \varepsilon \cdot \mu(r)] \tag{2.61}$$

$$\langle \mu(r) \rangle = 0$$

Here ε is a dimensionless small positive parameter characterizing the relative strength of index fluctuations. Equation 2.59 can now be rewritten as

$$\Delta\Psi(r,t) - \frac{1}{c^2}[1 + \varepsilon \cdot \mu(r)]\frac{\partial^2 \Psi(r,t)}{\partial t^2} = 0 \tag{2.62}$$

where $\langle n^2(r) \rangle$ has been incorporated into $1/c^2$.

The stationary random function is written in terms of its FT and $\mu(r)$, which is a random valued measure

$$\mu(r) = \int \exp\{ik \cdot r\}\mu(k)d^3k \tag{2.63}$$

The Laplace transformation (LT) of Equation 2.62, taking into account the initial conditions (2.60), is

$$\Delta\Psi(r,z) + \frac{z^2}{c^2}[1 + \varepsilon\mu(r)]\Psi(r,z) = j(r) \tag{2.64}$$

The FT of this equation is

$$\left[-k^2 + \frac{z^2}{c^2}\right]\Psi(k,z) + \frac{\varepsilon z^2}{c^2}\int \mu(k-k')\Psi(k',z)d^3k' = j(k) \tag{2.65}$$

2.3.8 Method of Diagrams for Green Function

Equations 2.62 and 2.64 are both of the type

$$(L_0 + \varepsilon L_1)\Psi = j \tag{2.66}$$

where L_0 is a nonrandom operator whose inverse $G^{(0)} = L_0^{-1}$, called the unperturbed propagator (or unperturbed Green's function), is known, and L_1 is a random operator. In the space (r) domain

$$L_0 = \Delta + \frac{z^2}{c^2}, \quad G^{(0)}(r,r';z) = \frac{\exp\{iz\,|r-r'|\}}{-4\pi\,|r-r'|}, \quad L_1 = \frac{z^2}{c^2}\mu(r) \tag{2.67}$$

acting as an integral convolution operator. In wave number (k) domain

$$L_0 = -k^2 + \frac{z^2}{c^2}, \quad G^{(0)}(k;z) = \frac{c^2}{z^2 - c^2 k^2}, \quad L_1 = \frac{z^2}{c^2}\mu(k-k') \tag{2.68}$$

acting as an integral convolution operator.

In r-domain L_1 is diagonal operator and L_0 is not; it is the converse in k-domain. The solution of Equation 2.66 is now formally expanded in powers of ε yielding

$$\Psi = (L_0 + \varepsilon L_1)^{-1}j = L_0^{-1}j - \varepsilon L_0^{-1}L_1 L_0^{-1}j + \varepsilon^2 L_0^{-1}L_1 L_0^{-1}L_1 L_0^{-1}j + \cdots \tag{2.69}$$

$(L_0 + \varepsilon L_1)^{-1} = G$ is called the perturbed propagator (or perturbed Green's function).

Let us represent the perturbation series for G with the aid of diagrams, which will be called bare diagrams to discriminate between them and other drossed diagrams to be introduced afterward. We make the following conventions:

1. The unperturbed propagator $G^{(0)}(r,r')$ is represented by a solid line $\overline{r \quad r'}$.
2. The random operator $-\varepsilon L_1$ is represented by a dot •.
3. Operators act to the right.

If so, we may write

$$G = \text{———} + \text{——•——} + \text{——•——•——} + \text{——•——•——•——} + \cdots \tag{2.70}$$

Let us write down explicity a few terms of the perturbation series in r-domain

$$G(r,r';z) = G^{(0)}(r,r';z) - \varepsilon \frac{z^2}{c^2} \int G^{(0)}(r,r_1;z)\mu(r_1)G^{(0)}(r_1,r';z)d^3r_1$$

$$+ \varepsilon^2 \frac{z^4}{c^4} \iint G^{(0)}(r,r_2;z)\mu(r_2)G^{(0)}(r_2,r_1;z)\mu(r_1)G^{(0)}(r_1,r';z)d^3r_1 d^3r_2 \qquad (2.71)$$

and in k-domain

$$G(k,k';z) = G^{(0)}(k;z)\delta(k-k') - \varepsilon \frac{z^2}{c^2}G^{(0)}(k;z)\mu(k-k')G^{(0)}(k';z)$$

$$+ \varepsilon^2 \frac{z^4}{c^4} \int G^{(0)}(k;z)\mu(k-k_1)G^{(0)}(k_1;z)\mu(k_1-k')G^{(0)}(k';z)d^3k_1 \qquad (2.72)$$

where $\delta(k - k')$ is Dirac's measure.

In order to help the interpretation of bare diagrams, it is sometimes useful to introduce subscripts under certain elements:

$$G(r, r'; z) = \underset{r \qquad r'}{\rule{1cm}{0.4pt}} + \underset{r \quad r_1 \quad r'}{\rule{1.5cm}{0.4pt}} + \underset{r \quad r_1 \quad r_2 \quad r'}{\rule{2cm}{0.4pt}} + \cdots$$

$$G(k, k'; z) = \underset{k \qquad k'}{\rule{1cm}{0.4pt}} + \underset{k \quad k_1 \quad k'}{\rule{1.5cm}{0.4pt}} + \underset{k \quad k_2 \quad k_1 \quad k'}{\rule{2cm}{0.4pt}} + \cdots \qquad (2.73)$$

If so, the dashed curve will connect the concrete points for which $\mu(r_1)$ and $\mu(r_2)$ (or $\mu(k - k_1)$ and $\mu(k_1 - k')$) are inside the integrals, that is,

$$\mu(r_1)\,\mu(r_2) \sim \underset{r_2 \qquad r_1}{\bullet\text{---}\bullet} \qquad (2.74)$$

or

$$\mu(k - k_1)\,\mu(k_1 - k') \sim \underset{k_1}{\bullet\text{---}\text{---}\bullet} \qquad (2.75)$$

We now give the physical interpretation of the perturbation expansion. The r-space diagrams correspond to multiple scattering of the wave at points r_1, r_2, $\ldots,r_N$. The k-space diagrams correspond to multiple interactions between Fourier components of the wave and of the random inhomogeneities; at each vortex of a diagram, a Fourier component k_p of the wave function interacts with a Fourier component $(k_{p+1} - k_p)$ of the random inhomogeneities, giving, as a result, a Fourier component $k_{p+1} \equiv k_{p+1} - k_p + k_p$ of the wave function. Both viewpoints are useful—the first one, particularly for single or double scattering, and the second one for multiple scattering, because of the wave vector conservation conditions.

In future description, we also need the expansion of the perturbed double propagator $G \otimes G^*$, that is, the tensor product of the perturbed propagator and its complex conjugate. In r-space

$$G \otimes G^* = G(r, r'; z) G^*(\eta, \eta'; z') \tag{2.76}$$

In k-space

$$G \otimes G^* = G(k, k'; z) G^*(k_1, k_1'; z') \tag{2.77}$$

This expansion can also be written in terms of diagrams:

$$G \otimes G^* = \quad + \quad + \quad + \quad + \cdots \tag{2.78}$$

If we make the convention that operators of the lower line are the complex conjugate of the usual ones, for example

$$\frac{\overset{k \bullet k'}{\underset{k_1 \bullet k_1'}{}}}{} = \varepsilon^2 \frac{z^2 z'^2}{c^4} G^{(0)}(k; z) \mu(k - k') G^{(0)}(k'; z) G^{(0)*}(k_1; z') \mu^*(k_1 - k_1') G^{(0)*}(k_1'; z')$$

$$\tag{2.79}$$

we can present the mean perturbed propagator following Reference 23.

2.3.9 An Exact Solution of 1D Equation

In this section, we study the one-dimensional (1D) equation:

$$\frac{\partial \Psi(x, t)}{\partial x} + \frac{1}{c}[1 + \varepsilon \mu(x)] \frac{\partial \Psi(x, t)}{\partial t} = 0 \tag{2.80}$$

where $\mu(x)$ is a real, centered, and stationary Gaussian random function with covariance function

$$\Gamma(x, x') = \langle \mu(x) \mu(x') \rangle \tag{2.81}$$

and the associated radiation problem

$$\frac{\partial \Psi(x)}{\partial x} - ik_0[1 + \varepsilon \mu(x)] \Psi(x) = \delta(x) \tag{2.82}$$

where $\delta(x)$ is Dirac's distribution at the origin. The wave number $k_0 = 2\pi/\lambda = 2\pi f/c$ is taken positive. Equations 2.80 and 2.82 can be treated simultaneously if we take the initial conditions

$$\Psi(x,0) = \frac{\delta(x)}{[1 + \varepsilon\mu(x)]} \qquad (2.83)$$

The LT of Equations 2.80 and 2.82 are then identical by introducing z and ck_0.

The underlying physical problem is the following: The monochromatic source of frequency $\omega = 2\pi f = ck_0$ is radiating into a semi-infinite 1D medium whose refractive index is $n(x) = 1 + \varepsilon\mu(x)$. Only propagation toward $(x > 0)$ is considered; and reflections are assumed to be negligible. Integration of Equation 2.82 gives

$$\Psi(x) = Y(x)\exp[ik_0x]\exp\left\{ik_0\int_0^x \varepsilon\mu(y)dy\right\} \qquad (2.84)$$

where $Y(x)$ is a Heaviside's step function [28].

To calculate now the mean value of Equation 2.84, the only random term is the second potential. For fixed x, $\int_0^x \varepsilon\mu(y)dy$ being a linear functional of the centered Gaussian random function $\mu(x)$, is a centered Gaussian random variable φ. If so, $\left\langle e^{ik_0\varphi}\right\rangle$ is the characteristic function of this random variable. As φ is Gaussian, that is,

$$\left\langle e^{ik_0\varphi}\right\rangle = e^{-\frac{1}{2}k_0^2 <\varphi^2>} \qquad (2.85)$$

we can now evaluate

$$\langle\phi^2\rangle = \left\langle \varepsilon^2\left|\int_0^x \mu(y)dy\right|^2\right\rangle = \varepsilon^2\int_0^x dy\int_0^y \Gamma(y - y')dy' \qquad (2.86)$$

and finally obtain

$$\langle\Psi(x)\rangle = Y(x)\exp[ik_0x]\exp\left\{-\frac{1}{2}k_0^2\varepsilon^2\int_0^x dy\int_0^y \Gamma(y - y')dy'\right\} \qquad (2.87)$$

The mean wave function is thus expressed in terms of the covariance function of the index of refraction. Higher-order moments such as $\langle \Psi(x)\Psi(x') \rangle$ are easily obtained, using the characteristic function of a multivariate Gaussian distribution [28]. We now introduce the covariance function

$$\Gamma(x - x') = \exp\left\{-\left|\frac{x - x'}{\ell}\right|\right\} \tag{2.88}$$

where ℓ is the range of index correlation. The mean wave function can now be calculated as

$$\langle \Psi(x) \rangle = Y(x)\exp[ik_0 x]\exp\left\{-\varepsilon^2 k_0^2 \ell^2 \left(\frac{x}{\ell} + e^{-\frac{x}{\ell}} - 1\right)\right\} \tag{2.89}$$

The dimensionless parameter which determines the behavior of the solution is $\varepsilon k_0 \ell$. There are two limiting approximations:

1. $\varepsilon k_0 \ell \ll 1$. It is a long wave length approximation ($\lambda \gg \ell$), and corresponds to weak interactions in quantum field theory. A uniform approximation for $\langle \Psi(x) \rangle$ is then

$$\langle \Psi(x) \rangle = Y(x)\exp[ik_0 x]\exp\left\{-\varepsilon^2 k_0^2 \ell x\right\} \tag{2.90}$$

As follows from (2.90), the initial excitation is damped with an extinction length

$$x_{ex} = \left(\varepsilon^2 k_0^2 \ell\right)^{-1} \tag{2.91}$$

Let us compare x_{ex} and the wavelength $\lambda \sim (k_0)^{-1}$

$$\frac{x_{ex}}{\lambda} \sim \frac{1}{\varepsilon^2 k_0 \ell} = \frac{1}{\varepsilon} \cdot \frac{1}{\varepsilon k_0 \ell} \gg 1 \tag{2.92}$$

The decaying is thus very slow; it is due to phase mixing and is not related to any dissipative mechanism. The mean wave function $\langle \Psi(x) \rangle$ can also be written as

$$\langle \Psi(x) \rangle = Y(x)\exp\left\{i\left(k_0 - i\varepsilon^2 k_0^2 \ell\right)x\right\} \tag{2.93}$$

The effect of randomness on the mean wave function, as follows from Equation 2.93, is simply a renormalization of the wave number. The renormalized wave number is now equal to $k = k_0 - i\varepsilon^2 k_0^2 \ell$, which has a small imaginary part (because $\varepsilon k_0 \ell \ll 1$). Below, we shall obtain this wave approximation as a sum of an infinite series extracted from the perturbation expansion of the mean propagator.

2. $\varepsilon k_0 \ell \gg 1$. It is a short wavelength approximation ($\lambda \ll \ell$) corresponding to strong interactions in quantum field theory. A uniform approximation for $\langle \Psi(x) \rangle$ is then

$$\langle \Psi(x) \rangle = Y(x) \exp[ik_0 x] \exp\left\{ -\tfrac{1}{2}\varepsilon^2 k_0^2 x^2 \right\} \tag{2.94}$$

The initial excitation is damped again, with an extinction length $x_{ex} = (\varepsilon k_0)^{-1} \sim \lambda/\varepsilon$; the damping decay is more rapid than in the proceeding case. This approximation is equivalent to a renormalization of the wave number, because x^2 appears in the second exponent in Equation 2.94.

Now we will present the FT of the exact mean function (2.94)

$$\langle \Psi(k) \rangle = \int\limits_{-\infty}^{\infty} \exp(-ik_0 x) \langle \Psi(x) \rangle \, dx$$

$$= \int\limits_{0}^{\infty} \exp[i(k_0 - k)] \exp\left(\varepsilon^2 k_0^2 \ell^2 x\right) \exp\left(-\varepsilon^2 k^2 \ell x\right) \exp\left\{ -\varepsilon^2 k_0^2 \ell^2 e^{-\frac{x}{\ell}} \right\} dx \tag{2.95}$$

Expanding the last exponential term in a uniformly convergent series and integrating Equation 2.95, yields

$$\langle \Psi(x) \rangle = \exp\left(\varepsilon^2 k_0^2 \ell^2\right) \sum\limits_{n=0}^{\infty} \frac{\left(-\varepsilon^2 k_0^2 \ell^2\right)^n}{n!} \frac{1}{ik - ik_0 + \varepsilon^2 k_0^2 \ell + n/\ell} \tag{2.96}$$

From Equation 2.96, it can be seen that $\langle \Psi(k) \rangle$ has the poles $k_n = k_0 + i\varepsilon^2 k_0^2 \ell + in/\ell$, which correspond to more and more damped partial waves in r-space. If $\varepsilon k_0 \ell \ll 1$, we can approximate $\langle \Psi(k) \rangle$ by the first partial wave ($n = 0$) that gives again Equation 2.93, apart from a factor $\exp\left(\varepsilon^2 k_0^2 \ell^2\right) \neq 1$.

2.3.10 Random Expansion at Short Wavelengths

Since optical waves are waves with short wavelengths even comparing with the inner scale of atmospheric turbulences (see Chapter 1), we will analyze the limiting case $\varepsilon k_0 \ell \gg 1$, presented above in the previous section. In this case, as was shown in Reference 45, the random refractive index behaves as a mere random value and not as a random function. This is easy to understand. Consequently, when ℓ is larger than the optical wave length, that is, $\ell \gg \lambda$, each realization (or sample) of the random index is a very slowly varying function, which can be approximated by a constant. At an intermediate level, between the general random function and random variable, we could try to approximate a random function by a linear function or a quadratic function with random variables as coefficients. We will use the Taylor's model, which is the short-wavelength approximation. The reader can find all other approximations in Reference 45. Thus, constructing a limited random Taylor expansion of the random function, we obtain a random equation in the following form:

$$\frac{\partial \Psi(x)}{\partial x} - ik_0[1 + \varepsilon \mu(x)]\Psi(x) = \delta(x) \tag{2.97}$$

where $\mu(x)$ is a random function, $\delta(x)$ is Dirac's distribution at the origin, and the wave number $k_0 = 2\pi/\lambda = 2\pi f/c$ is taken, as above, to be positive. Here λ is the wavelength, f is the radiated frequency and c is the velocity of light. We want to approximate the random function $\mu(x)$ by its random Taylor expansion introduced in Reference 28 in such a manner:

$$\mu(x) = \mu(0) + x\mu'(0) + \frac{x^2}{2}\mu''(0) + \cdots\cdots \tag{2.98}$$

where $\mu(0)$, $\mu'(0)$, $\mu''(0)$,... are not independent random variables. We cannot keep the covariance function $\exp\{-|x|/\ell\}$ because the corresponding random function is not mean square differentiable (see Reference 28). As we do not need to specify the covariance Γ, we shall only assume that it has derivatives of all orders at $x = 0$ and that $\Gamma(0) = 1$. This assumption leads to an approximation for $\Psi(x)$, as

$$\frac{\partial \Psi(x)}{\partial x} - ik_0\left[1 + \varepsilon \mu(0) + \varepsilon x\mu'(0) + \varepsilon \frac{x^2}{2}\mu''(0)\right]\Psi(x) = \delta(x) \tag{2.99}$$

It is solved for the mean wave function

$$\langle \Psi(x) \rangle = Y(x)e^{ik_0x}\left\langle \exp\left[ik_0\varepsilon x\left(\mu(0) + \varepsilon x\mu'(0) + \varepsilon \frac{x^2}{2}\mu''(0)\right)\right]\right\rangle \tag{2.100}$$

As $\mu(x)$ is a Gaussian random function, the multivariate distribution of $\mu(0)$, $\mu'(0)$, and $\mu''(0)$ is also Gaussian; it is thus determined by its second-order moment such as $\langle(\mu(0))^2\rangle$, $\langle(\mu'(0))^2\rangle$, $\langle\mu(0)\mu'(0)\rangle$, and so forth. They are easily calculated in terms of covariance function, for example,

$$\langle\mu(0)\mu'(0)\rangle = \lim_{h\to 0}\frac{\langle\mu(0)\mu(h)\rangle - \langle\mu(0)\mu(0)\rangle}{h} = \Gamma'(0) \equiv 0 \qquad (2.101)$$

because Γ is an even function.

The mean value in Equation 2.100 is easily related to the characteristic function of $\mu(0)$, $\mu'(0)$, $\mu''(0)$, and can be calculated in terms of Γ; this gives

$$\langle\Psi(x)\rangle = Y(x)e^{ik_0 x}\exp\left[-\frac{k_0^2\varepsilon^2 x^2}{2}\left(1 + \frac{x^2}{12}\Gamma''(0) + O(x^3)\right)\right] \qquad (2.102)$$

Let us compare this to the exact solution [45]

$$\langle\Psi(x)\rangle = Y(x)e^{ik_0 x}\exp\left[-\frac{k_0^2\varepsilon^2}{2}\int_0^x\int_0^y\Gamma(y-y')dydy'\right] \qquad (2.103)$$

Expanding the covariance function in power of x and integrating we get exactly the same result as in Equation 2.102. If the condition

$$\left|\frac{k_0^2\varepsilon^2}{\Gamma''(0)}\right| \gg 1$$

is satisfied, we can use the so-called random variable approximation [45]

$$\langle\Psi(x)\rangle = Y(x)e^{ik_0 x}\exp\left[-\frac{k_0^2\varepsilon^2 x^2}{2}\right] \qquad (2.104)$$

An equivalent condition is that the damping length $x_d = 1/k_0\varepsilon = \lambda/\varepsilon$ corresponding to this approximation should be much shorter than the range of random correlations $\ell = |\Gamma''(0)|^{-1/2}$. If it is satisfied, the wave cannot escape the region where the random index is properly approximated by a random variable. The "random variable" approximation is easily applied to any propagation equation because we only need to solve a partial differential equation with constant coefficients, and average afterwards. If we want a higher-order approximation, we must solve a partial differential equation with linear or quadratic coefficients. The case of linear coefficients can, in principle, be solved by means of a generalized Laplace transformation, but this is rather complicated.

Now we shall present the short-wave approximation for the scalar wave equation with point source

$$\Delta\Psi(r) + k_0^2[1 + \varepsilon\mu]\Psi(r) = \delta(r) \tag{2.105}$$

We assume μ to be a centered Gaussian random variable and $\langle\mu^2\rangle = 1$. Solving Equation 2.105, we get

$$\Psi(r) = \frac{\exp\{ik_0(1 + \varepsilon\mu)R\}}{-4\pi R}, \quad R = |r| \tag{2.106}$$

Taking the mean value of this wave function we find

$$\langle\Psi(r)\rangle = \frac{\exp(ik_0 R)\exp\left(-\frac{1}{2}\varepsilon^2 k_0^2 R^2\right)}{-4\pi R} \tag{2.107}$$

The damping due to phase mixing is thus exponential with a damping length

$$R_d = (\varepsilon | k_0 |)^{-1} > \lambda \tag{2.108}$$

The condition $R_d \ll \ell$ can be written as

$$\varepsilon | k_0 | \ell \gg 1 \tag{2.109}$$

It is thus a short wavelength condition ($\ell \gg \lambda$). The result of Equation 2.105 disagrees with a result derived by Tatarskii [32, 39] for $|k_0|\ell \gg 1$. His mean wave function

$$\langle\Psi(r)\rangle = \frac{\exp(ik_0 R)}{-4\pi R} \cdot \frac{1}{\left(1 + \varepsilon^2 k_0^2 R\ell\right)^{1/2}} \tag{2.110}$$

has not an exponential decrease, and a damping length

$$R_d = \left(\varepsilon^2 k_0^2 \ell\right)^{-1} \ll \lambda \tag{2.111}$$

His result is expressed as a certain integral over the solution of the Bourret equation (see Reference 29), and this integral is calculated by the method of stationary phase. The expansion of the solution of the Bourret equation used by Tatarskii is only valid for $|k_0|\ell \ll 1$. If the proper expansion is used, the result becomes identical with the one mentioned previously. This can also be checked on the 1D model for which the calculation is easier.

2.3.11 Exact Solution of the Scalar Wave Equation

Let us show how by following Reference 31 one can obtain the solution of the reduced scalar wave equation with random refractive index:

$$\Delta\Psi(r) + k_0^2[1 + \mu(r)]\Psi(r) = \delta(r) \tag{2.112}$$

where, once more, $\delta(r)$ is Dirac's distribution at the origin. Let us assume that the wave number k_0 has a small positive imaginary part and $\mu(r)$ is a centered Gaussian random function with covariance $\Gamma(r,r')$ that need not be stationary. In order to relate this equation to the heat equation, we introduce a new unknown function $\tilde{\Psi}(r,\theta)$ such that

$$\Psi(r) = -\frac{i}{k_0}\int_0^\infty \exp(ik_0\theta)\,\tilde{\Psi}(r,\theta)\,d\theta \tag{2.113}$$

Because of the positive imaginary part of k_0, this integral is convergent if $\tilde{\Psi}(r,\theta)$ is not increasing very fast at infinity. Equation 2.112 is now multiplied by k_0^2 and integrated by parts

$$k_0^2\Psi(r) = -\int_0^\infty \tilde{\Psi}(r,\theta)\frac{\partial}{\partial\theta}\exp(ik_0\theta)\,d\theta = \tilde{\Psi}(r,0) + \int_0^\infty \frac{\partial\tilde{\Psi}(r,\theta)}{\partial\theta}\exp(ik_0\theta)\,d\theta \tag{2.114}$$

Using Equations 2.113 and 2.114 we can write

$$\Delta\Psi(r) + k_0^2\Psi(r) + k_0^2\mu(r)\Psi(r) \equiv -\frac{i}{k_0}\int_0^\infty \exp(ik_0\theta)\left[\Delta\tilde{\Psi} + k_0^2\mu(r)\tilde{\Psi}\right]d\theta + \tilde{\Psi}(r,0)$$

$$+ \int_0^\infty \frac{\partial\tilde{\Psi}(r,\theta)}{\partial\theta}\exp(ik_0\theta)\,d\theta = \delta(r) \tag{2.115}$$

This is satisfied if we take

$$\frac{\partial \tilde{\Psi}(r,\theta)}{\partial \theta} = \frac{i}{k_0} \Delta \tilde{\Psi}(r,\theta) + ik_0 \mu(r)\tilde{\Psi}(r,\theta) \tag{2.116}$$
$$\tilde{\Psi}(r,0) = \delta(r)$$

Let us compare Equation 2.116 to the perturbed heat equation

$$\frac{\partial \tilde{\Psi}(r,\theta)}{\partial \theta} = \Delta \tilde{\Psi}(r,\theta) + V(r)\tilde{\Psi}(r,\theta) \tag{2.117}$$

and the Schrodinger equation

$$\frac{\partial \tilde{\Psi}(r,\theta)}{\partial \theta} = \alpha \Delta \tilde{\Psi}(r,\theta) + V(r)\tilde{\Psi}(r,\theta) \tag{2.118}$$

Equation 2.117 can be solved by functional integration for all functions $V(r)$ continuous and bounded from above, using the Wiener measure of the Brownian motion process. For Equation 2.118, there is no Wiener measure. It is well known that for the Schrodinger equation this solution through functional integration, given first by Feynman [21,23], is only a formal extension of the heat equation case. Fortunately, it can be shown that all equations such as Equation 2.118, where α a positive real part has can be rigorously solved with a complex Wiener measure.

This is the case here because $\mathrm{Re}(i/k_0) > 0$. Thus, the solution of Equation 2.118 is

$$\tilde{\Psi}(r,\theta) = \int_{\Omega} \exp\left[ik_0 \int_0^{\theta} \mu(\rho(\tau))\mathrm{d}\tau \right] \mathrm{d}W\left(\theta, r, \frac{i}{k_0}\right) \tag{2.119}$$

where Ω is the space of continuous function $\rho(\tau)$ such that $\rho(0) = 0$ and $\rho(\theta) = r$, and $\mathrm{d}W(\theta,r,i/k_0)$ is the complex Wiener measure corresponding to the complex heat equation

$$\frac{\partial \tilde{\Psi}(r,\theta)}{\partial \theta} = \frac{i}{k_0} \Delta \tilde{\Psi}(r,\theta) \tag{2.120}$$

Solution Equation 2.118 can also be written more explicitly as the limit of ordinary multiple integrals

$$
\tilde{\Psi}(r,\theta) = \lim_{n \to \infty} \left(\frac{4\pi i \Delta\tau}{k_0} \right)^{-\frac{3}{2}n} \int \cdots \int \exp\left\{ \frac{ik_0}{4\Delta\tau} \left[r_1^2 + (r_2 - r_1)^2 + (r - r_{n-1})^2 \right] \right\}
$$
$$
\times \exp\{ik_0\Delta\tau[\mu(r_1) + \mu(r_2) + \cdots + \mu(r_{n-1})]\} d^3 r_1 d^3 r_2 \cdots d^3 r_{n-1} \qquad (2.121)
$$

where $\Delta\tau = \theta/n$, $(4\pi i \Delta\tau/k_0)^{-(3/2)n}$ is the $3n$th power of the square root of $(4\pi i \Delta\tau/k_0)$, which has a positive real part. A formal proof of expression (2.121) was presented in Reference 45.

Now we will calculate the mean value of $\tilde{\Psi}(r,\theta)$ using Equation 2.119 by interchanging the functional integration. Finally, we get

$$
\langle \tilde{\Psi}(r,\theta) \rangle = \int_\Omega \left\langle \exp\left[ik_0 \int_0^\theta \mu(\rho(\tau)) d\tau \right] \right\rangle dW\left(\theta, r, \frac{i}{k_0} \right) \qquad (2.122)
$$

We shall now make use of the fact that $\mu(r)$ is a centered Gaussian random function. However, the Gaussian assumption can be dropped, because we actually only need to know the characteristic function of $\mu(r)$:

$$
F(\phi(r)) = \left\langle \exp i \int \phi(r)\mu(r) d^3 r \right\rangle \qquad (2.123)
$$

The following calculations are almost the same as those for the 1D model. For a fixed curve $\rho(\tau)$, a linear functional of $\mu(r)$

$$
\phi = \int_0^\theta \mu(\rho(\tau)) d\tau \qquad (2.124)
$$

is a centered Gaussian random value and $\langle \exp[ik_0\varphi] \rangle$ is its characteristic function

$$
\langle \exp[ik_0\phi] \rangle = \exp\left\{ -\frac{1}{2} k_0^2 \langle \phi^2 \rangle \right\} \qquad (2.125)
$$

where

$$\left\langle \phi^2 \right\rangle = \int\limits_0^\theta \int\limits_0^\tau \left\langle \mu(\rho(\tau))\,\mu(\rho(\tau')) \right\rangle \mathrm{d}\tau\,\mathrm{d}\tau' = \int\limits_0^\theta \int\limits_0^\tau \Gamma(\rho(\tau),\rho(\tau'))\,\mathrm{d}\tau\,\mathrm{d}\tau' \quad (2.126)$$

Turning back to the initial equation (2.122), we get

$$\left\langle \tilde{\Psi}(r) \right\rangle = -\frac{i}{k_0} \int\limits_0^\infty \mathrm{d}\theta \exp(ik_0\theta) \int\limits_\Omega \left\langle \exp\left[-\frac{1}{2}k_0^2 \int\limits_0^\theta \int\limits_0^\tau \Gamma(\rho(\tau),\rho(\tau'))\mathrm{d}\tau\,\mathrm{d}\tau' \right] \right\rangle$$

$$\times \mathrm{d}W\left(\theta, r, \frac{i}{k_0} \right) \quad (2.127)$$

This integral solves the problem. This functional integral can also be approximated for numerical purposes, for example, by multiple integrals:

$$\left\langle \tilde{\Psi}(r) \right\rangle_n = -\frac{i}{k_0} \int\limits_0^\infty \mathrm{d}\theta \exp(ik_0\theta) \left(\frac{4\pi i \Delta\tau}{k_0} \right)^{-\frac{3}{2}n} \int \cdots \int \exp\left\{ \frac{ik_0}{4\Delta\tau}\left[r_1^2 + (r_2 - r_1)^2 + (r - r_{n-1})^2 \right] \right\}$$

$$\times \exp\left\{ -\frac{1}{2}k_0^2(\Delta\tau)^2 \sum_{i,j=1}^{n-1} \Gamma_{i,j} \right\} \mathrm{d}^3 r_1 \mathrm{d}^3 r_2 \cdots \mathrm{d}^3 r_{n-1} \quad (2.128)$$

where $\Gamma_{i,j} = \Gamma(r_i, r_j)$ and $\Delta\tau = \theta/n$. The extension to higher order moments is straightforward, using characteristic functions of multivariate Gaussian distributions.

For a short-wave approximation, which is actual for optical wave propagation in the atmospheric links, we can evaluate integral (2.128). In this case, if the range of the covariance function is much longer than the wavelength, we can use a functional saddle point method to approximate the function

$$\exp\left[-\frac{1}{2}k_0^2 \int\limits_0^\theta \int\limits_0^\tau \Gamma(\rho(\tau),\rho(\tau'))\,\mathrm{d}\tau\,\mathrm{d}\tau' \right] \quad (2.129)$$

by a quadratic function of $\rho(\tau) - \rho_0(\tau)$, where $\rho_0(\tau)$ is the function that makes the exponent stationary. It is then possible to calculate exactly this approximate functional integral.

2.4 The Electromagnetic Wave Equation

As above, we consider the full electromagnetic wave equation with a random refractive index

$$\Delta \mathbf{E}(\mathbf{r}) - \nabla(\nabla \cdot \mathbf{E}(\mathbf{r})) + k_0^2[1 + \varepsilon \mu(\mathbf{r})]\mathbf{E}(\mathbf{r}) = \mathbf{j}(\mathbf{r}) \tag{2.130}$$

where $\mathbf{j}(\mathbf{r})$ is related to the actual current density $\mathbf{j}^*(\mathbf{r})$ by $\mathbf{j}(\mathbf{r}) = -i\omega\mu_0\mathbf{j}^*(\mathbf{r})$; ω is the angular frequency $\omega = 2\pi f$ and $\mu_0 = 4\pi \times 10^{-7}$ (Henry per meter) is the permeability of free space. This equation is not equivalent to the reduced scalar wave equation because of the term $\nabla(\nabla \cdot \mathbf{E}(\mathbf{r}))$, which is important when the refractive index changes much over a wavelength. We shall therefore only consider the case of long wavelengths such that $|k_0|\ell \ll 1$, and use the Bourret approximation [29]. This problem has already been treated by Tatarskii [32,39] but the results presented here do not agree. Taking the FT of Equation 2.130 we get

$$\left[(k_0^2 - k^2)\delta_{ij} + k_i k_j\right]\mathbf{E}_j(\mathbf{k}) + \varepsilon k_0^2 \int \mu(\mathbf{k} - \mathbf{k}')\mathbf{E}_i(\mathbf{k}'d\mathbf{k} = \mathbf{j}_i(\mathbf{k}) \tag{2.131}$$

The unperturbed propagator $G_{ij}^{(0)}(k)$ satisfies the following equation:

$$\left[(k_0^2 - k^2)\delta_{ij} + k_i k_j\right]G_{jl}^{(0)}(k) = \delta_{il} \tag{2.132}$$

This equation is easily solved:

$$G_{jl}^{(0)}(k) = \frac{1}{\left[k_0^2 - k^2\right]}\left(\delta_{ij} - \frac{k_i k_j}{k_0^2}\right) \tag{2.133}$$

The Bourret equation for the mean perturbed propagator $\langle G_{jl}(k)\rangle$ is

$$\langle G \rangle = \underline{\quad} + \overset{\frown}{\bullet\!\!-\!\!-\!\!-\!\!\bullet} \langle G \rangle \tag{2.134}$$

or

$$\langle G(k)\rangle = G^{(0)}(k) + G^{(0)}(k)\varepsilon^2 k_0^4\left[\int \Gamma(k - k')G^{(0)}(k')d^3k'\right]\langle G(k)\rangle \tag{2.135}$$

where $\Gamma(k)$ is the FT of the covariance function. After a few transformations, Equation 2.132 becomes [29,31]

$$\left[(k_0^2 - k^2)\delta_{ij} + k_i k_j - \varepsilon^2 k_0^4 \int \frac{\Gamma(k - k')}{k_0^2 - k'^2}\left(\delta_{ij} - \frac{k_i' k_j'}{k_0^2}\right) d^3 k'\right]\langle G_{jl}(k)\rangle = \delta_{il} \quad (2.136)$$

Let us now denote the tensor $T_{ij}(k)$ as

$$T_{ij}(k) = \int \frac{\Gamma(k - k')}{k_0^2 - k'^2}\left(\delta_{ij} - \frac{k_i' k_j'}{k_0^2}\right) d^3 k'$$

and assume that the covariance function is isotropic. Then the tensor $T_{ij}(k)$ is the convolution product of an isotropic tensor and an isotropic function; it is thus an isotropic tensor and can be written as

$$T_{ij}(k) = \chi(k)\delta_{ij} + \mu(k)\frac{k_i k_j}{k_0^2} \quad (2.137)$$

The Bourret equation for the mean propagator now becomes

$$\left[\left(k_0^2 - k^2 - \varepsilon^2 k_0^4 \chi(k)\right)\delta_{ij} + \left(1 - \varepsilon^2 k_0^2 \mu(k) k_i k_j\right)\right]\langle G_{jl}(k)\rangle = \delta_{il} \quad (2.138)$$

Let us now find the free oscillations which satisfy

$$\left[\left(k_0^2 - k^2 - \varepsilon^2 k_0^4 \chi(k)\right)\delta_{ij} + \left(1 - \varepsilon^2 k_0^2 \mu(k) k_i k_j\right)\right]\langle E_j(k)\rangle = 0 \quad (2.139)$$

There are two kinds of oscillations:

1. Transverse oscillations. Here $<\varepsilon>$ and **k** are perpendicular. The dispersion equation is

$$k_0^2 - k^2 - \varepsilon^2 k_0^4 \chi(k) = 0 \quad (2.140)$$

2. Longitudinal oscillations. Here $<\varepsilon>$ and $\mathbf{k}$ are parallel. The dispersion equation is

$$1 - \varepsilon^2 \left(k_0^2 \chi(k) + k^2 \mu(k) \right) = 0 \qquad (2.141)$$

Let us also find the renormalized wave number $K_\perp$ for transverse waves. We take for this purpose the correlation function $\exp(-R/\ell)$. After some straightforward manipulations, we find that for $k\ell \ll 1 (\ell \ll \lambda)$

$$\chi(K_\perp) = -\frac{2}{3}\ell^2(1 + 2iK_\perp\ell) + \frac{1}{3k_0^2} + O(\ell^4 K_\perp^2) \qquad (2.142)$$

The dispersion equation for transverse oscillations is solved for the renormalized wave number

$$K_\perp = \left[k_0^2 \left(1 - \varepsilon^2 k_0^2 \chi(k) \right) \right]^{1/2} \approx k_0 \left[1 - \frac{1}{6}\varepsilon^2 + \frac{1}{3}\varepsilon^2 k_0^2 \ell^2 (1 + 2ik_0\ell) \right] \qquad (2.143)$$

We compare this result to the corresponding formula for the scalar wave equation obtained in References 32, 39, 44 or deduced from Keller's result [30] with a covariance function of $\exp(-R/\ell)$:

$$K_\perp = k_0 \left[1 + \frac{1}{2}\varepsilon^2 k_0^2 \ell^2 (1 + 2ik_0\ell) \right] \qquad (2.144)$$

First of all, the imaginary part of for the electromagnetic wave equation has been reduced by factor $\sim(1/3)$ with respect to that in Equation 2.143, the damping length of the mean wave have thus increased by 50%. Second, due to the additional negative term $(\sim(1/6)k_0\varepsilon^2)$, the real part of $K_\perp$ is less than the real part of k_0 if $2k_0^2\ell^2 < 1$. As we assumed that $k\ell \ll 1 (\ell \ll \lambda)$, this is satisfied.

We conclude that the effective phase velocity of transverse waves increases at long wave lengths, instead of decreasing as is the case for the scalar wave equation. This needs some explanation. There are two wave modes actually in this medium: the transverse mode, whose phase velocity is approximately ω/k_0, and the longitudinal wave mode, whose phase velocity is much longer (infinite in the nonrandom case). Due to the term $k_i k_j E_j$ of Equation 2.131, the wave modes are coupled and part of the mean transverse wave has travels part of its way as a longitudinal wave. The traveling time thus decreased, and the phase velocity is increased. Without this coupling, it would be impossible to explain the increase of the phase velocity. As the additional term $(1/6)k_0\varepsilon^2$ does not depend on ℓ, it is possible that it corresponds rather to a diffraction effect by the scattering blobs (whose sizes are small compared to the wavelength), than to a volume scattering effect.

2.5 Propagation in Statistically Irregular Media

Below, we will assume that the mean refractive index is constant through space, but that its random part is not strictly stationary with respect to space translations. The correlation functions $\Gamma(x,x')$ are functions of $(x - x')$ but also of $(x + x')/2$. We shall assume that this additional space dependence has a scale of variations h, which is large compared to the wavelength. As there is no homogeneous turbulence in nature, this is a very common situation.

The FT $\mu(k)$ of such a slowly varying random function does not satisfy the wave vector conservation condition

$$\langle \mu(k_1)\mu(k_2)\cdots\mu(k_p)\rangle = 0, \quad \text{if} \quad k_1 + k_2 + \cdots + k_p \neq 0 \qquad (2.145)$$

and does not give rise to any singular terms in the perturbation series. All arguments based upon the extraction of the leading singular terms seem to disappear suddenly. We show, however, that if the condition $\varepsilon^2 K^4 l^3 h \gg 1$ is satisfied, in addition to the usual condition $K\ell \ll 1$, nothing is changed, because we have pseudo-singular terms. We assume that the additional space variation of the correlation functions $X(r_1, r_2, \ldots, r_n)$ is given by a factor $\exp[(is(r_1 + r_2 + \cdots + r_n/r_1)]$, where $\mathbf{s}$ is given vector. This is not of course the most general case, but it will be sufficient for our purpose. The scale of variation of this additional factor is $h = 1/|\mathbf{s}|$.

The Fourier transform of

$$\exp\left[\frac{is(r_1 + r_2 + \cdots + r_n)}{n}\right] \quad X(r_1, r_2, \ldots, r_n)$$

is

$$X(k_1 + \tfrac{s}{n}, k_2 + \tfrac{s}{n}, \ldots, r_n + \tfrac{s}{n}).$$

The wave vector conservation condition becomes thus

$$k_1 + k_2 + \cdots + k_n + s = 0 \qquad (2.146)$$

If we apply this to a connected diagram in k-space, such as we find that

$$k - k' = s \tag{2.147}$$

Because of the condition $h \gg \lambda$, which can also be written $|s| < K$, the wave vectors at the terminals of a connected diagram are almost equal. Instead of a squared unperturbed propagator, the terminals of ‿‿‿‿‿ introduce a factor

$$\frac{c^2}{z^2 - c^2 K^2} \frac{c^2}{z^2 - c^2 K'^2} = \frac{c}{2(K^2 - K'^2)}$$

$$\times \left[\frac{1}{K} \left(\frac{1}{z - cK} - \frac{1}{z + cK} \right) - \frac{1}{K'} \left(\frac{1}{z - cK'} - \frac{1}{z + cK'} \right) \right] \tag{2.148}$$

Let us find the corresponding contribution to the inverse LT. It is proportional to

$$\frac{1}{(K^2 - K'^2)} \left[\frac{\left(e^{-icKt} - e^{ic'Kt} \right)}{K} - \frac{\left(e^{-icK't} - e^{icK't} \right)}{K'} \right] \tag{2.149}$$

then, using the fact that $K - K'$ is small compared to K, we approximate Equation 2.149 by

$$\frac{1}{2K^2} \left[\frac{e^{-icKt} \left(1 - e^{ic(K-K')t} \right)}{K - K'} - \frac{e^{-icKt} \left(1 - e^{-ic(K-K')t} \right)}{K - K'} \right] \tag{2.150}$$

For $c(K - K') \ll 1$, we can make a Taylor expansion of Equation 2.150 and find

$$-\frac{ic}{2K^2} \left(te^{-icKt} + te^{icKt} \right) \tag{2.151}$$

This expression is not really singular, because it is only valid for $c|K - K'|t \ll 1$; this condition can also be written as $t \ll h/c$. If the damping time t_d, corresponding to the Bourret approximation [29] in the stationary case is small compared to h/c, then because the expression behaves exactly as a singular term; we call it a pseudo-secular term. As $t_d \sim (1/\varepsilon^2 cK^4 \ell^3)$, the condition $t_d \ll h/c$ can be written as

$$\varepsilon^2 K^4 \ell^3 h \gg 1 \tag{2.152}$$

because of $K\ell \ll 1$ and $\varepsilon^2 < 1$, h must be very large compared to the wavelength.

Conditions of optical wave propagation in the irregular turbulent atmosphere more detailed will be discussed in Chapters 3 and 4, and laser beams in Chapter 5, based on the stochastic approach and analyzing the phenomena based on the statistical functions and the corresponding statistical equations, instead of deterministic equations corrected only for description of optical wave propagation in free space.

References

1. Jenkis, F. A., and White, H. E., *Fundamentals of Optics*. New York: McGraw-Hill, 1953.
2. Born, M., and Wolf, E., *Principles in Optics*. New York: Pergamon Press, 1964.
3. Fain, V. N., and Hanin, Ya. N., *Quanty Radiophysics*, Moscow: Sov. Radio, 1965 (in Russian).
4. Akhamov, S. A., and Khohlov, R. V., *Problems of Nonlinear Optics*. Moscow: Fizmatgiz, 1965 (in Russian).
5. Lipson, S. G., and Lipson, H., *Optical Physics*. Cambridge: University Press, 1969.
6. Akhamov, S. A., Khohlov, R. V., and Sukhorukov, A. P., *Laser Handbook*, North Holland: Elsevier, 1972.
7. Marcuse, O., *Light Transmission Optics*, New York: Van Nostrand-Reinhold Publisher, 1972.
8. Kapany, N. S., and Burke, J. J., *Optical Waveguides*, Chapter 3. New York: Academic Press, 1972.
9. Fowles, G. R., *Introduction in Modern Optics*. New York: Holt, Rinehart, and Winston Publishers, 1975.
10. Grant, I. S., and Phillips, W. R., *Electromagnetism*, New York: John Wiley & Sons, 1975.
11. Plonus, M. A., *Applied Electromagnetics*. New York: McGraw-Hill, 1978.
12. Kong, J. A., *Electromagnetic Wave Theory*. New York: John Wiley & Sons, 1986.
13. Elliott, R. S., *Electromagnetics*: History, Theory, and Applications. New York: IEEE Press, 1993.
14. Kopeika, N. S., *A System Engineering Approach to Imaging*, Washington: SPIE Optical Engineering Press, 1998.
15. *Optical Fiber Sensors: Principles and Components, Handbook*, Vol. 1, J. Dakin and B. Culshaw Eds., Boston: Artech House, 1988.
16. Palais, J. C., Optical communications, in *Handbook: Engineering Electromagnetics Applications*, R. Bansal Ed. New York: Taylor & Francis, 2006.
17. Foldy, L. L., The multiple scattering of waves. I. General theory of isotropic scattering by randomly distributed scatterers, *Phys. Rev.*, vol. 67, pp. 107–119, 1945.
18. Dyson, F., The radiation theories of Tomonaga, Schwinger, and Feynman, *Phys. Rev.*, vol. 75, pp. 486–497, 1949.
19. Lax, M., Multiple scattering of waves, *Rev. Modern Phys.*, vol. 23, pp. 287–310, 1951.
20. Lax, M., Multiple scattering of waves. II. The effective field in dense systems, *Phys. Rev.*, vol. 85, pp. 621–629, 1952.
21. Furutsu, K., On the group velocity, wave path and their relations to the Poynting vector of E-M field in an absorbing medium, *J. Phys. Soc. Japan*, vol. 7, pp. 458–478, 1952.

22. Salpeter, E. E., and Bethe, H. A., A relative equation for bound-state problems, *Phys. Rev.*, vol. 84, pp. 1232–1239, 1951.

23. Furutsu, K., On the statistical theory of electromagnetic waves in a fluctuating medium (I), *J. Res. NBS (Radio Prop.)*, vol. 67D, pp. 303–323, 1963.

24. Matsubara, T., A new approach to quantum-statistical mechanics, *Prog. Theoret. Phys.*, vol. 14, pp. 351–361, 1955.

25. Martin, P. C., and Schwinger, J., Theory of many-particle systems (I), *Phys. Rev.*, vol. 115, pp. 1342–1349, 1959.

26. Schwinger, J., On the Green's functions of quantized fields, I, II, *Proc. Natl. Acad. Sci.*, vol. 37, pp. 452–455, 1951.

27. Twersky, V., Multiple scattering of electromagnetic waves by arbitrary configurations, *J. Math. Phys.*, vol. 8, pp. 569–610, 1967.

28. Buslaev, V. S., in *Topics in Mathematical Physics*, vol. 2, M. Sh. Birman Ed. New York: Consultants Bureau, 1968.

29. Bourret, R. C., Fiction theory of dynamical systems with noisy parameters, *Can. J. Phys.*, vol. 43, pp. 619–627, 1965.

30. Keller, J. B., Stochastic equations and wave propagation in random media, *Proc. Sympos. Appl. Math.*, vol. 13. American Mathematical Society, Providence, RI, 1964, pp. 145–147.

31. Chernov, L. A., *Wave Propagation in a Random Medium*. New York: McGraw-Hill, 1960.

32. Tatarskii, V. I., *Wave Propagation in a Turbulent Medium*. New York: McGraw-Hill, 1961.

33. Furutsu, K., On the Statistical Theory of Electromagnetic Waves in a Fluctuating Medium (II), *NBS Monograph*, No. 79, pp. 1–44, 1964.

34. Ishimaru, A., *Electromagnetic Wave Propagation, Radiation, and Scattering*. Englewood Cliffs, NJ: Prentice-Hall, 1991.

35. Zuev, V. E., Banah, V. A., and Pokasov, V. V., *Optics of the Turbulent Atmosphere*, Leningrad: Gidrometeoizdat, 1988.

36. Panofsky, H. A., and Dutton, J. A., *Atmospheric Turbulence: Models and Methods of Engineering Applications*. New York: John Wiley & Sons, 1984.

37. Andrews, L. C., and Phillips, R. L., *Laser Propagation through Random Media*, Bellingham, WA: SPIE Press, 1998.

38. Kraichnan, R. H., Dynamics of nonlinear stochastic systems, *J. Math. Phys.*, vol. 2, pp. 124–148, 1961.

39. Tatarskii, V. I., and Gertsenshtein, M. E., Propagation of wave in a medium with strong fluctuations of the refractive index, *JEFT*, vol. 17, pp. 548–563, 1967.

40. Klyatskin, V. I., *Statistical Description of Dynamical Systems with Fluctuating Parameters*. Moscow: Nauka, 1975.

41. Rytov, S. M., Kravtsov, Y. A., and Tatarskii, V. I., *Principles of Statistical Radiophysics*, Berlin: Springer, 1988.

42. Balesku, R., *Equilibrium and Nonequilibrium Statistical Mechanics*. New York and London: John Wiley and Sons, 1975.

43. Monin, A. S., and Yaglom, A. N., *Statistical Fluid Mechanics*. Cambridge, MA: MIT Press, 1971.

44. Charnotskii, M. I., Gozani, J., Tatarskii, V. I., and Zavorotny, V. U., in *Progress in Optics*, vol. 32, E. Wolf Ed. Amsterdam: Elsevier, 1993.

45. Blaunstein, N., Theoretical aspects of wave propagation in random media based on quantity and statistical field theory, *J. Electromagnetic Waves and Applications: Progress in Electromagnetic Research, PIER*, vol. 47, pp. 135–191, 2004.
46. Blaunstein, N., and Christodoulou, Ch., *Radio Propagation and Adaptive Antennas for Wireless Communication Links: Terrestrial, Atmospheric, and Ionospheric.* Hoboken, NJ: Wiley InterScience, 2007.

Chapter 3

Atmospheric Turbulence in the Anisotropic Boundary Layer

Victor V. Nosov

Contents

3.1 Overview

Physical properties of the turbulent atmosphere significantly affect the results of investigations in atmospheric optics. As was mentioned in Chapter 1, the correct knowledge of turbulent characteristics of the atmosphere is an important prerequisite for accurate prediction of propagation of short-wavelength (optical) radiation in the atmosphere.

This chapter presents the results of many years of experimental and theoretical investigations carried out by the author and his colleagues (see Bibliography at the end this chapter) to understand the main properties of atmospheric turbulence in the anisotropic boundary layer. The investigations were carried out in two directions.

First, we present investigations within the framework of the semiempirical theory of turbulence. This theory is statistical; it deals with statistical moments of turbulent fields.

Second, we throw light on the investigations of the local structure of turbulence, including issues of turbulence formation and evolution. These investigations make clear the mechanism of generation and existence of hydrodynamic turbulence, in which the role of stochasticity is significantly decreased. As will be seen from the results presented below, turbulence, which is usually considered as a purely stochastic phenomenon (chaos) in the international literature, is deterministic to a large extent.

3.1.1 Introduction in Semiempirical Theory of Turbulence

The theory of turbulence appeared in late nineteenth century and early twentieth century in pioneering papers by O. Raynolds, L. Richardson, J. Taylor, L. Prandtl, T. von Kármán, Lord Rayleigh, L. Keller, and A. Friedman (although the existence of two widely different types of flows referred to nowadays as laminar and turbulent, but this differentiation was noticed even in the first half of the nineteenth century). The further development of the theory is mostly connected with Russian investigators: A. N. Kolmogorov, A. M. Obukhov, A. S. Monin, A. M. Yaglom, G. S. Golitsyn, and others.

As is well known, the theory of turbulence deals with the description of fluid and gas flows based on equations of flow dynamics. The theory of turbulence is usually statistical, because of the high complexity of individual description of the fields of velocity, pressure, temperature, and other characteristics of turbulent flows. The complete statistical description of random hydrodynamic fields is given below by the characteristic functional, which satisfies equations with functional (variational) derivatives. Now there are no acceptable methods for providing solution to such equations. At the same time, for many practical problems, it is sufficient to determine only lowest-order statistical moments. Therefore, investigations in the theory of turbulence are traditionally based on the system of Reynolds equations as the result of averaging of hydrodynamic equations. However, in the Reynolds system of equations, the number of unknowns exceeds the number of equations. To close this system, some relationships between moments of hydrodynamic fields are usually stated. These relationships found from experiments or obtained from physical reasoning are referred to as semiempirical hypotheses, while the theory itself is called the semiempirical theory of turbulence.

It can be pointed out that the most theoretical papers on the dynamics of turbulent flows were and are devoted to methods of overcoming the difficulties associated with the problem of closure in the semiempirical theory of turbulence. These difficulties have not been fully overcome yet. Nevertheless, many important results were obtained in the theory of turbulence in two fields. One of it deals with the description of large-scale component of turbulence (comparable with general scales of the flow), while the other deals with small-scale components. In contrast to large-scale characteristics of turbulence, which depend significantly on the flow geometry and the character of external actions, small-scale characteristics at large values of the Reynolds number have a versatile character to a large extent (see, e.g., References 1–8).

The main hypotheses of the semiempirical theory of turbulence can usually be reduced to assignment of the relationship between the second moments of velocity and temperature pulsations (e.g., deviations from the average) and the average fields of velocity and temperature. The hypotheses are based on the analogy between the molecular and turbulent motions. In the most general form, the semiempirical hypotheses of the theory of turbulence were formulated by Monin in 1956 (see Reference 1). These hypotheses use the tensors of turbulence viscosity and turbulence temperature conductivity coefficients. The Monin hypotheses allow one to introduce the useful concepts of the isotropic and anisotropic boundary layers [9]. For plane-parallel flows, these tensors are isotropic (i.e., diagonal). The boundary layer with isotropic tensors is referred to as isotropic. If the temperature conductivity tensor is not isotropic, but the viscosity tensor can be considered isotropic, then the boundary layer is weakly anisotropic. At the same time, the concept of the isotropic boundary layer does not connect with the isotropy of hydrodynamic fields. In the isotropic layer, there is a preferred direction (e.g., a distance from the boundary plane); therefore, the fields are nonisotropic.

The hypothesis of similarity (i.e., closure) for small-scale components, proposed by Kolmogorov (1941) and Obukhov (1941) and based on the theory of dimension, allowed the formulation of the well-known Kolmogorov–Obukhov 2/3 law. The three-dimensional (3D) spectrum of turbulence corresponding to this law is called the Kolmogorov spectrum (see definitions in Chapter 1). In the inertial range of wave numbers, it has the 11/3 power decrease, while the one-dimensional (1D) spectrum of turbulence has the 5/3 power decrease. The 2/3 law is a very important achievement in the theory of turbulence. It is significant in many studies using statistical properties of turbulent flows; for example, it is applied in the theory of propagation of electromagnetic and acoustic waves in turbulent media in atmospheric optics and in other fields [1–8].

The semiempirical theory of turbulence for large-scale turbulence components (comparable with the scales of the flow as a whole) originates from the papers of Prandtl (1904, see Reference 1). Later on, the main ideas of the semiempirical theory of turbulence were developed by Obukhov [3] and Monin [1]. Now the semiempirical theory for large-scale components (with application of the 2/3 law) is called the Monin–Obukhov similarity theory. Within the framework of this theory, the methods for the calculation of turbulence characteristics in a temperature-stratified medium were developed and are discussed below. The methods allow calculation of the average dissipation rates of kinetic energy and temperature as well as calculation (based on the later) of the main parameters of spectra of atmospheric turbulence: the outer and inner scales of turbulence and the spectrum amplitude, which is usually referred to as the structure characteristic of fluctuations. In particular, in electrodynamic and optical applications, the turbulent atmosphere is traditionally described by the Kolmogorov–Obukhov theory. Turbulence is usually believed to have the Kolmogorov spectrum in the inertial range. In the energy and viscous ranges of wave numbers, different models are used, and parameters of these models are the outer L_0 and inner l_0 scales of turbulence (see detailed definitions in Chapter 1). One more parameter of the spectrum is its amplitude (or intensity) described by the structure characteristic of fluctuations of permittivity C_ε^2 or refractive index C_n^2.

The experience of the investigators has shown that in the atmosphere, the 2/3 law and the Monin–Obukhov theory of similarity are true over a rather smooth and uniformly heated extended underlying surface. This similarity corresponds to the isotropic boundary layer. Therefore, the existing theoretical methods for the calculation of characteristics of atmospheric turbulence assume the presence of the isotropic boundary layer and often give a large error under real conditions. The available optical models of turbulence, which usually include the calculated vertical profiles of C_n^2, the outer and inner scales, L_0, and l_0, are often constructed on the same assumption.

At the same time, many investigations, in particular, atmospheric-optical ones, often deal with the anisotropic boundary layer. For example, in the observational astronomy, optical instruments are often installed in mountain regions (i.e., the ground-based receiving telescopes are placed on mountain tops in order to diminish turbulent distortions). For turbulent flows in mountains, where we deal with the anisotropic boundary layer, one should not expect the constant Monin–Obukhov scale over the entire territory. Over the mountain terrain, stable vortex formations are often observed up to high altitudes [10]. In addition, the data of our measurements show that in mountains, the turbulence spectrum often deviates from the Kolmogorov one. The applicability of the model of isotropic layer (i.e., the similarity theory [1,2]) for mountains is not assessed yet. The turbulence models developed for the earth isotropic layer are usually inapplicable in mountains. Therefore, naturally, it is necessary to extend the semiempirical theory of turbulence to the anisotropic boundary layer. The extended theory should include, as a particular case, all statements of the semiempirical theory for the isotropic boundary layer. This extension can be reduced to the experimental check of the Monin semiempirical hypotheses directly for mountain conditions. Earlier, this check was not carried out in the needed volume. This is connected with the need to record (at any point of the mountain territory) experimental data for the large number of parameters.

Thus, Section 3.2 presents the results of the authors' many-year experimental studies of the properties of atmospheric turbulence in the anisotropic boundary layer [9,11–41]. The studies were initiated by practical needs of the observational astronomy. Therefore, the attention was accumulated at the optical properties of turbulence. For measurements, the small-size high-sensitivity acoustic meteorological system was developed at the V.E. Zuev Institute of Atmospheric Optics, SB RAS (Russia). The meteorological system has passed the complete set of meteorological tests and finally was certified.

It is shown in Section 3.2 that the similarity theory of turbulent flows can be extended to an arbitrary anisotropic boundary layer (see Reference 9). With the use of semiempirical hypotheses of the theory of turbulence, it was found theoretically and experimentally that the arbitrary anisotropic boundary layer can be considered as locally weakly anisotropic, in which the weakly anisotropic Monin–Obukhov similarity theory (with the nondiagonal tensor of turbulent temperature conductivity) is fulfilled locally (i.e., in some vicinity of every point in a layer). It was found that at the known characteristic temperature and velocity scales, averaging for the region of observation, the anisotropic boundary layer can be replaced with the isotropic one. This allows the use of the optical models of turbulence developed in Chapter 1 for the isotropic boundary layer.

Theoretical equations for the outer scale of turbulence in the anisotropic boundary layer have been derived and confirmed experimentally. Relationships between the outer scales have been determined by five different methods. They relate the Tatarskii outer scale of turbulent mixing and the parameters of different models of the turbulence spectrum presented also in Chapter 1.

As was also found in References 9, 11–38, at the known turbulent scales of velocity and temperature, the main parameter of turbulence in the anisotropic boundary layer is the variable Monin–Obukhov number (i.e., the ratio of the height to the Monin–Obukhov scale), which is different at every point of the layer. Therefore, the study of functional dependence of the turbulent scales of velocity and temperature on the Monin–Obukhov number can be considered as one of the top-priority problems of the semiempirical theory.

In the later studies within the framework of the semiempirical theory of turbulence [39–41], new results were obtained for the turbulent scales of the velocity and temperature fields observed under various climatic conditions and in different regions. The data confirm the earlier conclusion about the local applicability of the Monin–Obukhov similarity theory in the anisotropic atmospheric boundary layer and extend it to extreme temperature stratifications. In our experiments, we observed the Monin–Obukhov numbers in the range from +197 (high stability) to −102,819 (ultrahigh instability). Large (and extremely large) in the absolute value Monin–Obukhov numbers are characteristic only for the semiempirical theory in the anisotropic boundary layer. In the semiempirical theory for the isotropic layer, being a particular case of the anisotropic theory, absolute values of the Monin–Obukhov numbers usually do not exceed few units.

Section 3.3 briefly presents the results of our experimental studies of the processes of formation and breakdown of a Benard cell in air of a closed areas (room and pavilion) [80–82], as well as the results of the studies of the local structure of turbulence [83,85–87,90–108,110–143]. The main goal of the author to show difference in formation and breakdown of Benard cells in the closed space with respect to those created in the open irregular atmospheric layer filling by the turbulent structures.

Our investigations of the local structure of turbulence were started from the study of turbulence characteristics in closed local volumes (closed rooms and pavilions). Turbulence in closed rooms can be considered as a particular case of turbulence observed in the anisotropic boundary layer. The studies in this field have allowed us to obtain important results concerning coherent structures and, in general, the local structure of turbulence.

The data presented in Section 3.3 confirm the main scenarios of turbulence formation (stochastic scenarios of Landau–Hopf, Ruelle–Takens, Feigenbaum, Pomeau–Menneville). It is shown that temperature gradients cause the formation of Benard cell. It is found that the Benard cell breaks down by the Feigenbaum scenario. As this occurs, the main vortex in the cell breaks down into smaller ones as a result of a series of period-doubling bifurcations. Turbulence appearing as a result of the breakdown is shown to be coherent (inphase) and deterministic. The fractal character (local self-similarity) of the turbulence spectrum is found.

3.1.2 Local Structure of Turbulence

It should be emphasized that the semiempirical theory of turbulence now gives quite satisfactory answers to many practical questions. At the same time, the semiempirical theory usually fails to explain the physical mechanisms of turbulence formation and the local structure of turbulence. The data on physical mechanisms could be expected, to a large extent, for results of classical fluid dynamics. However, the studies in this field face the growing complexity of analytical equations when considering the hydrodynamic nonlinearity. Numerical solutions of hydrodynamic equations well take into account the nonlinearity, but, as the experience shows, for developed turbulent flows the physical interpretation of numerical data is difficult. In addition, the progress in this field till the middle of 2000s was significantly limited because of the absence of accessible high-performance computers.

As is known from classical mechanics of gases and hydro-mechanics, fluid and gas flows can be divided into two sharply different classes. The first class includes smooth flows varying in time only in connection with variation of active forces; they are referred to as laminar. The second class incorporates flows accompanied by chaotic pulsations of hydrodynamic fields both in time and in space; they are called turbulent. The presence of chaotic pulsations of the velocity field in turbulent flows leads to the sharp increase of mixing of the medium; therefore the intense mixing is often considered as the most characteristic feature of the turbulent motion.

In connection with the extreme complexity of individual description of turbulent hydrodynamic fields, the question on the local structure of developed turbulence usually does not arise, and statistical methods are forcibly used for the description of turbulent flows. At the same time, we can speak about the local structure of laminar and close-to-laminar flows. To study the local structure of these flows, it is possible to use nonstatistical analytical methods (see, e.g., Reference 42). In this case, we can speak about the local structure of large-scale ordered vortices.

Such vortices are often observed in the atmosphere. They have interesting properties and are usually referred to as coherent structures. In References 1, 2, it was defined a coherent structure as a nonrandom nonlinear superposition of large-scale turbulence components having high stability. This is likely the simplest definition of the coherent structure. The earliest definitions used the expansion of the instantaneous velocity field into the coherent large-scale and the random incoherent turbulent small-scale components (double expansion, see References 43, 44). Later, more complex expansions (e.g., triple) were applied. According to these definitions, turbulence consists of nonrandom coherent (large-scale) and purely random (small-scale) motions. Random motions are superimposed on coherent motions and (according to the double expansion) usually extend far beyond the boundaries of the coherent structure. In this case, coherent large-scale vortices are considered as simply independent main energy carriers and usually are not included in the structure of turbulence.

Coherent structures are actively studied (see, e.g., References 43–79). Near-wall small-scale turbulence, turbulent convection in the atmospheric surface layer in the presence of wind shear, "cloud streets" in the atmosphere, and "Langmuir circulation" in seas and lakes, and others are objects of investigation. It is shown that the main energy carriers in turbulent flows are large-scale ordered vortices, which affect significantly the formation of all flow characteristics. It was found that large-scale turbulent motions are deterministic, that is, are not random. Different methods of turbulence visualization (usually coloring of flows) are used in the investigations. However, the resolution of the used visualization methods is low. Therefore, only large-scale components of turbulent flows can be observed clearly, while small-scale inhomogeneities usually remain invisible.

Turbulence arising as a result of Benard cell breakdown satisfies all the attributes characterizing the appearance of chaos in typical dynamic systems [80–82]. These attributes usually include appearance of irregular long-lived spatial structures, whose form (or character) is determined by dissipative factors, local instability and fractal character of the phase space of these structures, and appearance of the central (at the zero frequency) peak in the spectrum. It is convenient to join these attributes by the single term "coherent structure," if we extend this already existent term and include small-scale components of cell breakdown into the composition of the coherent structure. We define the coherent structure as a compact formation including the long-lived spatial hydrodynamic vortex cell (arising as a result of long action of thermodynamic gradients) and the products of its discrete coherent cascade breakdown. For the extended understanding, the coherent structure is a solitary solution of flow dynamics equations (i.e., solitary wave). This is either a one-soliton topological solution or one soliton in a many-soliton solution. The coherent structure includes both large-scale and small-scale turbulence.

The frequency of the coherently breaking-down main vortex is the main attribute of the coherent structure. The dimensions of the coherent structure are unclear.

Flows external with respect to the main vortex can carry products of its breakdown to long distances, forming a long turbulent trace. The lifetime of the coherent structure is determined by the time of action of thermodynamic gradients. As a limiting case of the very weak local instability (when the parent structure is locally stable and does not break down, for example, in rather viscous media), the coherent structure can consist of only one long-lived parent structure. In this case, the parent structure is some configuration (usually, vortex configuration) of the laminar flow. Nonbreaking coherent structures, as ordinary nonsolitary waves, can be considered as varieties of laminar flows. Thus, the process of turbulence formation (origination) can be attributed to appearance of the coherent structure. The compact formation observed in our experiments includes the long-lived main energy-carrier vortex and the products of its simultaneous cascade discrete breakdown (coherent vortices with multiple frequencies). The frequency of the main vortex is determined by the dimensions of the convective Benard cell. According to our definition, this formation is the coherent structure.

The results obtained in Section 3.3 show that the known processes of transition of laminar flows into turbulent ones (Rayleigh–Benard convection, fluid flow past an obstacle, and others) can be considered as coherent structures (or sums of such structures). It is shown that the actual atmospheric turbulence can be considered as an incoherent mixture of different coherent structures with incommensurate frequencies of the main energy-carrying vortices. Irregularity (nonmultiplicity, incommensurability) of frequencies of the main vortices of different coherent structures leads to non-in-phase (incoherent) oscillations being breakdown products. If in one coherent structure the family of vortices (products of cascade breakdown) is coherent, then at the mixing of coherent structures with incommensurate frequencies of the main vortices the elements of one family are not in phase with elements of other family. Therefore, turbulence arising at the mixing of different coherent structures is naturally referred to as incoherent turbulence.

It was shown by us that experimental spectra of actually observed atmospheric turbulence are sums of spectra of individual coherent structures of different sizes (with different outer scales). At the same turbulence intensity, the curve corresponding to the inertial range of the 1D spectrum of Kolmogorov turbulence is the upper envelope of all the spectra of different coherent structures having different outer scales of turbulence. If the difference between the scales is not large, then the sum of the spectra of different coherent structures in the inertial range practically does not differ from the Kolmogorov 5/3 law (see also discussions carried out in Chapter 1). If the difference is large, then the sum of the spectra has a deep dip, which reveals one large structure with the 8/3 power decrease. In this case, turbulence is referred to as coherent. Thus, if coherent structures have close sizes and are well mixed, then the isotropy of turbulence described by the Kolmogorov spectrum is observed. If one of the coherent structures is much larger than others, then the anisotropy of turbulence described by the spectrum of coherent turbulence is observed.

As a result of the research missions in the 2000s under mountain and valley conditions, we have accumulated a large experimental database of near-surface measurements of the main turbulence parameters in various geographic regions and under various weather conditions. It follows from these data that extended zones of coherent turbulence are often observed in the open atmosphere, and one coherent structure among them has a decisive influence. Moreover, characteristic signs of rather large coherent structures (coherent turbulence) are present, to some or other extent, in the most part of the accumulated data. Incoherent Kolmogorov turbulence is usually observed only over extended areas with smooth and homogeneous underlying surface.

In the area with the decisive influence of one coherent structure (area of coherent turbulence), the values of the Kolmogorov and Obukhov constants (see also Chapter 1) can differ significantly from their values in the Kolmogorov turbulence. For this area, double expansion (into the large- and small-scale components) is performed. At its time, this expansion has served an experimental basis for earlier definitions of the coherent structure. In comparison with the incoherent Kolmogorov turbulence, the significant attenuation of optical radiation fluctuations occurs in the coherent turbulence.

As a result of our investigations of the local structure of turbulence, it became clear that coherent structures are important elements for understanding of the processes of turbulence formation and further evolution. That is why Section 3.3.7 lists briefly the established properties of individual coherent structures and properties of mixtures (sums) of different coherent structures. It flows from these properties that:

> *The coherent structure (in the extended definition), despite its complex inner structure, can be considered as a basis structural element or an elementary particle comprising the turbulence.*

This conclusion explains the inner structure of turbulence. It can be considered as a main result of our studies of the local structure of turbulence and discussed in detail in References 112–115.

The properties of coherent structures were found through the processing of accumulated experimental data by the methods of spectral analysis of random processes with the following formation of logical conclusions. However, these methods do not allow us to see coherent structures themselves and their mixtures. The possibility to see coherent structures is demonstrated in Section 3.3.8. The processes of turbulence origination and evolution were visualized in References 144–157. The visualization is carried out by constructing flow lines of medium motion in numerical solutions of Navier–Stokes equations for different boundary-value problems.

For example, in Reference 144, the numerical solution of coherent structures in closed rooms has been carried out. The results are presented for eight boundary-value problems: spectrograph pavilion and dome space of large astronomical telescopes (large solar vacuum telescope (LSVT) and big azimuthal telescope

(BAT)), cubic room, plane square cell, pipe of square section, near-wall turbulence (thermics), dome (hemisphere with viscous medium), open air over one heated spot. It was shown that solitary large vortices (coherent structures or 3D topological solitons) are observed indoor. In the case of identical boundary-value conditions, motion patters obtained by numerical simulation and independently observed experimentally by us (in air of the LSVT and BAT rooms) nearly coincide. The close agreement is also seen between the medium motion patterns inside other closed volumes obtained through numerical simulation by us and observed experimentally by other authors (including experimental data presented in the well-known album of Van Dyke [158]).

In the References 147–150, 152–155 the structure of turbulent air motions were numerically studied in closed volumes over irregularly heated surface. It was shown that solitary toroidal vortices (coherent structures or topological solitons) arise over an irregularly heated surface. The number of vortices and their inner structure depend on the shape and size of heated irregular spots. In the case of simple shapes (homogeneous heating in a cubic room, one heated circular spot), coherent turbulence arising as a result of coherent breakdown of vortices is observed in the volume. For complex shapes (thermal mottle), toroidal vortices become markedly deformed. Vortices can be elongated along the surface and have spiral flow lines.

> *In the process of evolution, the vortices mix noticeably. This yields the Kolmogorov (incoherent) turbulence.*

The results presented in References 147–150, 152–155 allow us to follow the evolution of the structure of turbulence formed over homogeneously and irregularly heated surfaces. For example, it is shown that the transition (through the chaotization period) from small thermics (toroidal mushroom-shaped vortices) to big stable vortices is inverse to the usually observed cascade coherent breakdown of large vortices into smaller ones.

> *Consequently, the formation of the near-wall turbulence (thermics) and its further evolution can serve an example of the inverse cascade of processes of energy transfer by the spectrum of motion scales.*

In References 95, 128, one can find the review of results concerning the properties of coherent structure. The comparison of the structure properties determined by us with the known results shows that our data far extend the ideas about coherent structures existent in the international literature. Our data are confirmed by the results of theoretical investigations of other authors, including data of numerical solutions of the Navier–Stokes equations. They are also confirmed by experimental results obtained by meteorological and optical methods. In general, our data are in harmonic agreement with data of other well-known approaches concerning various aspects of the problem of turbulence and generalize them.

As a result, the analysis of all the accumulated experimental data (and data of numerical calculations) on turbulence parameters in open air and in closed rooms shows that:

> *Laminar and turbulent flows can be considered as different phases of the same process of turbulence formation and evolution, which is the single process of formation and breakdown of topological solitons.*

Within the framework of the main conclusion, the non-Kolmogorov coherent turbulence and the Kolmogorov-incoherent turbulence are varieties (particular cases) of the single process of formation and breakdown of hydrodynamic topological solitons. These varieties are characterized by different intensities and different spatiotemporal distribution of energy sources of turbulence (thermodynamic gradients) at boundaries of continuous liquid medium.

3.2 Semiempirical Hypotheses of the Theory of Turbulence in the Anisotropic Boundary Layer

3.2.1 Main Principles of the Semiempirical Theory of Turbulence

As is well-known, the theory of turbulence is based on the description of fluid and gas flows by equations of flow dynamics. The complete statistical description of random hydrodynamic fields is given by the characteristic functional [1,2,4,5]. The characteristic functional contains information about the infinite set of field moments and satisfies the dynamic equations with functional derivatives. Now there are well-known acceptable methods for solution of such equations (see, e.g., Chapter 2). At the same time, for many practical problems, it is sufficient to determine only lowest-order statistical moments. That is why the studies in the theory of turbulence are traditionally based on the Reynolds system of equations being a result of averaging of the flow dynamics equations [1–7]. However, in the Reynolds system of equations, the number of unknowns exceeds the number of equations. This system is usually closed by specifying some relationships between moments of hydrodynamic fields. These relationships found from experiments or derived from physical reasoning (e.g., from dimensional considerations) are called semiempirical hypotheses of the theory of turbulence.

The main semiempirical hypotheses can usually be reduced to specification of the relationship between the second moments of pulsations (i.e., deviations from the average) of the velocity $\langle v_i' \cdot v_j' \rangle$ and temperature $\langle v_j' \cdot T' \rangle$ and the averaged fields of velocity $\bar{v}_i$ and temperature $\bar{T}$. These hypotheses are usually based on the analogy between the turbulent and molecular motions. Thus, the terms $\nu \cdot \partial \bar{v}_i / \partial x_j$ and $\chi \cdot \partial \bar{T} / \partial x_j$, presented in the averaged equations, are proportional to the components of the momentum and heat fluxes. Here, ν is the kinematic viscosity and χ is the temperature conductivity. They describe a medium without turbulence and are caused by molecular diffusion. In the turbulent medium, these components are supplemented with the terms $-\langle v_i' \cdot v_j' \rangle$, and $-\langle v_j' \cdot T' \rangle$. Therefore, these characteristics can be considered as components of the turbulent momentum and heat fluxes. Within the scope of the semiempirical theory, the structure of the dependence of turbulent momentum and heat fluxes on $\overline{v}_i$ and $\overline{T}$ is the same as in the case of purely molecular diffusion. In the general case of anisotropic turbulence, it was assumed that [1]

$$
\langle v_i' \cdot v_j' \rangle = \langle v_n' \cdot v_n' \rangle \delta_{ij}/3 - (K_{in}\Phi_{nj} + K_{jn}\Phi_{ni})/2, \quad \Phi_{ij} = \frac{\partial \overline{v}_i}{\partial x_j} + \frac{\partial \overline{v}_j}{\partial x_i}
$$

$$
\langle v_j' \cdot T' \rangle = -K_{Tji}\frac{\partial \overline{T}}{\partial x_i},
$$

(3.1)

where the repeating subscripts imply summation. The components K_{ij} of the symmetric tensor $\boldsymbol{K}$ in definitions (3.1) are called the coefficients of turbulent viscosity, while the components K_{Tij} of the tensor $\boldsymbol{K}_T$ have the meaning of the coefficients of turbulent temperature conductivity or coefficients of turbulent diffusion for a passive admixture, which is the potential temperature T (in the boundary layer, the ordinary and potential temperature may be considered identical). Hypotheses (3.1) replace 12 components of turbulent fluxes of momentum and heat with 27 new characteristics (by six components in symmetric tensors K_{ij} and Φ_{ij}, nine in tensor K_{Tij}, three derivatives $\partial \overline{T}/\partial x_j$ and three components in the sum $\langle v_n' \cdot v_n' \rangle$).

It was shown in References 1, 2, 4–8 that in the plane-parallel flows (between separated planes and in pipes) turbulent phenomena in the boundary layer are well described by the semiempirical hypotheses with application of only two scalar parameters K and K_T (which are called the coefficients of turbulent viscosity (turbulent exchange) and turbulent temperature conductivity, respectively). Turbulence in the boundary layer of the Earth's atmosphere can be considered as a particular case of the plane-parallel flow, if only we consider flows over extended surface area having a smooth, homogeneous (identical in the structure), and uniformly heated surface. As can be seen from Equation 3.1, for the plane-parallel flows the tensors $\boldsymbol{K}$ and $\boldsymbol{K}_T$ are isotropic ($K_{ij} = K \cdot \delta_{ij}$, $K_{Tij} = K_T \cdot \delta_{ij}$). In this connection, the boundary layer with the isotropic tensors $\boldsymbol{K}$ and $\boldsymbol{K}_T$ will be referred to as the isotropic boundary layer for brevity. If at least one of the tensors K and K_T is anisotropic, then the boundary layer is referred to as anisotropic [9].

In practice, however, it is necessary to specify these definitions. Thus, the isotropic boundary layer appears to be the more general concept than the boundary layer in plane-parallel flows. In contrast to plane-parallel flows, in the isotropic layer, the conditions that the horizontal derivatives and the vertical component of the average velocity are nonzero can take place in the general case. It is natural to call this case weakly isotropic, leaving the concept of the isotropic (or strongly isotropic) boundary layer only for plane-parallel flows. The analogous separation can also be done for the anisotropic boundary layer. Thus, if one of the tensors K or K_T is anisotropic, then the boundary layer can be called weakly anisotropic. If the both tensors K and K_T are anisotropic, then the layer can be called strongly anisotropic.

The concept of the isotropic boundary layer (for plane-parallel flows) is not connected with the isotropy of hydrodynamic fields themselves. In the isotropic layer, there is a preferred direction (distance from the boundary plane); therefore, the fields are not isotropic.

The components of the tensors K and K_T can be represented in the form of the products of the root-mean-square value of velocity pulsations by the components of the tensors of turbulence scales l_{ij} and l_{Tij} (i.e., the scales which are the average distances, turbulent formations can move to, keeping their individuality):

$$K_{ij} = \langle v'_n \cdot v'_n \rangle^{1/2} l_{ij}, \quad K_{Tij} = \langle v'_n \cdot v'_n \rangle^{1/2} l_{Tij},$$

For isotropic tensors K and K_T, the ellipsoids of the scales l_{ij} and l_{Tij} transform into spheres. In the general case, the tensors K_{ij} and K_{Tij} do not coincide. The variability of the temperature scales l_{Tij} is usually higher than that of the velocity scales l_{ij}. This can be seen in the case of free convection when there are no wind and no friction. Then turbulence receives energy from the energy of temperature instability, rather than from the energy of the averaged motion, and has the character of vertical thermal flows. Therefore, in the first approximation, the tensor K in Equation 3.1 can be believed isotropic (the tensor $\langle v'_i v'_j \rangle$ still remains anisotropic). Then, according to References 1, 7,

$$\langle v'_j \cdot v'_j \rangle = \langle v'_n \cdot v'_n \rangle \delta_{ij}/3 - K(\partial \overline{v}_i/\partial x_j + \partial \overline{v}_j/\partial x_i),$$
$$\langle v'_j \cdot T' \rangle = -K_{Tji}\partial \overline{T}/\partial x_i, \tag{3.2}$$

and the number of unknowns decreases down to 22.

The semiempirical hypotheses are actively used in the studies on the turbulent diffusion of passive admixtures, including the temperature diffusion. Hypotheses (3.2) are usually taken as a basis. The equations for the coefficients K and K_{Tij}, accepted now, are the results of generalization of experimental data obtained over approximately smooth surface (i.e., not in mountain regions). They take into account the action of thermal stratification. For the average wind velocity directed along the axis x_1, we have [1]

$$K_{Tji} = \beta_{ij}K_T, \quad K_T/K = \alpha, \quad \alpha = \mathrm{Pr}^{-1} \tag{3.3}$$

$$\beta_{33} = 1, \; \beta_{11} = 8.04, \; \beta_{22} = 4.21, \; \beta_{13} = -3.51, \; \beta_{31} = -0.49,$$
$$\beta_{12} = \beta_{21} = \beta_{23} = \beta_{32} = 0,$$
$$K = K(z) = æ \cdot V_* \cdot z / \varphi(\zeta), \quad \zeta = z / L, \quad K_T = K_T(z) = \alpha(z) \cdot K(z)$$

Here, $æ = 0.4$ is the Karman constant; z is the height above the underlying surface ($x_1 = x$, $x_2 = y$, $x_3 = z$); $\varphi(\zeta)$ is the universal similarity function specifying the type of stratification (see Figure 3.1); Pr is the turbulent Prandtl number. Here also the error of determination of the coefficients β_{ij} does not exceed 30% [1].

The similarity function depends on the stratification parameter $\zeta = z/L$, in which the scale of the length L is called the Monin–Obukhov scale (or the thickness of sublayer of dynamic turbulence). The scale L has the essential significance in the theory of thermally stratified atmosphere. It was introduced by Monin and Obukhov [3] from dimensional considerations and is described by the equation

$$L = V_*^2 / (\alpha \cdot æ^2 \cdot \beta \cdot T_*),$$
$$\beta = g / \overline{T}, V_*^2 = -\langle v_j' \cdot v_j' \rangle, \alpha \cdot æ \cdot V_* \cdot T_* = -\langle v_3' \cdot \overline{T}' \rangle, \tag{3.4}$$

where g is the gravity acceleration, $\overline{T}$ is the average value of the absolute temperature, V_* is the friction velocity (or the turbulent scale of velocity), and T_* is the

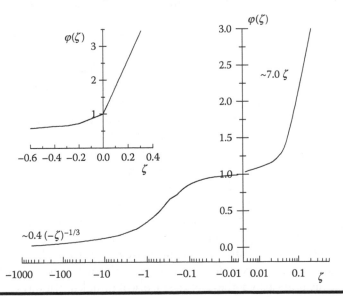

Figure 3.1 Universal similarity function $\varphi(\zeta)$, constructed by sewing of empirical values. The top left plot: data are taken from Reference 8 with known asymptotics [1]: $\varphi(\zeta) = 0.4 \, (-\zeta)^{-1/3}$, $\zeta < 0$, $|\zeta| \gg \zeta_0$; $\varphi(\zeta) = 7.0 \, \zeta$, $\zeta \gg \zeta_0$; $\zeta_0 = 0.05$. The error of measurement [1] of the coefficient of 0.4 was about 20%, while that for the coefficient of 7.0 was about 40%.

turbulent scale of the temperature field. For the neutral stratification, the parameter ζ coincides with the dynamic Richardson number Rf (Rf $= \zeta/\varphi(\zeta)$, Ri $=$ Rf$/\alpha$). Therefore, by analogy with the Richardson number, the parameter ζ is often called the Monin–Obukhov number. Experimental data for the ratio $\alpha(\zeta) = K_T/K$ show that at the neutral ($|\zeta| \leq 0.05$) and unstable ($\zeta < -0.05$) stratifications, the value of $\alpha(\zeta)$ is close to constant, $\alpha \approx 1.17$, Pr ≈ 0.85 [1,2]. However, at the stable stratification ($\zeta > +0.05$), it can decrease markedly (at high stability).

In Equations 3.3 and 3.4, the coefficient of turbulent viscosity $K(z)$ corresponds to the isotropic boundary layer, in which the characteristics V_*, T_*, and $\overline{T}$ are believed to be constant over all the layers. Therefore, the Monin–Obukhov scale L and the Monin–Obukhov number ζ (at a given height z) are numerical parameters of the turbulent flow over the entire considered temperature-stratified area of the Earth's surface. As can be seen from Equation 3.3, the tensor $\boldsymbol{K}_T$ is anisotropic, that is, boundary layer (3.2)–(3.4) is anisotropic as well. Since the anisotropy of boundary layer of turbulent diffusion (3.2)–(3.4) is caused only by the anisotropy of the temperature tensor K_{Tij}, while the other characteristics correspond to the isotropic layer, boundary layer (3.2)–(3.4) is weakly anisotropic.

Equations 3.2 through 3.4 are basic in the theory of similarity of turbulent flows in the atmosphere. This theory is usually referred to as the Monin–Obukhov similarity theory. Turbulent flows in mountain regions is of particular interest. Here, we could not expect the constant Monin–Obukhov scale over the entire territory. Over the mountain terrain, stable vortices are formed. Distortions of air flows from such rotor formations are observed up to high altitudes (from a mountain, for example, from 1 km-height up to 7–9 km [10]). At the same time, in atmospheric-optical studies, especially, in investigations on the influence of turbulence on the quality of optical images, we often have to deal with the anisotropic boundary layer in mountains (to decrease turbulent distortions, ground-based receiving telescopes are often installed on mountain tops). However, turbulence models developed for the isotropic boundary layer are usually inapplicable in mountains.

The applicability of the model of anisotropic layer (3.2)–(3.4) for mountains was not assessed yet. Therefore, it is interesting to experimentally check semiempirical hypotheses (3.1) or (3.2) directly for mountain conditions. Earlier this check was not carried out in the needed volume. This is connected with the need of recording (at every point of the mountain area of the surface) experimental data simultaneously for the large number of parameters.

In this subsection we established, following Reference 9, that the similarity theory of turbulent flows can be extended to an arbitrary anisotropic boundary layer. With the use of semiempirical hypotheses of the theory of turbulence, it was shown theoretically and experimentally that the arbitrary anisotropic boundary layer can be considered as locally weakly anisotropic. Excepting for the tensor of coefficients of turbulent temperature conductivity, the statements of the theory of isotropic layer (for plane-parallel flows) become true in the vicinity of every point of the layer. In the arbitrary boundary layer, the main parameter of turbulence is the variable Monin–Obukhov number. It was found that the anisotropic boundary layer can be replaced with the effective isotropic layer. Theoretical equations have been derived for the vertical outer scale of turbulence in the anisotropic boundary layer, and the agreement between experimental and theoretical values of the outer scale has been demonstrated.

3.2.2 Theoretical Equations for Dissipation Rates of Kinetic Energy ε and Temperature N in the Anisotropic Boundary Layer

Average values of the dissipation rate of the kinetic energy ε and the dissipation rate of temperature fluctuations N are important physical characteristics of the turbulent motion of the medium. They determine the intensity of velocity and temperature fluctuations. According to the Kolmogorov–Obukhov law, the structure functions of fluctuations of the longitudinal velocity $D_{rr}(r)$ and temperature $D_T(r)$ in the inertial range of scales r can be expressed through ε and N as

$$D_{rr}(r) = C_V^2 \cdot r^{2/3} \left(C_V^2 = C \cdot \varepsilon^{2/3} \right), \quad D_T(r) = C_T^2 \cdot r^{2/3} \left(C_T^2 = C_\theta \cdot \varepsilon^{-1/3} \cdot N \right). \quad (3.5)$$

The constants C and C_θ are called the Kolmogorov and Obukhov constants, respectively. Their values with the 10% error are [1,2]: $C = 1.9$, $C_\theta = 3.0$. The parameters C_V^2 [(m/s)$^2 \cdot$ cm$^{-2/3}$] and C_T^2 [deg$^2 \cdot$ cm$^{-2/3}$] are the structure characteristics of fluctuations of the longitudinal velocity and temperature.

The parameters ε and N can be expressed through statistical moments $\langle v_i' \cdot v_j' \rangle$, $\langle v_j' \cdot T' \rangle$ as follows [1,2,4]:

$$\varepsilon = -\langle v_i' \cdot v_j' \rangle \frac{\partial \overline{v}_i}{\partial x_j} + \langle v_3' \cdot T' \rangle \frac{g}{\overline{T}}, \quad N = -\langle v_j' \cdot T' \rangle \frac{\partial \overline{T}}{\partial x_j} \quad (3.6)$$

To represent ε and N as functions of derivatives of average hydrodynamic fields, it is necessary to replace the statistical moments with their semiempirical representations from hypotheses (3.1) or (3.2) in Equation 3.6.

However, for the arbitrary anisotropic boundary layer, being of the main interest, the application of the both hypotheses (3.1) and (3.2) is limited. Hypothesis (3.1), in which the tensors $\boldsymbol{K}$ and $\boldsymbol{K}_T$ should be considered different and anisotropic, is applicable for description of the arbitrary layer, but is characterized by uncertainty of the elements K_{ij} and K_{Tij}. Therefore, the equations resultant from Equation 3.6 can form the basis for the experimental study of the elements K_{ij} and K_{Tij} through measurement of other parameters in these equations. Hypothesis (3.2) is more detailed, but its application assumes the constant turbulent characteristics (V_*, T_*, L) in the entire layer and the approximately smooth underlying surface.

At the same time, in any boundary layer the underlying surface in a rather small vicinity of every point (locally) can be considered as approximately smooth.

Consequently, we can assume that in the arbitrary anisotropic boundary layer (including the layer over the mountain terrain) in some small vicinity of every observation point (locally) the conditions of applicability of the weakly anisotropic boundary layer of turbulent diffusion (3.2)–(3.4) (approximately smooth surface) take place. Then, according to Equations (3.2) through (3.4), in the vicinity of every observation point in the layer, variations of the turbulence characteristics are caused, as in the isotropic layer, mostly by variations of the three independent parameters V_*, T_*, and $\overline{T}$.

Using Equations 3.3 and 3.4, these parameters can be transformed into other three independent parameters with their inclusion into the Monin–Obukhov number: ζ, V_*, and T_*. The Monin–Obukhov number, as the stratification parameter, takes into account structural variations of the external energy influx, which transforms then into the turbulence energy, and, consequently, is a convenient characteristic. The independent parameters ζ, V_*, and T_* are functions of the radius vector of an observation point r. If we take some set of observation points in the boundary layer so that at the transition from one to other point (along some trajectory with the arc length $s(r)$), the number $\zeta(r)$ varies monotonically. Then with the replacement of independent variables $s(r)$ on $\zeta(r)$, we can consider the same dependences: $V_*(\zeta(r))$ and $T_*(\zeta(r))$. It should be noted that if the average temperature varies slightly in the boundary layer, then in place of the three parameters we have two independent parameters V_* and T_*, which can be replaced with ζ, V_* (or with ζ, T_*). Then along the trajectory, at which the number $\zeta(r)$ is monotonic, we can consider the function $V_*(\zeta(r))$ [the function $T_*(\zeta(r))$ is dependent on $V_*(\zeta(r))$] or the function $T_*(\zeta(r))$ [$V_*(\zeta(r))$ is dependent on $T_*(\zeta(r))$]. Thus, if two functions $V_*(\zeta)$ and $T_*(\zeta)$ are known along this trajectory (one function is sufficient at $\overline{T} \approx$ const), then the Monin–Obukhov number ζ becomes the single universal parameter, determining the characteristics of turbulence in the weakly anisotropic layer.

From the results of our measurements in the mountain boundary layer (see Section 3.2.3), it follows that the assumption about the local weak anisotropy of the arbitrary layer is fulfilled with a good accuracy. The arbitrary boundary layer, consequently, can be considered locally weakly anisotropic. In the arbitrary anisotropic boundary layer, all statistical characteristics of turbulence become functions of the Monin–Obukhov number, and every point in the layer is characterized by its own value of the Monin–Obukhov number.

According to the accepted assumption on the local weak anisotropy, semiempirical hypothesis (3.2) should be used in Equation 3.6, and K and K_T should be replaced with their representations from Equations 3.3 corresponding to the isotropic boundary layer ($K = \text{æ}V_*z/\varphi(\zeta)$, $K_T = \alpha K$). Then, as it follows from hypothesis (3.2):

$$V_*^2 = -\langle v_j' \cdot v_j' \rangle = K \cdot D^V, \quad D^V = \frac{\partial v_1}{\partial x_3} + \frac{\partial v_3}{\partial x_1}$$

$$\alpha \cdot \text{æ} \cdot V_* \cdot T_* = -\langle v_3' \cdot T' \rangle = K_T \cdot D^T, \quad D^T = \beta_{31}\frac{\partial T}{\partial x_1} + \beta_{32}\frac{\partial T}{\partial x_2} + \beta_{33}\frac{\partial T}{\partial x_3} \tag{3.7}$$

In Equation 3.7, the overbar of the average values of hydrodynamic fields was omitted from here for simplification of the description of the presented formulas. The direct experimental check of Equation 3.7 in the mountain anisotropic boundary layer (see below, Figures 3.10 and 3.11 according to Reference 9) has shown that the left-hand side of these equations is in a good agreement with the right-hand side.

Upon the substitution of semiempirical Equation 3.2 into Equation 3.6, taking into account Equation 3.7 and the condition of incompressibility, we get according to [9]:

$$\varepsilon = V_*^3 \cdot \mathfrak{x}^{-1} \cdot z^{-1} [\varphi(\zeta) + \varphi_V(\zeta) - \zeta], \tag{3.8}$$

$$\varphi_V(\zeta) = \mathfrak{x}^2 z^2 V_*^{-2} \varphi(\zeta)^{-1} \left[\Phi_{12}^2 + \Phi_{23}^2 + \frac{1}{2} \sum_i \Phi_{ii}^2 \right], \quad \Phi_{ij} = \frac{\partial v_i}{\partial x_j} + \frac{\partial v_j}{\partial x_i} \tag{3.9}$$

$$N = \alpha \cdot \mathfrak{x} \cdot V_* \cdot T_*^2 \cdot z^{-1} [\varphi(\zeta) + \varphi_T(\zeta)], \tag{3.10}$$

$$\varphi_T(\zeta) = T_*^{-1} \cdot z \cdot \left[(\beta_{13} - \beta_{31}) \frac{\partial T}{\partial x_1} + (\beta_{23} - \beta_{32}) \frac{\partial T}{\partial x_2} \right]$$

$$+ T_*^{-2} \cdot z^2 \cdot \varphi(\zeta)^{-1} \left[\eta_1 \left(\frac{\partial T}{\partial x_1} \right)^2 + \eta_2 \left(\frac{\partial T}{\partial x_2} \right)^2 + \eta_{12} \left(\frac{\partial T}{\partial x_1} \right) \left(\frac{\partial T}{\partial x_2} \right) \right], \tag{3.11}$$

$$\eta_1 = \beta_{11} - \beta_{13}\beta_{31}, \quad \eta_2 = \beta_{22} - \beta_{23}\beta_{32},$$

$$\eta_{12} = \beta_{12} + \beta_{21} - \beta_{23}\beta_{31} - \beta_{13}\beta_{32}, \quad \beta_{33} = 1.$$

Comparing representations (3.8) and (3.10) for ε and N obtained for the anisotropic boundary layer with the equations for ε and N in the isotropic layer [1,2,4], we can readily see that anisotropic ε and N differ from isotropic ones only by the presence of the functions $\varphi_V(\zeta)$ (for ε) and $\varphi_T(\zeta)$ (for N). These functions are added to the similarity function $\varphi(\zeta)$. Since the functions $\varphi_V(\zeta)$ and $\varphi_T(\zeta)$ are characteristics of the anisotropic layer, they can be called the anisotropy functions. According to definition (3.8), $\varphi_V(\zeta)$ characterizes the energy dissipation rate, and, therefore, it can be called the energy anisotropy function. The function $\varphi_T(\zeta)$ in definition (3.10) characterizes the dissipation rate of temperature fluctuations, and therefore it can be called the temperature anisotropy function.

In Equations 3.8 and 3.10, we still have to pass into the coordinate system, in which the axis $x_1 = x$ is directed along the average velocity of the horizontal wind. In this system, the transverse horizontal component of the average velocity is absent ($v_2 = 0$ along the axis $x_2 = y$), and the vector of average velocity $\mathbf{v}$ has the components $\mathbf{v} = (v_1, 0, v_3) = (u, 0, w)$. This fact usually allows us to believe that in some local vicinity of every point the transverse derivatives (along the axis x_2) are small in comparison with longitudinal (along the axis x_1) and vertical (along the axis $x_3 = z$) derivatives. These assumptions correspond to the often accepted symmetry conditions of the tensor Φ_{ij} in semiempirical hypothesis (3.1): $\Phi_{21} = \Phi_{12} = \Phi_{32} = \Phi_{23} = 0$. Consequently, the function $\varphi_V(\zeta)$ in Equation 3.9 can be written in the form

$$\varphi_V(\zeta) = 2\mathbf{x}^2 \cdot z^2 \cdot V_*^{-2} \cdot \varphi(\zeta)^{-1} \left[\left(\frac{\partial v_1}{\partial x_1} \right)^2 + \left(\frac{\partial v_3}{\partial x_3} \right)^2 \right], \tag{3.12}$$

If we additionally take into account the incompressibility condition $\partial v_n / \partial x_n = 0$, then from Equation 3.12, we obtain

$$\varphi_V(\zeta) = 4\mathbf{x}^2 \cdot z^2 \cdot V_*^{-2} \cdot \varphi(\zeta)^{-1} \cdot (\partial v_1 / \partial x_1)^2 \quad \text{or}$$
$$\varphi_V(\zeta) = 4\mathbf{x}^2 \cdot z^2 \cdot V_*^{-2} \cdot \varphi(\zeta)^{-1} \cdot (\partial v_3 / \partial x_3)^2. \tag{3.13}$$

The last equations should be fulfilled simultaneously. At the same time, according to hypothesis (3.2),

$$2K(\partial v_1 / \partial x_1) = \tau_1, \tau_1 = \langle v_n' \cdot v_n' \rangle / 3 - \langle v_1' \cdot v_1' \rangle,$$

$$2K(\partial v_3 / \partial x_3) = \tau_3, \tau_3 = \langle v_n' \cdot v_n' \rangle / 3 - \langle v_3' \cdot v_3' \rangle.$$

Upon the expression of the derivatives through τ_1, τ_3, K and substitution into Equation 3.13, we can find

$$\varphi_V(\zeta) = V_*^{-4} \cdot \varphi(\zeta) \cdot \tau_1^2, \varphi_V(\zeta) = V_*^{-4} \cdot \varphi(\zeta) \cdot \tau_3^2. \tag{3.14}$$

The requirement of the simultaneous fulfillment of equalities (3.14) corresponds to the assumption of isotropy of the tensor K_{ij} in semiempirical hypothesis (3.2). The experimental and theoretical (calculated by Equation 3.12) values of the function $\varphi_V(\zeta)$ in the anisotropic boundary layer are compared below in Section 3.2.3. However, the error of assumption on the isotropy of the tensor K_{ij} can be estimated tentatively already here from measurements of turbulent pulsations of the velocity components.

Figure 3.2 shows the results of measurements of standard (root-mean-square) deviations of turbulent pulsations of velocity components in the mountain anisotropic boundary layer in a wide range of values of the Monin–Obukhov numbers (from stable to very unstable local temperature stratifications, $-581 \le \zeta \le 0.3$, see Section 3.2.3).

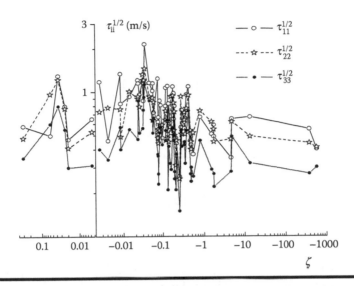

Figure 3.2 **Standard (root-mean-square) deviations of wind velocity fluctua-tions for all measurement sessions,** $\tau_{ii}^{1/2} = \overline{v_i'v_i'}^{1/2}$ **($i = 1,2,3$).** $\tau_{11} = \overline{v_1'v_1'}$**,** $\tau_{22} = \overline{v_2'v_2'}$**,** $\tau_{33} = \overline{v_3'v_3'}$**.**

As can be seen from Figure 3.2, the proportions between pulsations of different components are kept, on average, in the entire range of variability of the parameter ζ. Therefore, after averaging of experimental data over the entire range of variability of ζ, we obtain (in [m²/s²])

$$\langle \overline{v_1'v_1'} \rangle = 0.60, \langle \overline{v_2'v_2'} \rangle = 0.47, \langle \overline{v_3'v_3'} \rangle = 0.20,$$

which corresponds to $\langle \tau_1 \rangle = -0.17, \langle \tau_3 \rangle = 0.22$.

Then, two-way deviations of τ_1^2 and τ_3^2 in Equation 3.14 from their average value do not exceed 25%. Consequently, the unremovable error of measurement of the function $\varphi_V(\zeta)$ arising due to the assumption on the isotropy of the tensor K_{ij} in semiempirical hypothesis (3.2) also does not exceed 25%. The experience of measurement of turbulence characteristics in the atmosphere shows that this error is quite satisfactory.

It should be noted that since measurements of derivatives of the hydrodynamic fields in the atmosphere are usually accompanied by significant errors, for measurements of the energy anisotropy function $\varphi_V(\zeta)$ it is better to use Equation 3.12, which smooths measurement errors.

Equation 3.11 for the temperature anisotropy function $\varphi_T(\zeta)$ can also be simplified through the use of symmetry equalities of the tensor K_{Tij} in Equation 3.3 ($\beta_{12} = \beta_{21} = \beta_{23} = \beta_{32} = 0$):

$$\varphi_T(\zeta) = (\beta_{13} - \beta_{31}) \cdot T_*^{-1} \cdot z \cdot \left(\frac{\partial T}{\partial x_1}\right)$$

$$+ T_*^{-2} \cdot z^2 \cdot \varphi(\zeta)^{-1} \cdot \left[(\beta_{11} - \beta_{13}\beta_{31})\left(\frac{\partial T}{\partial x_1}\right)^2 + \beta_{22}\left(\frac{\partial T}{\partial x_2}\right)^2\right]. \quad (3.15)$$

In the coordinate system with the axis ox_1 directed along the average velocity of the horizontal wind, we obtain from Equation 3.15

$$\varphi_T(\zeta) = (\beta_{13} - \beta_{31}) \cdot T_*^{-1} \cdot z \cdot \left(\frac{\partial T}{\partial x_1}\right) + (\beta_{11} - \beta_{13}\beta_{31}) \cdot T_*^{-2} \cdot z^2 \cdot \varphi(\zeta)^{-1} \cdot \left(\frac{\partial T}{\partial x_1}\right)^2$$

$$(3.16)$$

Thus, as can be seen from Equations 3.13 and 3.16, the assumption on the local weak anisotropy of the arbitrary boundary layer reduces the calculation of the energy and temperature functions $\varphi_V(\zeta)$ and $\varphi_T(\zeta)$ (and the calculation of the dissipation rates ε and N) to determination of the turbulence scales V_*, T_*, the Monin–Obukhov number ζ, and two longitudinal derivatives $\partial T/\partial x_1$, $\partial v_1/\partial x_1$. In the isotropic boundary layer, the longitudinal derivatives are equal to zero (and $v_3 = 0$) and, consequently, $\varphi_V(\zeta) = \varphi_T(\zeta) = 0$.

3.2.3 Experimental Studies of Semiempirical Hypotheses in the Anisotropic Boundary Layer

In Section 3.2.2, the assumption of the local weak anisotropy of the arbitrary boundary layer was put forward. Based on this assumption, with the use of the semiempirical hypotheses of the theory of turbulence, the theoretical equations were derived for the main characteristics of turbulence, namely, the average dissipation rate of the kinetic energy ε and the average dissipation rate of temperature fluctuations N in the anisotropic boundary layer.

Below, based on data taken from Reference 9, the results of experimental verification of the obtained equations are presented for the dissipation rates ε and N, which, according to Equations 3.8 and 3.10, can be reduced to the results of experimental check of theoretical ideas for the energy and temperature anisotropy functions $\varphi_V(\zeta)$ and $\varphi_T(\zeta)$. The satisfactory agreement between the theory and the experiment is demonstrated.

Turbulent characteristics were measured in the atmospheric boundary layer in a mountain region. The mountains of the Baikal Astrophysical Observatory of the Institute of Solar-Terrestrial Physics SB RAS were located in the Irkutsk Region (Russia), on the shore of Lake Baikal. The region of measurements included the slopes and the top of the mountain where the Large Solar Vacuum Telescope

(LSVT) was installed on. The height of this mountain above the sea level is 680 m. The total length of the measurement path was about 3 km. The characteristics of atmospheric turbulence were mostly measured at a height of 2.7 m from the underlying surface. A total of five measurement sessions with different types of the regional meteorological situation were conducted. Measurements were carried out at 73 different points along the path. The measurements were carried out from October 12 till 28 2002 [9].

3.2.3.1 Measurement Instrumentation

For the measurements, we used the Meteo-2M mobile acoustic meteorological system developed in the Zuev Institute of Atmospheric Optics SB RAS (Russia). The operation principle of this station consists in the measurement of the speed of sound passed between two sensors on the measuring head.

The meteorological system employed four measuring "source–receiver" channels. The measurement frequency was determined by the speed of sound propagation in air and the reliability of instrumentation. For every channel, the output frequency was 10 Hz. The measured results were processed with the use of a two-stage procedure. The pre-processing, consisting the averaging of high-frequency measurements (with a frequency higher than 10 Hz), was performed on the base of the processor of the measuring head. It increased reliability of the instrumentation. The final postprocessing was performed on the base of the computer. The time of averaging was specified by the operator, and, depending on the measurement conditions, it was varied from 1 min to few days.

The meteorological system records 89 parameters (at the height of the center of the measuring head) and stored the measured data simultaneously in the form of a binary array and in the form of a text file. The main measured characteristics were: the average temperature of air (°C); average components of the wind velocity vector (m/s), including the absolute value of the averaged velocity vector (m/s), absolute value (in m/s) and the direction (in degrees) of the averaged horizontal velocity vector, absolute value (in m/s) and direction (upward-downward) of the averaged vertical component of the velocity vector; absolute value (in g/m^3) and relative (in %) humidity of air; atmospheric pressure (in mm-Hg); structure characteristics of temperature fluctuations C_T^2 (in deg$^2 \cdot$ cm$^{-2/3}$), a longitudinal component of the wind velocity C_V^2 (in (m/s)$^2 \cdot$ cm$^{-2/3}$), the acoustic refractive index $C_{n,a}^2$ (in m$^{-2/3}$), and the optical refractive index C_n^2 (in cm$^{-2/3}$). In addition, the system records standard deviations of temperature, components, and direction of the wind velocity vector, as well as coefficients of correlation, asymmetry, and the excess (between and for main measured parameters); total energy of turbulent motion (in m^2/s^2); moments of momentum and heat (in m^2/s^2); vertical fluxes of momentum and heat; dissipation rates of the kinetic energy ε (in m^2/s^3) and temperature fluctuations N (in deg^2/s); the turbulent Monin–Obukhov scales L (m), the scales of temperature T_* (°C) and wind velocity V_* (in m/s); Monin–Obukhov number ζ ($\zeta = z/L$), and other characteristics. The meteorological system also measures the correlation and structure functions, frequency spectra of fluctuations of temperature, and components of the wind velocity.

The meteorological system has passed the complete set of metrological tests, including tests in a thermal vacuum chamber, a thermal humidity chamber, and in a wind tunnel. Systematic errors of measurements are determined by calibration of the device. For the main averaged parameters, they are 0.3°C for temperature and 0.15 m/s for components of the wind velocity vector. However, the device sensitivity is much lower. Thus, for random parameters, sensitivity is 0.002°C for temperature and 0.03 m/s for the components of wind velocity. The systematic error of pressure measurements is 2 mmHg (sensitivity of 0.01 mmHg), and that for the relative humidity is 0.1%.

The passband of the sensitive element was determined by the sound transit time through the measuring head and is equal of 1.7 kHz. The upper boundary of the instrumental passband is determined by the output frequency and equals 10 Hz. The averaging, arising due to the time constant of the device, leads to the cutoff of high frequencies in the spatial spectrum of turbulence. That is why, the device is not sensitive to turbulent inhomogeneities, whose size, for example, is smaller than 10 cm in presence of the average wind velocity of 1 m/s. This time constant limits the possibilities of the experimental study of small-scale turbulence components. At the same time, it practically does not influence the accuracy of measurement of random characteristics of meteorological fields. Thus, the results of direct measurements of turbulence spectra showed that as random temperature and wind velocity are recorded, and the error introduced by the time constant of the device usually does not exceed 1%. This is caused by the insignificant contribution of the cut-off part of the spectrum to the total energy of fluctuations.

During the measurements of turbulent parameters, the averaging time, as known [1,2,4], was chosen from the condition that the scale of length of the averaged turbulent flow (average wind velocity multiplied by the averaging time) significantly exceeds the outer scale of turbulence and in the direction of average flow. Or, what is the same, the averaging time should exceed significantly the characteristic time correlation scale of the studied field. Then, the time average values are statistically stable. Measurements in the near-surface layer over a smooth surface are usually conducted with the averaging time no shorter than 100 s [1,2,4]. The corresponding length scale for the wind velocity of 1–10 m/s was of 0.1–1 km and exceeds the outer scale of turbulence. In the case of an uneven surface, the longitudinal outer scale of turbulence in the lower surface layer was obviously determined by the characteristic separation between surface irregularities. For the mountain terrain with the irregular surface, this near-surface separation is not large and can be estimated as tens of meters. Consequently, for the uneven surface, measurements in the near-surface layer can also be conducted with the averaging time of about 100 s. At measurements in a closed room, the outer scale of turbulence was limited by the room space. However, actually, it is much (5–10 times) smaller than the room space. For the averaging time of 100 s and the typical wind velocity of 0.05–0.5 m/s, the length scale is 5–50 m and exceeds the outer scale observed in the indoor environment. Therefore, in closed rooms, it is also possible to use the averaging with 100 s. In the current experiments, the averaging time at the measurement of turbulence parameters at one point was 2 min.

The relative error of measurement of the structure characteristics C_T^2, C_n^2, and C_V^2 is determined, first of all, by the device sensitivity and decreases with an increase of the average wind velocity, the averaging time, and the structure characteristics themselves. Thus, for example, under conditions of a weak turbulence $(C_n^2 = 5 \cdot 10^{-16}\ \mathrm{cm}^{-2/3})$ at the time of averaging ~ 2 min and a mean wind velocity of 0.5–10 m/s, the relative error of measurement of C_n^2 falls within the range 0.4%–14% (i.e., 0.4%, 7%, 14% for a wind velocity of 10, 1, and 0.5 m/s, respectively). The same errors are also characteristic of measurements of C_T^2 and C_V^2. The dissipation rates of kinetic energy ε and temperature N are measured based on Equation 3.5, which describes the Kolmogorov–Obukhov law. Consequently, the relative error of measurements of $\varepsilon(N)$ is practically a sum of errors in C_V^2 and the Kolmogorov constant C (C_T^2 and the Obukhov constant C_θ).

It should be noted that in later measurements we used the "Meteo3M" mobile ultrasonic meteorological system [159]. The measurement frequency in "Meteo3M" was increased and equal to 160 Hz. The sensitivity of "Meteo3M" was also increased: 0.002°C for temperature and 0.007 m/s for components of the wind velocity vector.

3.2.3.2 Requirements on the Duration of Observation Sessions

As is well known [1,2,4,7], during measurements of parameters of atmospheric turbulence in the isotropic boundary layer (plane-parallel flows over a homogeneous and homogeneously heated surface), a measurement session should necessarily be conducted in approximately the same meteorological situation, corresponding to a certain established pattern of turbulent motions. During field measurements, we should exclude the influence of diurnal variations, significant changes in the cloud situation, and, especially, cloud cover index variables. Then the stratification parameters controlling the near-surface turbulent meteorological situation over the territory (e.g., the Monin–Obukhov number or the Richardson number) are approximately constant, and the measured time average values are stable (depending only on the type of regional meteorological situation).

For measurements in the mountain anisotropic boundary layer, according to the results presented in Section 3.2.2, the requirement of the stable meteorological situation is not principal. In this boundary layer, all turbulence characteristics become functions of the Monin–Obukhov number, and every point in the layer is characterized by its own value of this number. However, to be certain, it is necessary to conduct a series of measurement sessions, in each of which the regional meteorological situation is approximately stable. From the comparison of these results, we can find the degree of influence of the meteorological situation.

In addition, as follows from Equations 3.7, 3.12, and 3.16, for the comparison of the experiment with the theory, it is necessary to measure different partial derivatives of meteorological fields. At the same time, the approximation of derivatives by difference relations usually leads to significant errors. These errors can be decreased considerably, if we use data obtained at the stable meteorological situation for measurement of the derivatives.

Thus, to conduct the necessary measurements, we should use the series of measurement sessions, in each of which the regional meteorological situation is approximately stable.

Since the known stratification parameters (Monin–Obukhov number or Richardson number) in the mountain boundary layer are different for every point of the layer, they cannot serve for indication of the general meteorological situation of the mountain region. The stratification parameters averaged over many points of the surface can likely be such indicators. However, in the process of measurements, these effective parameters are not known prior.

In this connection, during the measurements, the meteorological situation was controlled, first of all, by the stability of the solar radiation intensity, and by the cloud amount and cloud motion. During a measurement session, the sun should not be periodically covered by clouds or, if it was covered, should not be unclosed. This stability is usually observed in the region of the Baikal Astrophysical Observatory in October in the afternoon, starting from 13:00 to 14:00 local time. The duration of measurement intervals with the highly stable meteorological situation both for the open sun (scattered clouds located, as a rule, near the horizon) and for closed sun (overcast and dense clouds) is approximately identical and ranging from 3 to 5–7 h. However, unstable meteorological situations characterized by significant broken (or overcast, but not dense) clouds are observed more often (mostly, before noon). The duration of the open sun periods at broken clouds is usually shorter that 1–1.5 h.

In addition to the high-quality control of the cloud situation, the dynamics of variation of the regional turbulent meteorological situation for a particular time period can be estimated from measurements of the near-surface values of C_n^2 (or C_T^2, C_V^2) at one point. Preliminary measurements conducted in the afternoon at the same place have shown that in the case of permanently open sun at approximately the same cloud amount, the near-surface value of C_n^2 can vary (decrease 1.5–2 times) for 2–3 h due to the diurnal motion of the sun (the result is different for different geographic locations of the measurement point). At overcast or dense clouds, the same change takes longer time, approximately 2–4 h. The change of C_n^2 values appears to be more significant in the case of unstable meteorological situation and, when the sun is periodically open or closed by clouds or the density of clouds covering the sun changes.

It is clear that the duration of observation sessions should be minimized. If turbulent characteristics at one point are recorded for 5–7 min, including the measurement time of 2 min, the time for device setting-up procedures, and the time needed for transportation from one observation point to the next one, then at the highly stable meteorological situation lasting 3 h, it is possible to record about 30 points without marked errors for the measurement session. At not so stable meteorological situation, the measurement session should not last longer than 1–1.5 h.

For faster measurements and in connection with the need to move over a vast territory, we used a cross-country car; pre-measurement and post-measurement procedures were minimized. Measures were provided to avoid the influence of foreign thermal sources on the measurement results (during the measurements, the car engine was turned off, equipment and people were at a sufficient distance on the leeward).

For the period of research mission (October 12–28, 2002), five measurement sessions were carried out. In addition, preliminary observations were conducted to monitor the stability of regional turbulent meteorological situations. The first measurement session was conducted on October 17 before noon (sunny morning, wind speed 1–2 m/s, temperature from −1°C to 0°C, and humidity 49%–61%); the second session took place on October 17 in the afternoon (overcast conditions, white haze, wind speed 0.3–5 m/s, temperature from −0.4°C to +1.7°C, humidity 46%–62%); the third session took place on October 18 in the afternoon (sunny day, wind speed 1–8 m/s, temperature from 0°C to +3.7°C, and humidity 35%–63%); the fourth and fifth sessions were conducted on October 22 in the afternoon (overcast conditions, wind of 1–3 m/s, temperature −0.9°C to +1.1°C, humidity 37%–57%). In the first three sessions, the measurements were conducted along the entire path at different long sections, whereas in the fourth and fifth sessions the measurements were conducted only on the top of the LSVT mountain with a more detailed resolution, and in the fifth session the vertical measurements in the lowest 5-m layer were carried out. The first and third measurements were carried out at the fine sunny weather in similar stable meteorological situations, while the second, fourth, and fifth measurements were carried out at the stably closed sun.

3.2.3.3 Characteristics of the Measurement Path

The measurement path was chosen so it was possible to combine the measurement of the necessary turbulence parameters and the measurements of the near-surface fields of the structure characteristic C_n^2. The values of C_n^2 are needed for the prediction of image quality in the Large Solar Vacuum Telescope (LSVT). That is why the measurement path included the slopes and the top of the LSVT-mountain.

According to the goal of the measurements, the path at the slopes of the LSVT mountain should be in the LSVT observation sector, along the projection of the optical path onto the underlying surface, that is, strictly to the south from the telescope (the telescope was oriented from the north to the south). The real route by the southern slope of the LSVT mountain differs from the north–south direction by approximately 20°C to 35°C. Analogous deviations of real routes from desirable ones appear at observations on the northern slope and measurements in the canyon between two approximately parallel ridges going down to the Lake Baikal.

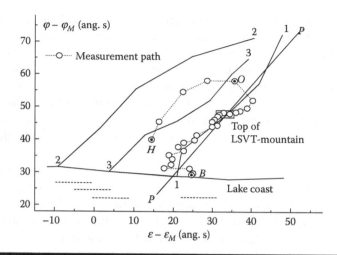

Figure 3.3 Map of the measurement path: φ is latitude, ε is longitude; $(\varphi_M, \varepsilon_M)$ are the latitude and longitude of the reference point *M*. The path includes *HO*, *OB*, and *HOB* routes. Orientation of the ridges and the canyon bottom: *11*—ridge of the LSVT-mountain; *22*—ridge of the neighboring mountain; *33*—canyon bottom; *PP*—reference vertical plane.

The entire path is divided into two sections: *HO* and *OB* (see Figure 3.3). The point *H* is located near the field office of the LSVT administration. The section *HO* connects the points *H* and *O* and passes on the slope of the mountain neighboring to the LSVT mountain. The point *O* is at the line of the canyon bottom at the conditional point of junction of the LSVT mountain and the neighboring mountain. At this conditional point, the canyon bottom goes up and the canyon terminates. The section *OB* connects the points *O* and *B*.

From the point *O*, the path passes by the northern (rear) slope of the LSVT mountain, rises to the top, and then goes down by the southern (front) slope of the LSVT mountain to the Lake Baikal to the point *B*. The point *B* is located on the lake coastal line, just near the water. All the coordinates in Figure 3.3, including the coordinates of the observation points at the measurement path are referred to the coordinates of some reference point *M*, which lies in the water area of the lake several hundreds of meters from the lake coast near the point *B*.

In the cases when it is necessary to study the behavior of turbulence characteristic along the entire path trajectory (usually at a height of 2.7 m above the surface), it is convenient to use the central ground angle α as an independent argument. Then, the path sections corresponding to observation sessions 1–3 are sewed in one route *HOB*. In the angle α, the one arc corresponds to approximately 31 m on the geoid surface. Thus, the central angle α increases at the motion along the trajectory of the compound path. Sometimes, the *HOB* route (sessions 1–3) complements with the route on the top of the LSVT mountain (session 4, with the more detailed resolution).

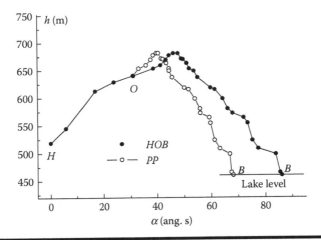

Figure 3.4 **Vertical profile of the measurement path:** h **is height above the sea level,** α **is the central ground angle. Closed dots correspond to the** *HOB* **route, while open dots correspond to the** *OB* **route along the reference plane** *PP* **(projections of points of the** *OB* **route onto the** *PP* **plane).**

Figure 3.4 shows the recorded vertical profile of the measurement path (about the sea level). For comparison, the figure shows the vertical profile of the *OB* path section, where every point is projected onto the auxiliary reference vertical plane *PP*.

This projection allows us to assess the path zigzagging and maximal deviations from the fixed direction specified by the *PP* plane, as well as deviations from the direction of telescope orientation (north–south). As can be seen from Figure 3.4, the actual vertical profile of the LSVT-mountain corresponding to the projection onto the *PP* plane appears to be sharper than the profile of the measurement path.

Deviations of the vertical profile of the measurement path from the averaged vertical profile of the LSVT mountain are 5–10 m (both up and down). The underlying surface along the measurement path is covered by thin (mostly pine) forest 7–15 m high. At some sections, the forest is absent, and rocks somewhere covered by low grass are observed.

3.2.3.4 General Results of Measurement of Turbulence Characteristics in the Mountain Boundary Layer

It was found experimentally that the turbulence in the mountain boundary layer is characterized by significant anisotropy. Even at the stable regional meteorological situation, depending on the location of the measurement point, the structure characteristics C_T^2, C_V^2, and C_n^2 can vary by more than two orders of magnitude. The strong influence is exerted by surface areas with the developed heat exchange and high heat capacity (e.g., the influence of the Lake Baikal surface). Near these areas, C_T^2, C_V^2, and C_n^2 are practically independent of the type of meteorological situation.

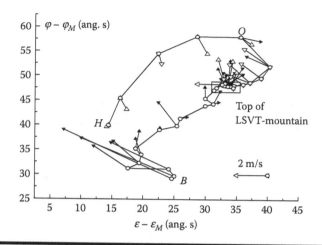

Figure 3.5 **Wind map of the measurement path at the stable meteorological situation. Arrows with open triangles correspond to session 1, arrows with closed triangles are for session 3. The measurement path is shown by the solid curve (*HO, OB, HOB* routes).**

Large inhomogeneities of terrain and human-made objects create stable rotor perturbations of air flows. Figures 3.5 and 3.6 depict the wind maps of the observation area. As can be seen from Figure 3.5, air flows on the slopes of two parallel ridges are directed to the canyon bottom (between the ridges). At close meteorological situations, the stability of these flows was observed (near point *O*).

At the top of the LSVT mountain, wind flows from the lake reflect from a height (26 m) of the LSVT building, and, as can be seen from Figure 3.6, stable rotor flows take place (especially, near points 8, 10, 11, 14 in Figure 3.6). Near the centers of these vortex formations, we observed low values of the velocity vector, increased pressure, and decreased humidity.

In the vicinity of the centers, the local temperature stratification can alternate from very unstable to stable. This can be seen, for example, from Figure 3.7, which shows the field of recorded values of the Monin–Obukhov number ζ at the top of the LSVT-mountain at a height of 2.7 m from the surface. Thus, at the same, as in Figure 3.6, point 8 the local very unstable stratification is observed ($\zeta = -388$). However, just near, at point 11 (several tens of meters from point 8), the Monin–Obukhov number already corresponds to the local stable stratification ($\zeta = +0.3$).

As follows from our findings, the values of the Monin–Obukhov number ζ measured at the same height of 2.7 m vary at the same observation point at the alternation of the type of regional meteorological situation. The main turbulence characteristics, namely, average dissipation rates of kinetic energy ε and temperature N, as well as turbulent scales of temperature T_* and velocity V_*, also vary considerably in the mountain boundary layer.

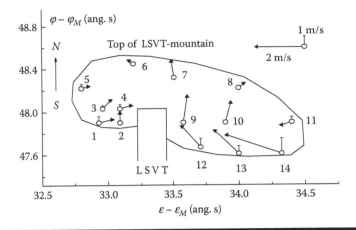

Figure 3.6 **Wind map of the top of the LSVT-mountain (field of 3D vectors of mean wind at the mountain top at a field near LSVT, whose boundary is shown by the solid curve): φ is latitude, ε is longitude, $(\varphi_M, \varepsilon_M)$ are the latitude and longitude of the reference point M. Observation session 4, measurements at a height of 2.7 m from the underlying surface. Vertical bars show the vertical component of the mean 3D wind vector.**

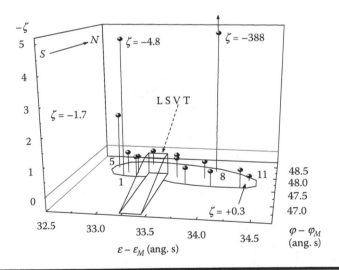

Figure 3.7 **Field of values of the Monin–Obukhov number ζ on the top of the LSVT mountain. Observation session 4, measurements at a height $z = 2.7$ m from the underlying surface. The values of ζ widely different from the stratification boundaries $|\zeta| = 0.05$ are shown separately; for other points: $-0.34 < \zeta < -0.05$.**

The average temperature T (°C) is also variable. From measurements of the dissipation rate ε by the equation $l_0 = \nu^{3/4}\varepsilon^{-1/4}$, where $\nu = 1.3 \cdot 10^{-5}$ m²/s is the kinematic viscosity of air at 0°C, we have obtained the values of the Kolmogorov inner scale of turbulence l_0. The values of the scale l_0 fall in a range from 0.3 to 1.2 mm (average of 0.64 mm). The Kolmogorov inner scale is in the inverse dependence with the intensity of fluctuations of the air flow velocity. The smaller the scale, the larger the velocity fluctuations, which increase near rather large obstacles.

3.2.3.5 Experimental Verification of Local Weak Anisotropy of the Mountain Boundary Layer via the Anisotropy Functions

According to the results of Section 3.2.2, for measurements in the mountain anisotropic boundary layer, the requirement of the stable regional meteorological situation is not principal. In this boundary layer, all turbulence characteristics become functions of the Monin–Obukhov number.

Figures 3.8 and 3.9 depict the experimental results for the scales of temperature T_* and velocity V_* for all observation sessions as functions of the Monin–Obukhov number. In every session, the regional meteorological situation is described by its own set of the Monin–Obukhov numbers, varying in the entire range $-581 \leq \zeta \leq 0.3$. As can be seen from Figures 3.8 and 3.9, the joining of all observation sessions in one does not leads to the significant spread of the data.

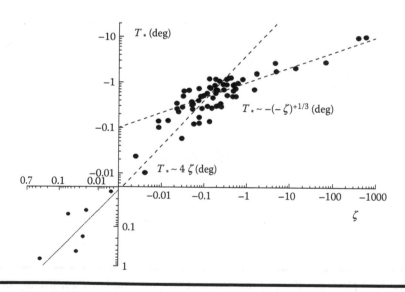

Figure 3.8 Turbulent scale of temperature field T_* in the mountain boundary layer for all measurement sessions as a function of the Monin–Obukhov number ζ.

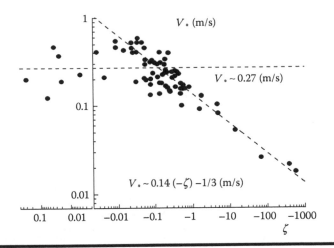

Figure 3.9 **Turbulent scale of the velocity field V_* in mountain boundary layer for all measurement sessions as a function of the Monin–Obukhov number ζ.**

All the data stably group around certain smoothed dependence curves shown in Figures 3.8 and 3.9. Some spread of the points observed in the zone of stable stratification and at $(\zeta > +0.05)$ appears permanently in the most measurements of different turbulent characteristics in the atmosphere (taken from different authors). Thus, Monin and Yaglom [1,2] explained this fact by the intermittence of turbulence under stable conditions and, consequently, by the insufficient ordinary averaging time. As it follows from our measurements, in the mountain boundary layer, the experimental results, as functions of the universal parameter—the Monin–Obukhov number—can be joined regardless of the type of regional meteorological situation (at least, for meteorological situations observed during the period of measurements).

Figures 3.10 and 3.11 show the results of comparison of the semiempirical theory with the experiment for the functions

$$D^T = \beta_{31} \cdot \partial T / \partial x_1 + \beta_{33} \cdot \partial T / \partial x_3, \quad D^V = \partial v_1 / \partial x_3 + \partial v_3 / \partial x_1$$

in the mountain boundary layer (see Equation 3.7, in which at the mean wind velocity along the axis x_1 the horizontal transverse derivative $\partial T / \partial x_2$ can be neglected).

These figures also depict the experimental values of the derivatives $\partial T / \partial z$ and $\partial u / \partial z$. In the case of the isotropic boundary layer (when ζ is fixed, there are no longitudinal derivatives and $D^T = \partial T / \partial z$, $D^V = \partial u / \partial z$), Equation 3.7 is fundamental in the semiempirical theory of turbulence (in the similarity theory) and are reliably confirmed by experiments [1,2,4].

In the anisotropic layer, all components of equalities (3.7) are functions of the number ζ varying at an arbitrary shift of the observation point. As can be seen from

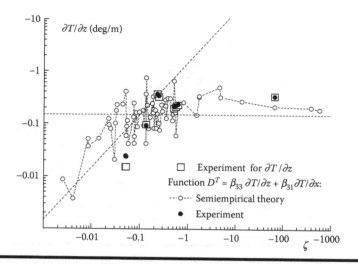

Figure 3.10 Comparison of experimental and theoretical values of the vertical derivative (with respect to *z*) of the mean temperature of air $\partial T/\partial z$ in the mountain boundary layer. Open squares are for experimental values of $\partial T/\partial z$ from measurements in the lower 5-m layer near LSVT (observation session 5). Open and closed circles are, respectively, for the semiempirical theory and experiment for the function D^T in Equation 3.7. Straight lines are asymptotics of the theoretical function D^T at $|\zeta| \to 0$ and $\zeta \to -\infty$.

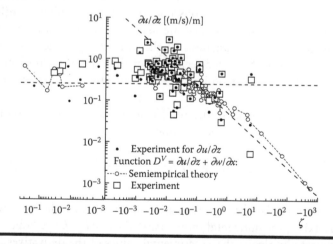

Figure 3.11 Comparison of experimental and theoretical values of the vertical derivative (with respect to *z*) of the longitudinal component (*u*) of the mean wind velocity $\partial u/\partial z$ in the mountain boundary layer. Closed circles are experimental values of $\partial u/\partial z$ in all the measurement sessions. Open circles and open squares are, respectively, the semiempirical theory and experiment for the function D^V [in Equation 3.7]. Straight lines are asymptotics of the theoretical function D^V at $|\zeta| \to 0$ и $\zeta \to -\infty$.

Figures 3.10 and 3.11, in the wide range of variation of the Monin–Obukhov number ζ, the semiempirical theory is in satisfactory agreement with the experiment. The consideration of longitudinal derivatives improves this agreement.

It can be seen from Equations 3.7 and 3.3 that for the anisotropic layer the theoretical functions D^T and D^V can be represented in the form

$$D^T = T_*(\zeta) \cdot \varphi(\zeta)/z, \quad D^V = V_*(\zeta) \cdot \varphi(\zeta)/(\varkappa z),$$

where the scales $T_*(\zeta)$ and $V_*(\zeta)$ are functions of ζ (see Figures 3.8 and 3.9). The same representations are also valid for the isotropic layer, but the scales T_* and V_* in this case are constant.

As can be seen from the measurements (see Figures 3.10 and 3.11), inside the anisotropic layer there are local areas, in which we can neglect longitudinal derivatives in the functions D^T, D^V in comparison with the vertical ones. This means that in these areas the conditions of the isotropic layer take place for the functions D^T and D^V. The plane-parallel flow over an extended part of the surface can be considered as an extension of some small local area with the isotropic conditions. Consequently, in extended isotropic layers, the constant scales T_* and V_* are not arbitrary, but as can be seen from the theoretical reason for D^T and D^V, are determined by the particular value of ζ.

Figures 3.12 and 3.13 show the results of comparison of the semiempirical theory with the experiment for the energy and temperature anisotropy functions $\varphi_V(\zeta)$ and $\varphi_T(\zeta)$. As can be seen from Figures 3.12 and 3.13, the theory is in the satisfactory agreement with the experiment. The theoretical values of these functions

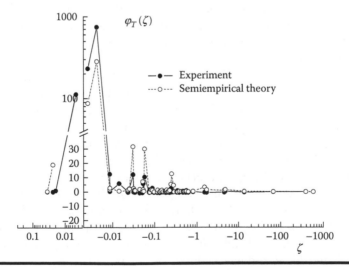

Figure 3.12 Comparison of experimental and theoretical results for the temperature anisotropy function.

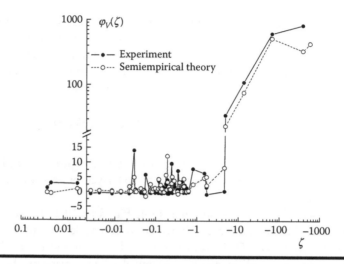

Figure 3.13 **Comparison of experimental and theoretical results for the energy anisotropy function.**

appear to be close to the experimental values in a wide range of variability of the Monin–Obukhov number (from stable to very unstable local temperature stratifications, $-581 \leq \zeta \leq 0.3$).

The anisotropy functions have maxima in different ranges of variability of the Monin–Obukhov number ζ. If $\varphi_T(\zeta)$ is mostly concentrated in the range $|\zeta| \lesssim 0.1$, then $\varphi_V(\zeta)$ concentrates in the range $-1 \gtrsim \zeta \gtrsim -1000$. Beyond these ranges, the both functions are close to zero, while at the maxima they achieve values close to 1000. Despite the low accuracy of measurement of the derivatives of hydrodynamic fields, the agreement is observed at the variation of the functions $\varphi_V(\zeta)$ and $\varphi_T(\zeta)$ by more than three orders of magnitude. Therefore, it cannot be a consequence of experimental errors.

Since the functions $\varphi_V(\zeta)$ and $\varphi_T(\zeta)$ determine the dissipation rates ε and N, it follows from Figures 3.12 and 3.13 that the anisotropic boundary layer influences the energy (ε) and temperature (N) turbulence characteristic in the significantly nonsymmetrical way. The appearance of maxima of the anisotropy functions in different ranges of variation of the Monin–Obukhov number ζ is associated with the corresponding behavior of the scales $T_*(\zeta)$ and $V_*(\zeta)$ in these areas. As can be seen from the smoothed empirical dependences for the scales $T_*(\zeta)$ and $V_*(\zeta)$ in Figures 3.8 and 3.9, at small values of $|\zeta|$ ($|\zeta| \rightarrow 0$) $T_*(\zeta) \rightarrow 0$, and at large negative values of ζ ($\zeta \rightarrow -\infty$) $V_*(\zeta) \rightarrow 0$. Therefore, if in these areas the derivatives $\partial T/\partial x_1$ (at $|\zeta| \rightarrow 0$), $\partial v_1/\partial x_1$ or $\partial v_3/\partial x_3$ (at $\zeta \rightarrow -\infty$) are limited (and nonzero), then $\varphi_T(\zeta) \rightarrow \infty$ at $|\zeta| \rightarrow 0$ and $\varphi_V(\zeta) \rightarrow \infty$ at $\zeta \rightarrow -\infty$ (due to normalization to the scales T_* and V_*) according to Equations 3.16 and 3.12.

In the range of variation of the Monin–Obukhov number $-0.1 \gtrsim \zeta \gtrsim -1$, in which the both anisotropy functions are simultaneously close to zero, Equations 3.8

and 3.10 for the anisotropic dissipation rates ε and N coincide with the equations for the isotropic dissipation rates. Consequently, in this range of ζ, the conditions of the isotopic layer take place in the anisotropic boundary layer.

Thus, it follows from the results of the measurements in the mountain boundary layer, the assumption on the local weak anisotropy of the arbitrary boundary layer is fulfilled with a good accuracy. Consequently, the arbitrary boundary layer can be considered as locally weakly anisotropic. This means that, by introducing the anisotropy functions $\varphi_V(\zeta)$ and $\varphi_T(\zeta)$, we can extend the similarity theory for the isotropic dissipation rates to the arbitrary anisotropic boundary layer.

3.2.4 Effective Isotropic Layer

In Section 3.2.3, it is shown that in a wide range of variation of the Monin–Obukhov number (at the range of $-0.1 \gtrsim \zeta \gtrsim -1$), the conditions of isotropic layer take place in the mountain boundary layer. It is interesting to answer the question: is it possible to replace an arbitrary boundary layer with some effective isotropic boundary layer? This will allow researchers to use simple semiempirical relations valid in the isotropic layer for description of the anisotropic layer.

As follows from Equations 3.8, 3.10, 3.13, and 3.16, in the arbitrary boundary layer, the dissipation rates ε and N depend on five parameters: V_*, T_*, T, $\partial u/\partial x$, and $\partial T/\partial x$. We denote $\partial u/\partial x = V^X(\zeta)$ and $\partial T/\partial x = T^X(\zeta)$ and will assume that the form of the functions $V^X(\zeta)$ and $T^X(\zeta)$ is known. Then the equations for ε and N can be represented in the form

$$\varepsilon = V_*^3 \cdot \text{\ae}^{-1} \cdot z^{-1} \cdot [\varphi(\zeta) - \zeta + \varphi_V(\zeta, V_*, V^X(\zeta))],$$

$$N = \alpha \cdot \text{\ae} \cdot V_* \cdot T_*^2 \cdot z^{-1} \cdot [\varphi(\zeta) + \varphi_T(\zeta, T_*, T^X(\zeta))], \qquad (3.17)$$

$$\zeta = z \cdot \alpha \cdot \text{\ae}^2 \cdot g \cdot T_* \cdot V_*^{-2} \cdot T^{-1},$$

where all arguments in the functions φ_V and φ_T are written in the explicit form. Setting the values of the parameters V_*, T_*, T in system (3.17), we can find the values of the left-hand sides ε, N, ζ. Converting the problem, we can find V_*, T_*, T from known ε, N, ζ.

System of Equation 3.17 corresponds to the anisotropic layer. If we take $\varphi_V = 0$, $\varphi_T = 0$ in Equation 3.17, then system (3.17) describes the isotropic layer. Upon the introduction of new variables $V_{*\text{eff}}$, $T_{*\text{eff}}$, and T_{eff} instead of V_*, T_*, and T, we get

$$\varepsilon = V_{*\text{eff}}^3 \cdot \text{\ae}^{-1} \cdot z^{-1} \cdot [\varphi(\zeta) - \zeta], \quad N = \alpha \cdot \text{\ae} \cdot V_{*\text{eff}} \cdot T_{*\text{eff}}^2 \cdot z^{-1} \cdot \varphi(\zeta),$$

$$\zeta = z \cdot \alpha \cdot \text{\ae}^2 \cdot g \cdot T_{*\text{eff}} \cdot V_{*\text{eff}}^{-2} \cdot T_{\text{eff}}^{-1}. \qquad (3.18)$$

Equate the left-hand sides of systems (3.17) and (3.18) to each other, which is equivalent to taking the values of ε, N, and ζ from the actual mountain layer in

Equation 3.18. Upon the solution of system of Equation 3.18, we find the values of the parameters $V_{*\text{eff}}$, $T_{*\text{eff}}$, and T_{eff} corresponding to the effective isotropic layer.

The problem can be simplified, if we assume that the relative changes of the absolute temperature are small and take the absolute temperature equal to its average values over all measurement sessions $T = T_{\text{eff}} = 273.7$ K (average value $\langle T \rangle = +0.5°C$ was deviated in different sessions by 1–2°C). Then, upon expression of one variable through other from the third equation of system (3.18) and substitution of this variable into the first two equations, we obtain the system of two nonlinear equations with two unknowns $V_{*\text{eff}}$ and $T_{*\text{eff}}$.

Figures 3.14 and 3.15 depict the results of comparison of the turbulent scales of temperature and velocity of the anisotropic boundary layer T_* and V_* with the effective scales for the isotropic layer $T_{*\text{eff}}$ and $V_{*\text{eff}}$. It follows from these figures that the effective turbulent scales of temperature and velocity $T_{*\text{eff}}$ and $V_{*\text{eff}}$ appear to be close to constants in a wide range of variation of the Monin–Obukhov number ($-581 \leq \zeta \leq +0.3$). As is well-known, these scales should be constant in the isotropic layer. The data of Figures 3.14 and 3.15 suggest that the anisotropic boundary layer can be replaced with the effective isotropic boundary layer.

The effective scales of the isotropic layer averaged over all observation points (denoted by brackets) turn out to be $\langle T_{*\text{eff}} \rangle = -0.07°C$, $\langle V_{*\text{eff}} \rangle = 0.05$ m/s. These values of the scales at $z = 2.7$ m and $T_{\text{eff}} = 273.7$ K correspond to the Monin–Obukhov number of the effective isotropic layer $\langle \zeta_{\text{eff}} \rangle = -0.5$. These values of the three parameters, $\langle T_{*\text{eff}} \rangle$, $\langle V_{*\text{eff}} \rangle$, and $\langle \zeta_{\text{eff}} \rangle$, characterize completely the effective isotropic layer corresponding to the real mountain boundary layer. The value of $\langle \zeta_{\text{eff}} \rangle$ falls within the range with the conditions of the isotropic layer ($-0.1 \gtrsim \zeta \gtrsim -1$) and can serve an indicator of the general meteorological situation over the

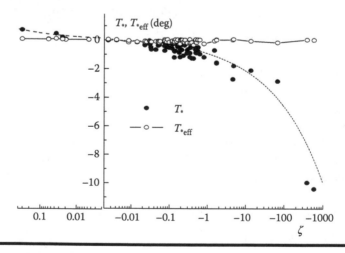

Figure 3.14 Comparison of the turbulent scale of temperature of the anisotropic boundary layer T_* with the effective scale for the isotropic layer $T_{*\text{eff}}$: $T_* \sim 4\zeta$ deg, $\zeta > \zeta_*$ ($\zeta_* = -0.125$); $T_* \sim -(-\zeta)^{+1/3}$ deg, $\zeta < \zeta_*$; $\langle T_{*\text{eff}} \rangle = -0.066°C$.

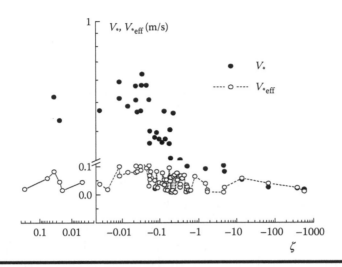

Figure 3.15 Comparison of the turbulent scale of velocity (friction velocity) of the anisotropic boundary layer V_* with the effective scale for the isotropic layer V_{*eff}: $V_* \sim 0.27$ m/s, $\zeta > \zeta_* (\zeta_* = -0.125)$; $V_* \sim 0.14 (-\zeta)^{-1/3}$ m/s, $\zeta < \zeta_*$; $\langle V_{*eff} \rangle = 0.049$ m/s.

mountain region under study during the experiments. Thus, if we find the values of the parameters $\langle T_{*eff} \rangle$, $\langle V_{*eff} \rangle$, and $\langle \zeta_{eff} \rangle$, characterizing the effective isotropic layer, then we can use Equation 3.18 ignoring the anisotropy functions $\varphi_V(\zeta)$ and $\varphi_T(\zeta)$.

The possibility of the replacing the anisotropic boundary layer with the effective isotropic layer allows us to use the optical models of turbulence developed for the isotropic boundary layer. For this, it is necessary to know the characteristic temperature $\langle T_{*eff} \rangle$ and velocity $\langle V_{*eff} \rangle$ scales average for the region of observation. If these scales are known, then, by use of Equation 3.18, we can reconstruct the regional average parameters, $\langle \varepsilon_{eff} \rangle$, $\langle N_{eff} \rangle$, and $\langle \zeta_{eff} \rangle$, corresponding to the effective isotropic layer. Finally, by the known equations of similarity theory [1,2,4,178], we can find the near-surface values of the optical parameters of turbulence C_n^2, L_0, and l_0, which usually [4,178] serve as main initial parameters for construction of vertical profiles (optical models of turbulence). Thus, the replacement of the anisotropic boundary layer with the effective isotropic layer assumes only that at every point of the underlying surface, it is possible to calculate the necessary vertical profile with the use of the calculated regional average near-surface value. In this case, no requirements are imposed on the terrain of the underlying surface, and therefore it should be taken into account separately.

For the first time, the mountain anisotropic boundary layer in References 179, 180 was replaced with the effective isotropic layer with allowance for the terrain of the underlying surface. It was shown that the dependence of jitter of astronomical images on the zenith angle is determined by the structure of terrain of the underlying surface at the observation site. Thus, we terminated the many-year discussion of astronomers about significant deviations from the well-known Tatarskii secant law [4]. The principle of interchangeability in the semiempirical theory of turbulence is formulated in

Reference 30. Using this principle, we can reduce the terrain variations to the corresponding variations of near-surface values of optical parameters of turbulence. This gives us the possibility of considering near-surface values locally deviating from the regional average value. The joint application of the replacement of the anisotropic layer with the effective isotropic one and the principle of interchangeability has allowed us to give recommendations on selection of optimal, from the viewpoint of the image quality, places for location of new ground-based telescopes, and recommendation on improvement of the image quality in already existing telescopes [31–35].

3.2.5 Outer Scale of Turbulence in the Anisotropic Boundary Layer

As is well known, the outer scale of turbulence L_0 can be determined in different ways. For example, Tatarskii [4] determines the vertical outer scale L_0^T from the condition of equality of the average squared difference of random temperature values at two points z_1 and z_2 to its systematic difference (scale of turbulent mixing). This condition gives

$$C_T^2 \mid z_1 - z_2 \mid^{2/3} = (dT/dz)^2 \mid z_1 - z_2 \mid^2;$$
$$L_0^T = \mid z_1 - z_2 \mid /(\alpha \cdot C_\theta)^{3/4} = \{C_T^2 /[\alpha \cdot C_\theta \cdot (dT/dz)^2]\}^{3/4}, \tag{3.19}$$

where, as before, $\alpha = \mathrm{Pr}^{-1} \approx 1.17$; C_θ is the Obukhov constant. We can determine the outer scale L_0^D from deviation of the structure function of temperature fluctuations from the 2/3 law. In the space of Fourier transforms, this scale corresponds to the scale L_0^V determined from deviations of the 1D spatial or temporal frequency spectra from the 5/3 law (see also information on turbulence spectral dependence in Chapter 1). There are also scales, which are parameters in different theoretical models of the energy range of the 3D spectrum of fluctuations (e.g., the von Kármán outer scale L_0^K). For practical needs, it is interesting to find the relations between these scales, to obtain the theoretical ideas about their applicability in the anisotropic boundary layer, and to compare the theoretical and experimental results in the unified manner.

For the von Kármán model of the 3D spectrum of turbulence, the structure function $D(r)$ and 1D spectral density $V(k)$ are described by the equations [4]:

$$D_\nu(r) = 2 \cdot a_\nu^2 \cdot [1 - 2^{1-\nu} \cdot \Gamma^{-1}(\nu) \cdot (r/r_0)^\nu \cdot K_\nu(r/r_0)];$$
$$V_\nu(k) = \Gamma^{-1}(\nu + 1/2) \cdot \Gamma^{-1}(\nu) \cdot \pi^{-1/2} \cdot a_\nu^2 \cdot r_0 \cdot (1 + k^2 r_0^2)^{-\nu-1/2}, \tag{3.20}$$

where r_0 is some spatial scale (correlation radius); a_ν^2 is variance; and K_ν is the MacDonald function. Considering, for example, temperature fluctuations, we should take $\nu = 1/3$ and account for $r_0^{-1} = k_0 = 2\pi/L_0^K$, where L_0^K is the von Kármán outer scale.

Expanding $D_\nu(r)$ and $V_\nu(k)$ at $\nu = 1/3$ in power series of r/r_0 and k_0/k, respectively, and at $\nu = 4/3$ (assuming $r_0 = r_1$, $k_0 = k_1$) in power series of r/r_1 and k_1/k, we finally get

$$D_{1/3}(r) = \alpha_0 r^{2/3} - \alpha_1 r^2 + O((r/r_0)^{8/3}), \quad V_{1/3}(k) = \beta_0 k^{-5/3} - \beta_1 k^{-11/3} + O((k_0/k)^{17/3}),$$

$$D_{4/3}(r) = \alpha_2 r^2 + O((r/r_1)^{8/3}), \quad V_{4/3}(k) = \beta_2 k^{-11/3} + O((k_1/k)^{17/3}).$$

Here, α_0, α_1, β_0, and β_1 are the positive constants dependent on $a_{1/3}^2$ and r_0; α_2 and β_2 depend on $a_{4/3}^2$ and r_1. The parameters $a_{1/3}^2$, r_0, and $a_{4/3}^2$, r_1 can be related to each other, if we impose the conditions $\alpha_0 = C_T^2$, $\alpha_2 = (dT/dz)^2$ on α_0 and α_2. These conditions follow from definition (3.19). They allow us to find the relation between the Tatarskii scale L_0^T and other scales.

Determine the outer scale L_0^V $\left(k_* = 2\pi/L_0^V\right)$ from the condition of intersection of $V_{1/3}(k)$ and $V_{4/3}(k)$ at the point k_*, at which the relative deviation of $V_{1/3}(k)$ from the dependence $\beta_0 k^{-5/3}$ (corresponding to the inertial range) is equal to the preset value δ_V. Analogously, we can determine the outer scale L_0^D $\left[L_0^D = r_*/(\alpha C_\theta)^{3/4}\right]$ from the condition of intersection of $D_{1/3}(r)$ and $D_{4/3}(r)$ at the point r_*, at which the relative deviation of $D_{1/3}(r)$ from the inertial interval (dependence $\alpha_0 r^{2/3}$) is equal to δ_D. The deviations δ_V and δ_D turn out to be related. Thus, at $|\delta_V| \ll 1$ we have $|\delta_D| \approx 1.14 |\delta_V|^{3/4}$.

Thus, we have four outer scales determined in different ways: L_0^T, L_0^K, L_0^V, and L_0^D. At small deviations δ_V and δ_D, all these scales are related by linear dependences (with awkward expressions for the coefficients). For example, at $\delta_V = 0.3$ ($\delta_D \approx 0.37$), we obtain the following expressions of the scales through the Tatarskii scale:

$$L_0^V \approx 7.3 L_0^T, L_0^D \approx 0.72 L_0^T, \quad L_0^K \approx 12.4 L_0^T \tag{3.21}$$

(or through the von Kármán outer scale: $L_0^V \approx 0.6 L_0^K$, $L_0^D \approx 0.06 L_0^K$, $L_0^T \approx 0.08 L_0^K$). With allowance for the known relation between the von Kármán L_0^K and exponential outer scales L_0 (usually, $L_0 = 0.54 L_0^K$ or $L_0^K \approx 1.85 L_0$ [175–177]). Equations 3.21 establish the relations between five outer scales determined by five different methods: L_0, L_0^T, L_0^K, L_0^V, and L_0^D.

As follows from definitions (3.5), $C_T^2 = C_\theta \varepsilon^{-1/3} N$. We can substitute Equations 3.8 and 3.10 for ε and N in the anisotropic layer into this equation. The vertical derivative dT/dz can be expressed from Equation 3.7, where $D^T = -0.49$ $\partial T/\partial x + \partial T/\partial z$. Upon the substitution of C_T^2 and $\partial T/\partial z$ into definition (3.19), we find the equation for the Tatarskii outer scale generalized for the case of an arbitrary anisotropic layer:

$$L_0^T = \mathfrak{æ} \cdot z \cdot [\varphi(\zeta) + \varphi_T(\zeta)]^{3/4} [\varphi(\zeta) + \varphi_V(\zeta) - \zeta]^{-1/4} \cdot |\varphi(\zeta) + 0.49 z \cdot T_*^{-1} \cdot \partial T/\partial x|^{-3/2}.$$

$$\tag{3.22}$$

Assuming here $\varphi_T(\zeta) = 0$, $\varphi_V(\zeta) = 0$, $\partial T/\partial x = 0$, we obtain the well-known equation for the isotropic layer

$$L_0^T = \mathfrak{x} \cdot z \cdot \varphi(\zeta)^{-3/4} \cdot [\varphi(\zeta) - \zeta]^{-1/4}. \tag{3.23}$$

In the isotropic layer, the simpler equation [4] $L_0^T = \mathfrak{x} \cdot z/\varphi(\zeta)$ is also used. It differs insignificantly from Equation 3.23 in the limiting cases of very unstable and very stable stratification.

Compare the theory with the experiment. For this, we use different methods for obtaining of experimental values of the vertical outer scale.

One of such methods is the substitution of measured values of C_T^2 and $\partial T/\partial z$ into definition (3.19) (conditionally, this method can be referred to as "by the Tatarskii definition"). As can be seen from Figure 3.10, the experimental values of $\partial T/\partial z$ were determined from measurements in the lower 5-m layer (observation session 5, a total of six points for $\partial T/\partial z$), and they are relatively few. Therefore, for a more complete comparison, we present other independent methods allowing reconstruction of experimental values of the outer scale. These methods can be proposed from the results of measurement of temporal frequency spectra of temperature fluctuations.

Figure 3.16 shows sampled frequency spectra of temperature $W(f)$ obtained in our measurements at different values of the Monin–Obukhov number ζ. As follows from Figure 3.16, all the spectra are characterized by the presence of the 5/3-inertial frequency range f, in which $W(f) \sim f^{-5/3}$, and saturation in the range of low frequencies, which is well described by the von Kármán model (see also Chapter 1).

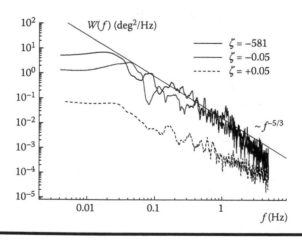

Figure 3.16 Experimental nonnormalized spectra of temperature fluctuations. The upper curve in the low-frequency range corresponds to the very unstable stratification, while the lower curve corresponds to the stable stratification.

We apply the von Kármán model of spectrum (3.20) for determination of the von Kármán outer scale L_0^K from stable characteristics of the spectrum. These characteristics include the value of the spectrum at the lower boundary of the recorded frequency range (denote it as $W(0)$) and the value of the coefficient w_* of $f^{-5/3}$ in the inertial range $[W(f) = w_* \cdot f^{-5/3}]$. We use the equation [4] $V(k) = v\, W_{\exp}(kv)$, where v is the absolute value of the average wind velocity vector ($v = \mathbf{v}$). This equation relates the 1D spatial spectrum p $V(k)$ determined by Equation 3.20 with the temporal frequency spectrum $W_{\exp}(\omega)$ being the ordinary 1D Fourier transform of the correlation function ($\omega = 2\pi f$). Taking into account that $W(f)$ is the transformation by positive frequencies and $W(f) = 4\pi \cdot W_{\exp}(2\pi f)$, we find two methods for determination of the von Kármán scale L_0^K from characteristics of the spectra:

$$(1)\ L_0^K = 4.8[W(0) \cdot v/C_T^2]^{3/5}, \quad (2)\ L_0^K = (v/f)\{[W(0)/W(f)]^{6/5} - 1\}^{1/2}. \quad (3.24)$$

The second method for the frequencies of the inertial range simplifies and gives (2) $L_0^K \approx v \cdot [W(0)/w_*]^{3/5}$. We can conditionally refer to the first of these methods as "from spectra by saturation" and the second one as "from spectra by the 5/3 law."

Figure 3.17 shows the results of comparison of experimental and theoretical results for the Tatarskii outer scale L_0^T in the mountain boundary layer. With the use of experimental values of the von Kármán scale L_0^K obtained from spectra based

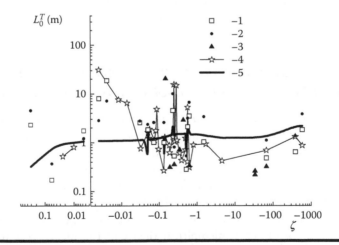

Figure 3.17 **Comparison of experimental and theoretical results for the Tatarskii outer scale of turbulence L_0^T in the mountain anisotropic boundary layer: (1) experiment (from spectra by the 5/3 law); (2) experiment (from spectra by saturation); (3) experiment (by Tatarskii definition); (4) semiempirical theory for the anisotropic layer; (5) semiempirical theory for the isotropic layer.**

on methods (3.24), the coefficient of the von Kármán scale into the Tatarskii scale (3.21) was applied. This coefficient is applicable to any boundary layer.

The comparison of the scales L_0^T measured by three methods (by Tatarskii definition, from spectra by saturation, and from spectra by the 5/3 law) shows that in the anisotropic boundary layer, the agreement between the experiment and the semiempirical theory is satisfactory (3.22).

For comparison, we used the data of all observation sessions (including session 5 of high-altitude observations). Therefore, because of the clear linear height dependence, the theoretical scales scale of both the isotropic $[L_0^T = \mathscr{a}z/\varphi(\zeta)]$ and anisotropic layers (3.22) demonstrate jumps at some values of ζ (where the height z differs from its permanent value of 2.7 m). As can be seen from Figure 3.17, for this ζ the experimental data also show jumps. In addition, in the region with insignificant anisotropy ($-0.1 \gtrsim \zeta \gtrsim -1$), the theoretical values of the anisotropic and isotropic outer scales, as expected, turn out to be close (curves 4 and 5 have jumps at practically coinciding points ζ).

In the region of very unstable local stratification, the anisotropic outer scale is smaller than the isotropic one. As follows from (3.22), this decrease is caused by the factor $[\varphi(\zeta) + \varphi_V(\zeta) - \zeta]^{-1/4}$, in which the values of $(\varphi_V(\zeta) - \zeta)$ are large. The both scales (anisotropic and isotropic) decrease in the region of weakly stable stratification. The marked difference between them (anisotropic scale larger than isotropic one) is observed in the range of dynamic turbulence (neutral stratification). The increase of the anisotropic scale is connected with the growth of the function $\varphi_T(\zeta)$ in this range. However, if we take into account the longitudinal derivative $\partial T/\partial x$ in theoretical Equation 3.22, then the increase of the anisotropic outer scale becomes limited. As can be seen from Figure 3.17, this improves the agreement between the theory and the experiment.

3.2.6 Further Development of the Semiempirical Theory of Turbulence in the Anisotropic Boundary Layer

In the later studies within the scope of the semiempirical theory of turbulence [39,40], new results have been presented for the turbulent scale of the velocity and temperature fields measured under various climatic conditions and different regions.

The numerical regularization algorithm for the reconstruction of spatial derivatives of the average temperature from long series of experimental data obtained in the B atmospheric anisotropic boundary layer has been developed [41]. The algorithm is based on the solution of the heat conduction equation with allowance for the relationships of the semiempirical theory of turbulence. The determined average derivatives have allowed to find the experimental dependence of the turbulent Prandtl number on the Monin–Obukhov number. It was shown that in the atmospheric boundary layer in the rather wide range of the Monin–Obukhov number, the turbulent Prandtl number only slightly differs from unity. A decrease of the inverse turbulent Prandtl number at strong instability was noticed.

3.2.6.1 Turbulent Scales of the Monin–Obukhov Similarity Theory in the Anisotropic Boundary Layer

The turbulent scale of velocity V_* (i.e., the friction velocity) and temperature T_* are important characteristics of turbulence in the Monin–Obukhov similarity theory [1,2]. In the case of plane-parallel flows (that is, in the case of an isotropic boundary layer), the velocity and temperature scales are connected with the vertical derivatives of the average horizontal flow velocity u and the average absolute temperature T as [1]

$$\mathrm{d}u/\mathrm{d}z = V_*/(\mathrm{æ} \cdot z), \quad \mathrm{d}T/\mathrm{d}z = T_*/z,$$

where, as in Equation 3.7, $\mathrm{æ} = 0.4$ is the von Kármán constant, z is the height above the underlying surface. These equations follow from Equations 3.7, in which the horizontal derivatives of the average meteorological fields (with respect to x_1 and x_2) can be neglected at the average wind velocity along the axis x_1. These equations were taken as primary semiempirical hypotheses (for neutral stratification in the isotropic boundary layer), and their complication has led to semiempirical hypotheses (3.1)–(3.3) in the anisotropic boundary layer [1,4].

In Sections 3.2.1 through 3.2.3, devoted to the semiempirical hypotheses of the theory of turbulence in the atmospheric anisotropic boundary layer, it was shown theoretically and experimentally that the theory of similarity of turbulent flows can be extended to the arbitrary anisotropic boundary layer, which can be considered as locally weakly anisotropic. In particular, the weak anisotropy of the mountain boundary layer was checked experimentally.

In this boundary layer, all turbulence characteristics become functions of the stratification parameter and of the parameter ζ ($\zeta = z/L$, L is the Monin–Obukhov scale). During the last two decades, the parameter ζ was determined as the Monin–Obukhov number. This name reflects the essence of the parameter characterizing the local temperature stratification and the names of the authors of the similarity theory, who introduced the Monin–Obukhov scale (as a thickness of sublayer of dynamic turbulence [3]) into the corresponding theoretical framework. For the neutral stratification, the Monin–Obukhov number coincides with the dynamic Richardson number. As was shown in Reference 9, in the anisotropic boundary layer the Monin–Obukhov number, varying from one point to other within the layer, characterizes variations in the structure of the external influx of energy, which transforms then into the energy of turbulence, and it can be considered as the main parameter of turbulence in this layer. In this case, the turbulent scales V_* and T_* become functions of the Monin–Obukhov number ζ.

In the near-surface layer, in the case of strong instability ($\zeta \ll -1$), it is possible to find simple asymptotic dependences of the turbulent scale V_* and T_* on the Monin–Obukhov number ζ from Equations 3.2 through 3.4. In this case, the absolute value of the longitudinal horizontal derivative of the average temperature is usually much smaller than the absolute value of the vertical derivative. The vertical derivative itself in the strong instability is close to a negative constant [9], that is, $dT/dz \approx \text{const} < 0$, and $\zeta \ll -1$ (see Figure 3.10). For observations near the underlying surface, the height z can be believed approximately constant ($z \approx \text{const}$). In this case, variations of the Monin–Obukhov number ζ are mostly caused by variations of the scale L. Taking into account the known asymptotics of the similarity function $\varphi(\zeta)$ (Figure 3.1), we can, following References 1, 8, find from Equations 3.2–3.4:

$$T_* = c_1 \cdot |\zeta|^{1/3}, \quad V_* = c_2 \cdot |\zeta|^{-1/3}; \quad \zeta \ll -1, \tag{3.25}$$

where c_1, c_2 are unknown constants to be determined from experiment.

In the zone of strong stability ($\zeta \gg 1$), the influence of the $\alpha(\zeta)$, $\alpha = \text{Pr}^{-1}$ (i.e., α is inverse turbulent Prandtl number) becomes marked. The exact equation for this function is unknown nowadays. This is why the question on theoretical equations for the scales V_* and T_* at strong stability remains open.

The results of research expeditions during the 2000s in mountain and valley regions have been accumulated in References 39, 40 in the form of the large experimental database of near-surface measurements of turbulent characteristics under various meteorological conditions. The accumulated data significantly extend the domain of applicability of the results [9]. In particular, new data were obtained for the turbulent scales of velocity field V_* and temperature T_* [39,40]. The range of observed Monin–Obukhov numbers was extended significantly. In one of the measurement sessions (Lake Baikal, 2006), the ultrastrong instability with the highest Monin–Obukhov number, $\zeta = -102819.2$ ($T_* = 53$ degrees, $V_* = 3.2 \cdot 10^{-3}$ m/s), was observed. Totally, the range of the observed Monin–Obukhov numbers available for the theoretical analysis are changed over a wide range of $-102,819 \leq \zeta \leq +197$. This range includes both the ultrastrong stability (when $\zeta \gg 1$) and ultrastrong instability (when $\zeta \ll -1$) situations.

Figures 3.18 and 3.19 depict experimental data for the scales of temperature T_* and velocity field V_* as functions of the universal parameter determined above as the Monin–Obukhov number ζ. The data plotted in Figures 3.18 and 3.19 were obtained in the mountain boundary layer at different observed time, in different geographic regions, and in different regional meteorological situations.

The dashed lines in Figures 3.18 and 3.19 in the range $\zeta < 0$ show theoretical dependences (3.25) for turbulent scales lie at the range of unstable stratification. The recorded data stably group around these dependences. The experiments for $\alpha(\zeta)$ [1,2] show that at $\zeta < 0$ we can take $\alpha = \alpha_0$. Then, at negative ζ, the parameter $T_* ((\alpha/\alpha_0)T_* = T_*)$ is plotted as an ordinate in Figure 3.18. The relative error of data shown in Figures 3.18 and 3.19 mostly does not exceed 40%. This error corresponds to the maximal error (two to three times higher) [1,2] of measurements of the structure characteristics of temperature C_T^2 and velocity C_V^2 fluctuations in

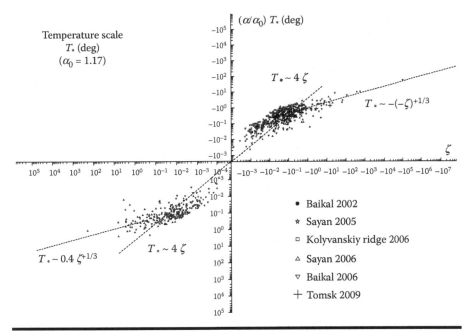

Figure 3.18 Turbulent scale of temperature T_* in the mountain anisotropic boundary layer. Rage of variation of the Monin–Obukhov number: $-102{,}819 \leq \zeta \leq +197$.

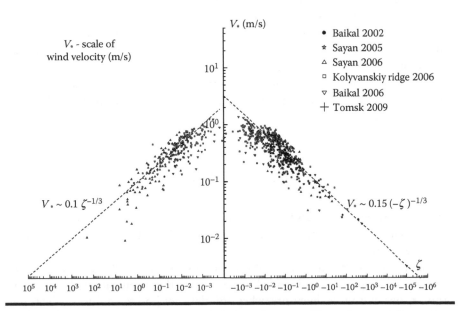

Figure 3.19 Turbulent scale of velocity V_* in the mountain anisotropic boundary layer. Range of variation of the Monin–Obukhov number: $102{,}819 \leq \zeta \leq +197$.

the actual atmosphere. This is in agreement with the similarity theory, in which $C_T^2 \sim T_*^2$, $C_V^2 \sim V_*^2$ [1,2]. Consequently, the errors of C_T^2, and C_V^2 are approximately twice larger comparing with the errors of T_*, and V_*.

As it follows from Figures 3.18 and 3.19, the data of all measurement sessions are in a good agreement with each other, as functions of the Monin–Obukhov number, and they can be joined regardless the type of the regional meteorological situation. Thus, the conclusions drawn in Reference 9 (see also Sections 3.2.2 and 3.2.3.5) are additionally confirmed.

Based on References 39, 40, it is possible to state that the locally weak anisotropy of the mountain boundary layer is observed in measurements carried out in different seasons, in different mountain regions, and along paths with different height changes.

3.2.6.2 Turbulent Prandtl Number in the Anisotropic Boundary Layer

It is known [1,2] that two geometrically similar flows of incompressible fluid are also mechanically similar, if their Reynolds numbers coincide. In the case of temperature-inhomogeneous fluid, for the flows to be similar, the coincidence of several dimensionless characteristics (e.g., similarity criteria) is required. One of these characteristics is the Prandtl number $\mathrm{Pr} = \nu/\chi$ (ν is kinematic viscosity, χ is the temperature conductivity), which depends only on the type of fluid and is a characteristic of the medium. Namely, for air, $\mathrm{Pr} \approx 0.7$.

In the turbulent medium, in averaged motion equations (Reynolds equations [1,2]) the terms $v \cdot \partial \overline{v}_i / \partial x_j$ and $\chi \cdot \partial \overline{T} / \partial x_j$ ($\overline{v}_i$ and $\overline{T}$ are average velocities and temperature) were complemented, respectively, with the moments of pulsations of random velocity and temperature, respectively, $\langle v_i' \cdot v_j' \rangle$ and $-\langle v_j' \cdot T' \rangle$. In the semiempirical theory of turbulence, the relations between these moments and spatial derivatives of the average velocity and average temperature are determined (see Equation 3.2):

$$
\begin{aligned}
\langle v_i' \cdot v_j' \rangle &= -K \cdot (\partial \overline{v}_i / \partial x_j + \partial \overline{v}_j / \partial x_i), \\
\langle v_p' \cdot T' \rangle &= -K_T \cdot \beta_{pk} \cdot (\partial \overline{T} / \partial x_k); \quad i,j,k,p = 1,2,3; i \neq j,
\end{aligned}
\tag{3.26}
$$

where the doubled subscripts imply summation. Here, K and K_T are the coefficients of turbulent viscosity (turbulent exchange) and turbulent temperature conductivity, respectively (see Equation 3.3).

The turbulent Prandtl number Pr_T is the ratio of these coefficients: $\mathrm{Pr}_T = K/K_T$ (here the subscript T in Pr_T is added to distinguish the turbulent number from the ordinary one, but in the literature, when this does not lead to confusion, the turbulent Prandtl number is often designated identically to the ordinary number Pr). In the theory, the inverse turbulent Prandtl number α, $\alpha = \mathrm{Pr}_T^{-1} = K_T/K$ is often used.

Analogously to the ordinary (molecular) number Pr, the turbulent Prandtl number Pr_T depends only on the nature of turbulence in fluid. Early experimental data for the inverse number α obtained in measurements of the average velocity and temperature in pipes and channels have shown [1,2] that α insignificantly differs from unity (from 1.1 to 1.4). The average value of α is $\alpha \approx 1.17$ ($Pr_T \approx 0.85$). This value is observed at the neutral ($|\zeta| \leq 0.05$) and unstable ($\zeta < -0.05$) stratifications, where ζ is the Monin–Obukhov number ($\zeta = z/L$, z is height, L is the Monin–Obukhov scale). However, under conditions of stability ($\zeta > +0.05$) α can decrease markedly [1,2]. At the same time, the data of later experiments reported in the second edition of the monographs [1,2], are contradictory (even negative values of α were observed).

In our paper [41], we have undertaken an attempt to find the dependence of the turbulent Prandtl number Pr_T on the Monin–Obukhov number ζ based on the measurements of turbulence characteristics in the atmospheric anisotropic boundary layer. The measurements were conducted in 2006–2010 in mountain regions of the Baikal Astrophysical Observatory (Listvyanka village, Irkutsk Region) and Sayan Sun Observatory (Mondy village, Buryatiya) of the Institute of Solar-terrestrial Physics SB RAS. For measurements, we used the mobile ultrasonic meteorological system [159]. The meteorological system records random values of wind velocity components, temperature, and other meteorological parameters with a frequency of 160 Hz and stores them in the computer memory. As a result, a file of digital values with the length N (length of the sample of recorded values) is created. For the two-minute averaging interval usually used by us (with the time interval $\Delta t = 0.00626$ s between values), the sample length is $N = 19150$.

To reconstruct the inverse turbulent Prandtl number α from accumulated measured data, it is possible to use the second equality in semiempirical Equations 3.26. Upon convergence with the set of some weighting coefficients c_p ($p = 1, 2, 3$), it takes the form

$$\Omega_M(c_p) = -\alpha \cdot K \cdot \Omega_D(c_p),$$

$$\Omega_M(c_p) = c_p \cdot \langle v'_p \cdot T' \rangle, \quad \Omega_D(c_p) = c_p \cdot \beta_{pk} \cdot (\partial \overline{T}/\partial x_k), \quad K = \varkappa \cdot V_* \cdot z/\varphi(\zeta), \tag{3.27}$$

where $x_1 = x$, $x_2 = y$, $x_3 = z$. Since the accuracy of setting of semiempirical numerical coefficients β_{pk} in Equation 3.27 is not high and the measurements of turbulent characteristics in the atmosphere are also usually accompanied by significant errors, the use of the weighting coefficients c_p in Equation 3.27 allows us to smooth, to some extent, the errors of semiempirical Equation 3.26.

As can be seen from Equation 3.27, for the determination of α, it is needed to know mixed moments between pulsations of velocity and temperature $\langle v'_p \cdot T' \rangle$, the coefficient of turbulent viscosity K, spatial derivatives of average temperature $\partial \overline{T}/\partial x_k$ ($k = 1, 2, 3$). The first two parameters can be easily calculated from the accumulated measured data. However, the calculation of spatial derivatives of the average temperature $\partial \overline{T}/\partial x_k$ is a sophisticated problem.

To calculate the derivatives of the average temperature from accumulated measured data, a numerical algorithm was developed in Reference 41. The algorithm employs the heat-conduction equation in a moving medium

$$\partial T(t,x)/\partial t + v_k(t,x) \cdot [\partial T(t,x)/\partial x_k] = \chi \cdot \Delta_x T(t,x),$$

where $T(t, x)$ and $v_k(t, x)$ are the random temperature and flow velocity components.

In the atmosphere, the dissipative right-hand side of this equation is usually a small addition to the terms in the left-hand side of the equation. The Taylor hypothesis of frozen turbulence is based on this fact [1,2,4]. The applicability of the frozen turbulence hypothesis in the atmosphere is now beyond question. It is confirmed experimentally (down to very small wave numbers), and there are also many indirect confirmations obtained from the comparison of theoretical conclusions based on the hypothesis with the experiment.

The heat-conduction equation should work at all time instants t_i, $1 \leq i \leq N$. Then for every i of the range $2 \leq i \leq N - 1$, we can construct the system of three algebraic equations for the three neighboring time points

$$t_{i-1} = t_i - \Delta t, t_i, t_{i+1} = t_i + \Delta t$$

to find three random unknowns at one central time point t_i:

$$y_k = \partial T(t_i,x)/\partial x_k \quad (k=1,2,3).$$

The system of algebraic equations has the form

$$v_{jk} \cdot y_k = f_j, \tag{3.28}$$

$$v_{jk} = v_k(t_j,x), y_k = \partial T(t_i,x)/\partial x_k,$$
$$k = 1,2,3; j = i-1,$$
$$i, i+1; 2 \leq i \leq N-1;$$
$$f_j = -\partial T(t,x)/\partial t \big|_{t=t_i} + F_{ji},$$
$$F_{ji} = \chi \cdot \Delta_x T(t_i,x) + sgn(j-i) \cdot \{v_{ik} \cdot [\partial^2 T(t,x)/\partial t \, \partial x_k]$$
$$- [\partial^2 T(t,x)/\partial t^2] + \chi[\Delta_x \, \partial T(t,x)/\partial t]\} \big|_{t=t_i} \cdot \Delta t + O([\Delta t]^2),$$
$$F_{ji} = o(\Delta t, [\partial T(t,x)/\partial t] \big|_{t=t_i}).$$

Due to the smallness of the discretization range Δt and the term $\chi \cdot \Delta_x T(t_i, x)$ in the heat-conductivity equation, the function F_{ji} in Equation 3.28 can be neglected in the calculations. In this case, the solution of system (3.28) becomes simpler.

The average spatial derivatives of temperature can be found through averaging of the obtained arrays (of length $N - 2$) of random derivatives y_k.

It should be noted that due to the presence of experimental errors in the coefficients (v_{jk}) and the right-hand side (f_j), the solution of system (3.28) (and also the solution of Equation 3.27) is an ill-posed inverse problem. In such problems, the absolute value of the inverse operator exceeds unity. This leads to a wide spread in the values of the obtained solutions. As shown in Reference 160, ill-posed problems are solved through introduction of the regularization parameter, whose proper choice can decrease the absolute value of the inverse operator.

From tensor's theory follows that the inverse operator for system (3.28) is proportional to the inverse determinant of the matrix of coefficients $\det^{-1}(v_{jk})$ (and $\det(v_{jk}) \sim [\Delta t]^3$). To decrease it, we can introduce the regularization parameter. However, the data of reconstruction of the vertical derivative of the average temperature (see Figure 3.20) show that there is not such needed.

Figure 3.20 shows the results of comparison of the vertical derivatives of average temperature measured directly in the atmosphere [9] (see Figure 3.10) and reconstructed from solution of the problem (3.28). As it follows from Figure 3.20, the measured and reconstructed derivatives are in a good agreement with each other (and with the semiempirical theoretical asymptotic of the average vertical derivative at $\zeta \to -\infty$) despite the fact that data of direct measurements of the derivatives in the atmosphere, and the data used for reconstruction, correspond to different regions, different meteorological conditions, and different seasons. Moreover, the period between obtaining these data is several years.

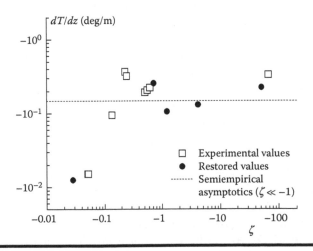

Figure 3.20 Vertical derivative of the average temperature $dT(t, x)/dz$. Closed circles are for solution of problem (3.28). Open squares are experimental data [9] obtained in atmospheric measurements. Straight line is semiempirical theoretical asymptotic of the derivative at $\zeta \to -\infty$.

As can be seen from Figure 3.20, the measured and reconstructed derivatives fall within the ranges of neutral and unstable stratifications ($\zeta < 0$), including the ultrastrong unstable stratification ($\zeta \ll -1$). In the stability range ($\zeta > 0$), no measurements of the derivatives in the atmosphere were conducted [9], and that is why we restrict the following consideration to the range of negative ζ (neutral and unstable stratifications, $\zeta < 0$).

Thus, the reconstructed (from solution of problem (3.28)) and independently measured average derivatives are in agreement with each other. Therefore, it is more important to introduce the regularization algorithm for solution of Equation 3.27. The action of this algorithm (for reconstruction of α from Equation 3.27) consists in achieving of some selected criterion by the solution. The equality

$$\Omega_M(c_p)/\Omega_M(c_p') = \Omega_D(c_p)/\Omega_D(c_p'), \tag{3.29}$$

is used as a regularization criterion, where c_p' ($p = 1, 2, 3$) is other set of weighting coefficients, different from c_p. This equality follows from Equation 3.27.

The algorithm based on the smoothing properties of the sets of weighting coefficients (c_p and c_p') in criterion (3.29) was applied as regularization algorithm in Reference 41. For this purpose, every element in these sets was considered as an independent random parameter uniformly distributed in the range (0, 1) (with the average value of 1.2). For every random realization of the weighting coefficients, after solution of system (3.28), the left-hand and right-hand sides of equality (3.29) were calculated. Since the errors of calculation of Ω_M in Equation 3.29 (dependent on the errors of moment calculation) appear to be much smaller than the errors of calculation of Ω_D (dependent on errors of reconstruction of the derivatives in system (3.28) and on errors of setting of the semiempirical coefficients β_{pk}), the accuracy of calculation of the left-hand side in equality (3.29) is generally higher than that of the right-hand side. Therefore, the iterative procedure based on the search for coincidence (with a preset accuracy) of the left-hand and right-hand sides of equality (3.29) allows us to obtain, for some determined particular random realization of the sets of c_p and c_p', the values of $\Omega_M(c_p)$ and $\Omega_D(c_p)$ rather accurate by criterion (3.29). Then, coming back to Equation 3.27, we obtain the value of α.

Figure 3.21 shows the dependence of the inverse turbulent Prandtl number α vs. the Monin–Obukhov number ζ, as judged from measurements of the turbulence characteristics in the anisotropic atmospheric boundary layer.

The straight line in Figure 3.21 corresponds to $\alpha = 1.17$ obtained in the earlier measurements [1,2]. As can be seen from Figure 3.21, in a wide range of the Monin–Obukhov number ($0 < -\zeta \lesssim 4$, neutral and weakly unstable stratification), the Prandtl number is close to unity. These results in a good agreement with the data obtained in References 1, 2. At the same time, at the strong instability ($\zeta = -51.1$, i.e., $\zeta \ll -1$) the value of α decreases.

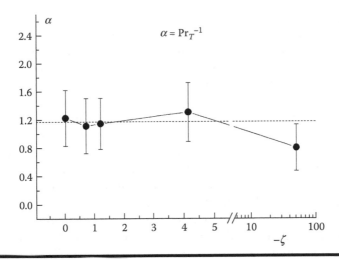

Figure 3.21 **Inverse turbulent Prandtl number α in the atmosphere as a function of the Monin–Obukhov number ζ.**

3.2.7 Main Results

Thus, in Section 3.2, we presented the data of long-term experimental and theoretical studies of properties of atmospheric turbulence in the anisotropic boundary layer. These data should be taken into account for the correct prediction of characteristics of optical radiation in the atmosphere. The main results can be formulated as follows:

1. It was found that the Monin–Obukhov theory of similarity of turbulent flows can be extended to an arbitrary anisotropic boundary layer. With the use of the semiempirical hypotheses of the theory of turbulence, it was shown theoretically and experimentally that the arbitrary anisotropic boundary layer can be considered as locally weakly anisotropic. In this layer, the weakly anisotropic similarity theory is locally true (in some vicinity of every point in the layer).

2. At the known turbulent scales of velocity and temperature, the variable Monin–Obukhov number (ratio of the height to the Monin–Obukhov scale), different at every point of the layer, has been shown to be the main parameter of turbulence in the anisotropic boundary layer.

3. It was demonstrated that the semiempirical theory of turbulence for the isotropic layer is a particular case of the semiempirical theory for the anisotropic layer. In this particular case, both anisotropy functions are simultaneously equal to zero. In the isotropic boundary layer, the absolute values of the Monin–Obukhov number usually do not exceed few units.

4. In the anisotropic boundary layer, there is a range of variation of the Monin–Obukhov number, in which the conditions of the isotropic layer take place (both anisotropy functions in this range are close to zero simultaneously).
5. It was found that at the known characteristic scales of temperature and velocity average for the observation region, the anisotropic boundary layer can be replaced with the isotropic one. This provides the possibility of using the optical models of turbulence developed for the isotropic boundary layer.
6. Theoretical equations have been derived for the outer scale of turbulence in the anisotropic boundary layer; they have been confirmed experimentally. The relations have been found between the outer scales determined by five different methods; they connect the Tatarskii outer scale of turbulent mixing and parameters of different models of the turbulence spectrum.
7. In the atmospheric boundary layer, the turbulent Prandtl number has been shown to be close to unity in a rather wide range of the Monin–Obukhov number (neutral and weakly unstable stratification). The decrease of the inverse turbulent Prandtl number at the strong instability was noted.
8. New results have been presented for the turbulent scales of velocity field and temperature measured at different times under different climatic conditions, and in different regions. The data confirm the conclusion on the local applicability of the Monin–Obukhov similarity theory in the anisotropic atmospheric boundary layer and extend it to extreme temperature stratifications. In our experiments, we observed the Monin–Obukhov numbers in the range from +197 (strong stability) to −102,819 (ultrastrong instability). The large (and extremely large), in the absolute value, Monin–Obukhov numbers are characteristic only of the semiempirical theory in the anisotropic boundary layer. The data of all measurement sessions are in a good agreement with each other as functions of the Monin–Obukhov number, and they can be joined regardless of the type of regional meteorological situations. Therefore, the study of functional dependences of the turbulent scales of velocity and temperature on the Monin–Obukhov number can be considered as one of the top priority problems of the semiempirical theory of turbulence.

3.3 Coherent Structures in the Turbulent Atmosphere

In Section 3.2, we presented the results of the studies carried out within the scope of the semiempirical theory of turbulence in the anisotropic atmospheric boundary layer. These studies were naturally continued in the study of turbulence characteristics in closed local volumes (namely, in closed rooms). Turbulence in closed rooms can be considered as a particular case of turbulence observed in the anisotropic boundary layer.

The studies in this field have allowed to obtain important results on coherent structures and, in general, on the local structure of turbulence. In particular, it was found that coherent structures are important elements for understanding of the processes of formation and further evolution of turbulence. The brief review of these studies of the local structure of turbulence is discussed in Section 3.3.1. Section 3.3.2 (based on data published in References 80–82, 128) presents the results of experimental study of the processes of Benard cell formation and breakdown in air to show similarity of the processes occurring in the air in room and pavilion and in the open atmospheric areas consisting local boundary layers.

3.3.1 Brief Review of the Studies of the Local Structure of Turbulence

We start our explanations of the local structure of turbulence from the study of turbulence characteristics in closed volumes (room and pavilion), because the processes occurring in the closed areas are similar with those observed in open atmospheric boundary layers. The results of measurement of turbulence characteristics in the pavilion of astronomical spectrograph of the large solar vacuum telescope (LSVT) have been interesting for such a comparison [80,81].

It has been shown that the Benard cell is formed as a result of temperature gradients. Our data confirm the main scenarios of turbulence formation (Landau–Hopf, Ruelle–Takens, Feigenbaum, Pomeau–Menneville stochastization scenarios). It was found that the breakdown of the Benard cell follows the Feigenbaum's scenario. As this takes place, the main vortex in the cell breaks down into smaller ones as a result of the series (about 10) period-doubling bifurcations. It was shown that the arising turbulence is coherent and deterministic. The fractal character (local self-similarity) of the turbulence spectrum was discovered.

Historically, the theoretical study of stability in the motion of fluid between two spaced planes (the existence of Benard cells follows this scenario) has allowed the formulation of several scenario of turbulence formation (origination). In this case, the turbulence was called incipient turbulence. Based on this historical definition, we can say that random motions of air in closed rooms are an example of incipient turbulence.

It is convenient to join the studied properties of the processes of Benard cell formation and breakdown by the term "coherent structure," if we extend this, already existing, term and include small-scale components into the coherent structure (see Section 3.3.5). We define coherent structure as a compact formation including the long-lived spatial hydrodynamic vortex cell (arising as a result of long action of thermodynamic gradients) and products of its coherent cascade breakdown. The coherent structure satisfies all attributes characterizing the appearance of chaos (turbulence) in typical dynamic systems. Our results show that the known processes of transition of laminar flows into turbulent ones (Rayleigh–Benard convection, fluid flow around obstacles, and others) can be considered as coherent structures (or sums of such structures).

The appearance of coherent structures in a closed room (in the form of observed Benard cell) follows from experiments reported in our papers [80,81]. These results are very important for detection of coherent structures (in the extended definition of a coherent structure). Therefore, the experiment and its results are described in Section 3.3.2 with necessary additions following from consideration of the results of experiment from the viewpoint of formation of coherent structures.

In general, the data show that the coherent structure in its extended definition can be considered as a structure element of hydrodynamic turbulence. This structure element is a vortex formation arising as a result of transition of the energy of thermodynamic perturbation into the energy of motion of a solid medium. The vortex can have different shapes and sizes, and under instability conditions, it breaks down into smaller vortices in the cascade and coherent way. From the mathematical point of view, the structure element of turbulence is a soliton-like solution of flow-dynamic equations. It is either a solitary 3D topological one-soliton solution or one soliton in a many-soliton solution.

Coherent structures in open air (Section 3.3.6) were considered with application of data of other measurements carried out during 2000s (see, e.g., References 9, 80–82, 128). All these measurements used the mobile ultrasonic meteorological system recording more than one hindered of atmospheric parameters with the sufficient accuracy (see Section 3.2).

It was shown that the actual atmospheric turbulence can be considered as a result of mixing of different coherent structures. It follows from the measurement data that extended areas, in which one coherent structure has the decisive influence, are often observed in open air. These areas are referred to areas of coherent turbulence, because turbulence in a single coherent structure is coherent and deterministic. Since the atmospheric coherent turbulence differs from the incoherent Kolmogorov turbulence, the values of the Kolmogorov constant C and the Obukhov constant C_θ in the Kolmogorov–Obukhov law are refined. It is shown that the error of determination of these constants can achieve 93%. This is one of the reasons for large errors in measurements of turbulent characteristics of the atmosphere. The effect of attenuation of phase (refractive) fluctuations of the propagating optical radiation is shown to be observed in coherent turbulence.

Section 3.3.7 briefly lists the determined various properties of single coherent structures and the properties of mixtures (sums) of different coherent structure. Twenty-eight such properties are determined and formulated by now.

The properties of the coherent structure were revealed through processing of the accumulated experimental data by the methods of spectral analysis of random processes with the following formulation of logical conclusions. However, these methods do not allow us to see coherent structures and their mixtures. The possibility of seeing coherent structures (visualizing them by flow lines in numerical solutions of the Navier–Stokes equations) is demonstrated in Section 3.3.8. Numerical solutions of the Navier–Stokes equations are obtained now for several tens of various boundary-value problems (see, e.g., References 144–157).

The results of numerical calculations show that large solitary vortices (coherent structures, topological 3D solitons) are observed in closed rooms. The cascade breakdown of these vortices gives rise to coherent turbulence. It is shown that solitary toroidal vortices (coherent structures or topological solitons) arise also above irregularly heated surfaces. The number of vortices and their internal structure depend on the shape and size of thermal inhomogeneities. For complex heating shapes, toroidal vortices become deformed. In the process of evolution, vortices mix markedly. This produces the Kolmogorov (incoherent) turbulence.

The transition from small thermics (toroidal mushroom-like vortices) to large stable vortices through the period of chaotization was observed. This transition is inverse to the usually observed cascade coherent breakdown of large vortices into smaller ones. Consequently, the formation of near-wall turbulence (thermics) and its further evolution can serve as an example of the inverse cascade process of energy transfer by the spectrum of motion scales. Numerical calculations confirm the experimental conclusion formulated by us that the mixing of coherent structures with close sizes (and close frequencies of the main vortices) yields Kolmogorov turbulence.

3.3.2 Incipient Convective Turbulence: Benard Cells

Theoretical models predicting the existence of Benard cells were developed a long time ago. As follows from the theory, for appearance of cells (the convective motion breaks down to), the temperature gradient between the opposite planes is required. Depending on the gradient value, the cells can take the shape of hexagonal prisms (with the axis along the gradient vector). At the center of such prisms, fluid moves upward along the prism axis (in parallel to the gradient direction), and at the edges it moves downward ($d\nu/dT < 0$, ν is kinematic viscosity), or vice versa ($d\nu/dT > 0$). In more complex (than two-spaced planes) spatial areas, the cells can take shapes different from hexagonal, for example, longitudinal and longitudinal-transverse cylinders (often with waists) and others. The experiments with model objects (vessels) with the use of water, oil, and liquid helium as carriers confirm the fact of appearance of convective Benard cells [1,2,161,162]. For air inside large closed rooms, such experiments were not conducted before. This is connected with the fact that such experiments require small-size sensors sensitive to motion of weak air flows.

As was shown in References 1, 2, 161, 162, for Benard cells (stationary periodic motion) to appear, the Rayleigh number Ra should be larger than the critical number Ra_{cr}. By definition, $Ra = g \cdot \beta \cdot h^3 \cdot (T_0 - T_h)/(\nu \cdot \chi)$, where T_h and T_0 are respectively the air temperatures at the top and bottom of a layer with the height h, g is the acceleration due to gravity, β is the thermal expansion coefficient ($\beta = 1/T_0$), and χ is the temperature conductivity of air. Upon the substitution of the parameter values [80,81] ($T_0 = 285.1$ K, $T_h = T_0 + h \cdot d\bar{T}/dh$, $h = 5$ m, $d\bar{T}/dh = -0.145°/m$, $\nu = 1.3 \cdot 10^{-5}$ m²/s, $\nu = 0.7\chi$), we obtain $Ra = 1.3 \cdot 10^{10}$. The critical Rayleigh numbers Ra_{cr}, according to References 1, 2 fall within the range $Ra_{cr} = 657–1708$. It is seen from here that recorded Ra far exceeds the critical value $(Ra \gg Ra_{cr})$. That is why there exist stationary periodic motions.

Historically, the theoretical study of stability of fluid motion between two spaced planes (the existence of Benard cells follows from) has allowed the formulation of several scenarios of turbulence formation (origination). In this case, the turbulence was called incipient. In particular, it was found [1,2,42] that under conditions of instability and convective motions in closed volumes, the arising turbulence is undeveloped. One part of the fluid energy is contained in regular (laminar) motions (vortices in Benard cells), while another part is contained in turbulent motions.

Based on this historic definition, it can be outlined that random air motions in closed rooms are examples of incipient turbulence.

Statistical characteristics of incipient undeveloped turbulence in air are practically not investigated [1,2,4]. Therefore, it is interesting to reveal some characteristic features of undeveloped turbulence distinguishing it from developed turbulence. Here, we consider the structure and correlation functions of temperature fluctuations (since they are the simplest).

Figures 3.22 and 3.23 depict the results of comparison of the experimental data for main statistical characteristics of air temperature fluctuations in a closed room and in open air.

For the closed room, the data correspond to the measurements at point 5 inside the spectrograph pavilion [80,81] ($\zeta = -4.5$, $C_n^2 = 1.6 \cdot 10^{-14}$ cm$^{-2/3}$, $h = 1.1$ m, $\bar{T} = 11.95°$, $v = 0.09$ m/s, $v = |v|$ is the absolute value of average velocity). For open air, typical experimental data were taken. They were obtained separately in summer measurements over an approximately smooth underlying surface at dry fine weather ($\zeta = -3.8$, $C_n^2 = 6.5 \cdot 10^{-16}$ cm$^{-2/3}$, $h = 3.1$ m, $\bar{T} = 24.56°$, $v = 0.86$ m/s).

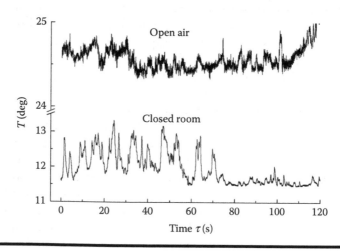

Figure 3.22 **Random temperature *T* realizations in a closed room and in open air.**

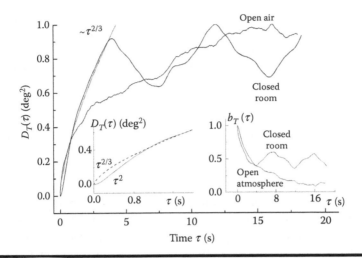

Figure 3.23 **Normalized structure $D_T(\tau)$ and correlation $b_T(\tau)$ (bottom right inset) functions in a closed room and in open air (bottom left inset shows the initial part of $D_T(\tau)$).**

As can be seen from Figure 3.22, which shows 2-min realizations of random temperature, for open air, the random fluctuation process is close to stationary. Under conditions of closed room, the process of fluctuations is clearly divided into two intervals with different turbulent conditions, and the conditions alternate each other in a discrete step. The step demonstrates the appearance of a new stationary motion with new characteristics from the old one. This phenomenon is called bifurcation of stability alternation [1,2].

It follows from Figure 3.23 that the structure functions of temperature fluctuations $D_T(\tau)$ in both developed and undeveloped turbulence at small separations τ are Kolmogorov: $D_T \sim \tau^{2/3}$. However, at large separations, they become significantly different. The correlation coefficient of fluctuations b_T inside a closed room, in contrast to open air, has a series of local maxima. The maxima are large enough. A local minimum of the structure function D_T in the correlation coefficient corresponds to every such maximum (their arguments coincide). At very small τ, the structure function $D_T(\tau)$ in a closed room has a square area (Figure 3.23, bottom left insert) more extended than in open air, which corresponds to increased values of the inner scale of turbulence.

3.3.3 Models of Temperature Fluctuation Spectra in Incipient Turbulence

For problems of optical radiation propagation under conditions of undeveloped turbulence, theoretical models of temperature fluctuation spectra are needed, first of all, for the spatial 3D spectrum $\Phi_T(\varkappa)$.

As was shown in Reference 4, the temporal frequency spectra of temperature fluctuations $W_T(f)$ in open air are satisfactorily described by the von Kármán model. Spectra of Kolmogorov developed turbulence have the extended inertial range, in which $W_T \sim f^{-5/3}$ (see also Chapter 1). In this range, the energy is transferred from vortices with scales to smaller ones.

Figure 3.24 depicts the smoothed temporal frequency spectra of temperature fluctuations W_T in a closed room and in open air.

As follows from Figure 3.24, in comparison with the open air, the spectra in the closed room decrease much faster in the inertial range. In addition, in this range, there are only some short frequency sections (located stepwise), in which the turbulence can be considered as Kolmogorov ($W_T \sim f^{-5/3}$). These sections are observed between steps of the spectral function at frequencies corresponding to local maxima of the correlation function of fluctuations (or minima of the structure function). If we smooth the steps, then the experimental spectra of undeveloped turbulence have some characteristic parts with the fast power-law decrease. Thus, if $W_T \sim$ const in the extended energy range, then with the increase of the frequency (in the inertial range) first $W_T \sim f^{-8/3}$, and then $W_T \sim f^{-12/3}$. Consequently, in incipient turbulence, the energy transfer from large vortices to smaller ones is insignificant, that is, vortices are slightly blurred. With the further increase of the frequency, in the viscous range, where the spectral density is close to the noise level, the decrease of the spectrum becomes more slowly, $W_T \sim f^{-2/3}$. The analogous behavior of the spectra is also observed at other measurement points in the pavilion. At some points, the stepwise form of the spectrum is even more pronounced.

To construct the theoretical model $\Phi_T(\text{æ})$ of the temperature spectrum of incipient turbulence, we can use the von Kármán model with the decrease in the

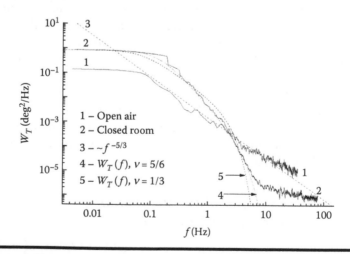

Figure 3.24 **Smoothed temporal frequency spectra of temperature fluctuations W_T in a closed room and in open air.**

inertial range corresponding to Figure 3.24. This approximate model was obtained in References 80, 81 (see also Chapter 1):

$$\Phi_T(\mathfrak{x}) = A_0 \cdot C_T^2 \cdot (6.6\mathfrak{x}_0)^{2(\nu-1/3)} \cdot \left(\mathfrak{x}^2 + \mathfrak{x}_0^2\right)^{-(\nu+3/2)} \cdot \exp\left(-\mathfrak{x}^2/\mathfrak{x}_m^2\right),$$
$$A_0 = 0.033, \ \mathfrak{x}_0 = 2\pi/L_0^K, \ \mathfrak{x}_m = 5.92/l_0 \tag{3.30}$$

where L_0^K and l_0 are the von Kármán outer and inner scales of turbulence. For developed turbulence, $\nu = 1/3$. Then, $\Phi_T(\mathfrak{x}) \sim \mathfrak{x}^{-11/3}$ in the inertial range. In the incipient turbulence, according to Figure 3.24, we should take $\nu = 5/6$, which gives $\Phi_T(\mathfrak{x}) \sim \mathfrak{x}^{-14/3}$ in the most part of the inertial range. The further, faster decrease of the spectrum is described by the exponential factor in Equation 3.30. The parameters of this spectrum L_0^K, and l_0 for $\nu = 5/6$ and $\nu = 1/3$ are given in Table 3.1. As can be seen from Table 3.1, the inner scale of incipient turbulence l_0 (average $l_0 = 1.9$ cm for $\nu = 5/6$, and $l_0 = 2.7$ cm for $\nu = 1/3$) is an order of magnitude larger than the inner scale in open air (0.7–4 mm [1,2,4,9]).

The spectrum $\Phi_T(\mathfrak{x})$ at $\nu = 5/6$, which can be considered as the spectrum of incipient turbulence, is in a better agreement with the experiment than at $\nu = 1/3$ (spectrum of developed turbulence, $\nu = 1/3$, but with parameters L_0^K and l_0 for the

Table 3.1   Parameters of Spectra of Incipient Turbulence

$\nu = 5/6$			$\nu = 1/3$	
L_0^K (cm)	l_0 (cm)	*Observation Point*	L_0^K (cm)	l_0 (cm)
33.2	1.2	1	83.0	1.8
47.4	2.3	2	101.6	3.5
19.8	1.2	3	53.9	1.2
27.0	1.8	4	67.5	2.8
62.6	2.3	5	125.1	4.1
18.5	1.6	6	46.4	2.1
46.4	2.3	7	126.5	2.9
18.0	1.84	8	36.0	2.8
60.3	2.3	9	139.2	3.9
18.1	1.8	10	38.9	1.8
30.4	2.3	11	76.1	2.9
25.4	2.3	12	54.3	3.2

incipient turbulence, Table 3.1). It can be easily seen from comparison of curves 4 and 5 in Figure 3.24. The areas under these curves differ the experimental one, respectively, by 15% ($\nu = 5/6$) and 28% ($\nu = 1/3$). However, model (3.30) with $\nu = 1/3$ is widely used. It allows the solutions of problems of wave propagation in developed turbulence to be easily extended to the case of incipient turbulence. Therefore, model (3.30) with $\nu = 1/3$ is often preferable.

The maximal error of approximation of actual spectra by Equation 3.30 falls on the range of very high frequencies (viscosity range). Therefore, in problems of wave propagation, in which the viscous interval plays the important role, the model more detailed than that given by Equation 3.30 should be used. In problems in which the main contribution is given by fluctuations of the wave phase (shifts of optical beams, image jitter, and others), the viscous range is not a significant contributor, and therefore approximation (3.30) is applicable.

Figure 3.25 presents the diagram of size distribution of the outer scale of turbulence L_0^K in the vertical plane inside the pavilion. Here, the same periodicity is observed for the inner scale of turbulence l_0 (see Table 3.1). The smaller inner scale corresponds to the smaller outer scale. As is known, the outer and inner scales determine the maximal and minimal sizes of inhomogeneities. As follows from Figure 3.25 and Table 3.1, the spatial periodicity of the size of inhomogeneities of the temperature field (chessboard-like structure) takes place in the pavilion.

The areas with decreased size of the outer scale can be called turbulent locks (or foci). In these areas, the intensified breakdown of large-scale averaged flow into smaller spatial components is observed. The intensity of random temperature variations characterized by the parameter C_n^2 (periodic) in the foci decreases on average.

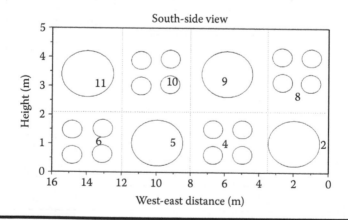

Figure 3.25 Distribution of values of the von Kármán outer scale of turbulence L_0^K in the vertical plane passing through the center of the pavilion and the east-west line (according to Table 3.1). Circles of larger diameter correspond to larger values of L_0^K. Digits show numbers and positions of observation points.

3.3.4 Scenarios of Stochastization of Convective Flows

Compare the results of measurements in the pavilion with the known data on turbulence formation from laminar flows (scenario of stochastization).

As known [1,2], the main stochastization scenarios include the Landau–Hopf, Ruelle–Takens, Feigenbaum, and Pomeau–Menneville scenarios. In the following will be shown that all the main scenarios are confirmed in the incipient convective turbulence.

1. *Pomeau–Menneville scenario*: It is known [1,2,42] that, with the increase of distance (Reynolds number) in laminar flows in pipes, small turbulent areas, in which the flow is not laminar, arise first. These areas are called turbulent locks (or spots). With an increase of the distance, the locks become longer, finally joining in one solid turbulent jet. Turbulent locks are also observed in the experiments by other schemes [42]. The appearance of locks leads to intermittent alternation of laminar and turbulent flows. The appearance of turbulence through intermittence is called the Pomeau–Menneville scenario [1,2,163].

   It follows from our measurements that turbulent locks and intermittence (and, as can be seen from Figure 3.22, the corresponding bifurcations of stability alternation) exist also in the periodic flows in the Benard cell. Areas with decreased spatial components (decreased outer L_0^K and inner l_0 scales) play the role of locks. Locks appear to be frozen in the structure of Benard cell and interchange with areas of large scales L_0^K and l_0. Consequently, the data confirm the Pomeau–Menneville scenario.

2. *Landau–Hopf scenario*: Incipient turbulence in Benard cell is a convenient model, with which we follow the process of breakdown of energy-carrying vortices into smaller ones. Actually, the only energy-carrying vortex in the Benard cell is the toroidal vortex of average motions [80,81]. Its size is determined by the size of the room, where it is formed. In open air, it is rather difficult to measure the size of the main energy-carrying vortex, because it depends on the climate-forming factors. The outer scale of turbulence, which itself is a product of breakdown, is usually taken as this vortex.

   Figure 3.26 depicts the correlation coefficient b_T and the selected nonsmoothed frequency spectrum W_T of temperature calculated from the data of measurements in the pavilion. The correlation coefficient b_T was calculated from different selected estimates [164–166]. However, they give the coinciding results and demonstrate local maxima of b_T. The correlation function can be calculated with an arbitrarily small error at the large sample size N (variance of the b_T estimate is proportional to $1/N$). In our case, $N = 19,139$, and therefore the 95% confidence boundaries of determination of the function b_T shown in Figure 3.26 are ± 0.014. This confidence range is much smaller than the value of b_T maxima.

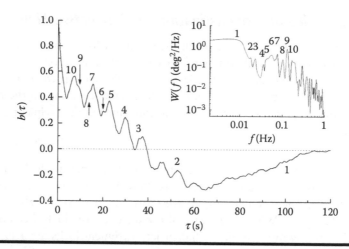

Figure 3.26 **Correlation coefficient b_T and nonsmoothed frequency spectrum W_T (top right inset) in the pavilion. Digits are numbers of maxima of b_T and W_T corresponding to each other.**

The selected spectrum W_T is calculated without smoothing with a rectangular spectral window. As known [164,165], this window acts as a slit with the width $\sim 2/T$ ($T = 120$ s). This resolution is quite sufficient to find maxima of the spectrum W_T being in one-to-one correspondence with the maxima of b_T. The arguments of b_T and W_T maxima are connected by the relation $\tau_k f_k = 1$, $k = 1, 2,...$ (the values of τ_1, f_1 usually determine characteristic scales of decrease of the functions b_T and W_T). From the relation $2\pi R_1 = v \tau_1 = v/f_1$, we can easily reconstruct the diameter $2R_1$ of the main energy-carrying vortex. In the pavilion, it is equal to 294.4 cm ($v = 9$ cm/s, $f_1 = 0.00973$ Hz) at point 5.

From comparison of spectra W_T shown in Figures 3.24 and 3.26, we can see that at the standard smoothing of the spectrum by a wide spectral window (variance of the smoothed spectrum in Figure 3.24 is 1% of the variance of the selected spectrum in Figure 3.26) actual maxima of the spectra disappear. Therefore, to calculate the frequencies of spectral maxima (harmonics), we should use data of Figure 3.26. However, a rectangular spectral window has large side lobes leading to oscillations, especially at high frequencies. We can dispose of these lobes by applying any of widely used nonrectangular windows (the difference between them is small), for example, the Welch window [166]. In comparison with the rectangular window, this window decreases the variance approximately two times, increasing the width of the frequency band also approximately two times. The increase of the band is acceptable, because it turns out to be smaller than the average width of the spectral maxima, which can be seen from the data for W_T in Figure 3.26.

To improve the sample estimate, we can apply additionally a digital threshold filter, which removes weak (lower than the average value shown in Figure 3.24) harmonics in the spectrum. These harmonics can be interpreted either as a high-frequency side lobes not fully suppressed by the Welch window or as a weak transient process of breakdown of the energy-carrying vortex into smaller ones. It is shown below (see Figure 3.29) that the structure of weak harmonics is fractal. It does not coincide with the structure of side lobes. That is why the side lobes turn out to be practically suppressed. The level of the used threshold filter in the form of the smoothed spectrum W_T (with 1% variance) was chosen from the detailed analysis of selected spectra at many measurement points in the pavilion. Thus, the threshold filter separates main maxima (or first-order harmonics) from the spectrum.

Figure 3.27 shows the frequencies of first-order harmonics (arguments of the maxima) in the W_T fluctuation spectrum. Based on the above analysis, they can be interpreted as frequencies of stable vortices observed in the temperature field. The vortex frequencies f_n appear to be multiple to the frequency of the main energy-carrying vortex $f_1 = 0.00973$ Hz. Upon normalization to f_1, they are integer natural numbers ($n = 1,2,...$):

$$f_n/f_1 = 1, 6, 8, 11, 13, 17, 20, 66, 68, 77, 90, 93, 109, 113, 117, 120,$$
$$127, 130, 133, 136, 144, 150, 152, 157, 162, 164, 167, 175, 179, 183,...$$

Multiple frequencies are the exact result of discrete breakdown of the main energy-carrying vortex into smaller ones. The multiplicity means also that the phases of different harmonics (oscillations) are rigidly connected (matched).

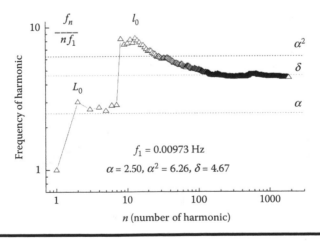

Figure 3.27 **Frequencies of stable harmonics (vortices) f_n in the spectrum of W_T fluctuations; α, α^2, δ are Feigenbaum constants.**

In this case, the oscillations themselves are usually called coherent (inphase). It should be noted that the function $f_n/(nf_1)$ shown in Figure 3.27 is resistant to variations of the level of the threshold filter. Variations of this function appear to be insignificant at variations of the filter level. From the frequencies f_n, using the relation $2\pi R_n = v/f_n$, $n = 1, 2,...$, following from the law of conservation of momentum of a liquid particle at remote edges of the breaking down vortex and its daughter vortices (Heisenberg equality [1,2]), we can calculate the corresponding vortex radii R_n (in cm):

$$R_n = 147.21, 24.52, 18.39, 13.38, 11.32, 8.66, 7.36, 2.23, 2.16, 1.91, 1.63, 1.58,$$
$$1.35, 1.30, 1.26, 1.23, 1.16, 1.13, 1.11, 1.08, 1.02, 0.98, 0.97, 0.94, 0.91, 0.90,$$
$$0.88, 0.84, 0.82, 0.80,...$$

The vortex diameters are shown in Figure 3.28.

From the comparison of the data in Figure 3.28 and Table 3.1 (point 5, $v = 5/6$), we can see that the second frequency $6f_1$ corresponds approximately to the outer scale L_0^K ($2R_1 = 294$ cm, $2R_2 = 49$ cm), while the frequencies $127f_1$ and $130f_1$ (at which the inertial range terminates) correspond to the inner scale l_0 ($2R_{17} = 2.3$ cm).

In the viscous range and in a part of the inertial range, vortices being products of breakdown of large vortices (their frequencies are multiple to the lower frequencies) are observed. For example,

$$f_n/f_2 = 11, 15, 20, 24, 25,... (n = 8, 11, 16, 21, 22,...);$$
$$f_n/f_3 = 15, 17, 18, 19, 36,... (n = 16, 20, 21, 23, 52,...).$$

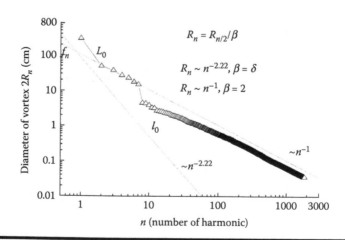

Figure 3.28 Diameters of stable harmonics (vortices) $2R_n$ in the W_T fluctuation spectrum.

The processes observed inside the pavilion are stationary. Therefore, both the formation of the main vortex in the Benard cell (due to temperature gradient) and its breakdown into smaller vortices occur permanently (replicate, possibly, at the same time). As we can see, the result is the limiting N-periodic flow with the frequencies f_n, $n = 1, 2, \ldots, N$. The transition of a small perturbation into the stable periodic flow follows from the solution of Landau equations, and the appearance of time-periodic flows is referred to as the normal Hopf bifurcation. The Landau–Hopf scenario describes the formation of turbulence as a series of normal bifurcations, giving rise to the limiting N-periodic flow ($N \gg 1$) with, generally speaking, and incommensurate frequencies.

However, the incommensurability of frequencies (phases) usually does not take place (as can be seen from our data). Therefore, with the main idea retained, it is believed now that normal bifurcations form serial subharmonics [1,2,167]. It can be easily seen that our results confirm the Landau–Hopf scenario.

3. *Ruelle–Takens scenario*: The Ruelle–Takens scenario can be considered as the refined Landau–Hopf scenario. The difference consists in the number of normal bifurcations, after which the flow can be considered as turbulent. According to this scenario, the turbulence appears (a strange attractor appears) already after three normal bifurcations [1,2,48,49,168]. This means that already the 3-period flow ($N = 3$) is turbulent. Then the convective flow becomes turbulent after formation of the main vortex in the Benard cell and two events of its breakdown.

4. *Feigenbaum scenario*: The Feigenbaum scenario describes the appearance of turbulence (strange attractor) as a result of the infinite series of period-doubling bifurcations [167]. These bifurcations show themselves only at a change of the value of some control parameter μ, for example, Reynolds, Rayleigh numbers, and others. As is well known, the Feigenbaum scenario follows from the versatility of arrangement of periodic points $x_0, \left(x_1^0, x_1^1\right)$, $\left(x_2^0, x_2^1, x_2^2, x_2^3\right)$, ... of 2^m multiple cycles. In the plot of $x(\mu)$, these points x_m^k correspond to tree branches, which double nonsymmetrically at critical bifurcation points μ_m. For both x_m^k and the parameter μ_m, the following asymptotic relations are valid ($m \gg 1$, α and δ are Feigenbaum constants):

$$\frac{x_m^k - x_m^{k+2^{m-1}}}{x_{m+1}^k - x_{m+1}^{k+2^m}} = \begin{cases} -\alpha, 0 \le k < 2^{m-1} \\ -\alpha^2, 2^{m-1} \le k < 2^m \end{cases}, \quad \sigma_m(\mu) = \frac{\mu_{m-1} - \mu_m}{\mu_m - \mu_{m+1}} = \delta,$$
$$\alpha = 2.503, \delta = 4.669.$$

Here, the parameter x, due to versatility, is usually understood as the main parameter characterizing the nonlinear dynamic system. For example, coordinates measured in [m] or ordinary velocity [m/s], as in the case of Navier–Stokes equations, or normalized [Hz] velocity.

The equality $\sigma_m(\mu) = \delta$ is asymptotic. However, it is shown in Reference 167 that it is applicable already after two-three period-doubling events (accurate to several percent). The good predictability of the theory just is the consequence of the high rate of convergence δ ($\delta = 4.67$). For approximate estimates, this equality can be used at $m \geq 0$. Actually, it can be easily seen that the equation $\sigma_m(\mu) = \delta$ has the solution $\mu_m = c\delta^{-m} + \mu_\infty$, $c = \mathrm{const}$. Assuming in this solution $m = 0$ and $m = 1$, we obtain the system of equations for determination of unknown constants c, μ_∞. From here, $c = \mu_0 - \mu_\infty$, $\mu_\infty = \mu_0 + (\mu_1 - \mu_0) \cdot \delta/(\delta - 1)$. For the logistic equation $x_{m+1} = \mu \cdot x_m \cdot (1 - x_m)$ considered in Reference 167, as known, we can take $\mu_0 = 1$ and $\mu_1 = 3$. Then, $\mu_\infty = 3.54508$. This value differs only slightly from the accurate value $\mu_\infty = 3.56994$ found in Reference 167. It should be noted that μ_m can both increase ($c < 0$) and decrease ($c > 0$) with an increase of m depending on the sign of c (at $\mu_\infty \geq 0$).

In the above case, the value of the control parameter μ is fixed (the dimensions of the pavilion, the temperature gradient, and other parameters are preset), and therefore we can observe only the result of breakdown of the main vortex. However, due to the self-reproduction of harmonics, if doubling bifurcations are present, the harmonics being results of these bifurcations can be observed.

Let, for example, the main parameter of the system x is a shift measured in length units. Then, x_0 ($m = 0$) can be identified with the main vortex radius R_1. The following parameters (x_1^0, x_1^1) should be the result of x_0 breakdown (nonsymmetric bifurcation at $m = 1$). Their role can be played only by two following largest radii R_2 and R_3 (the others are too small). Then, as follows from the breakdown diagram [167], we can take R_4, R_5 or R_5, R_6 as a pair x_2^0, x_2^2 ($m = 2$) (the first element in these pairs is approximately equal to a half of R_2, while the second is equal to a half of R_3). Upon the substitution of these parameters into the Feigenbaum relation, we obtain $(R_2 - R_3)/(R_4 - R_5) = 2.97$, $(R_2 - R_3)/(R_5 - R_6) = 2.30$. The other couple x_2^1, x_2^3 ($m = 2$) should be taken from obvious products of breakdown of vortices with radii R_2, R_3 (their frequencies are multiple to f_2, f_3). The first elements of the breakdown products are characterized by R_8, R_{16} (see above). From the Feigenbaum relation, we can find $(R_2 - R_3)/(R_8 - R_{16}) = 6.11$. Thus, despite we are at the beginning of the bifurcation tree ($m = 1, 2$), our data give values close to required α and α^2 for the absolute value of the left-hand side of the Feigenbaum relation. It should be noted that R_n satisfies the relation $R_n = R_{n/2}/\beta$ (Figure 3.28). In the inertial and viscous ranges, $\beta \approx 2$. However, $\beta \approx \delta$ in the energy range and in the inertial range near l_0, in this case the equality $R_n = R_{n/2}/\delta$ coincides with the Feigenbaum similarity relation for Fourier harmonics of the parameter x [1,2,167].

The Feigenbaum constants α, α^2, and δ are most pronounced in the plot of harmonic frequencies. As can be seen from Figure 3.27, with an increase of n the normalized frequency $y_n = f_n/(n \cdot f_1)$ experiences two large stepwise changes. The first step is observed near the outer scale of turbulence L_0^K (as a result, the frequency y_n

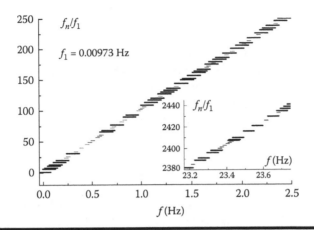

Figure 3.29 Devil's self-similarity staircase (normalized frequencies of the first- and second-order harmonics f_n/f_1). Long dashes are for the first-order harmonics, and short dashes are for the second-order harmonics. Devil's staircase in the range of 23.1–23.7 Hz (bottom right inset).

saturates to the level of α), and the second step is near the inner scale l_0 (y_n saturates to the level of δ, omitting the level of α^2). These steps correspond to the large steps of the function f_n/f_1, which can be interpreted as an analog of the "devil's staircase" (see Figure 3.29). The presence of the constants α, α^2, and δ in Figure 3.27 and fulfillment of the similarity relations confirm the Feigenbaum scenario.

Since $y_m \to \delta$ and $\sigma_m(\mu) \to \delta$ at $m \gg 1$, the bifurcation values of the control parameter μ_m can be related to the harmonic frequencies f_m and vortex radii R_m. They are different parameters of the same process of turbulence formation. Actually, for positive ε, δ_1, and δ_2, as soon as $m^* > \delta_1$ and $m > \delta_2$, we should have $|y_m - \delta| < \varepsilon$ and $|\sigma_{m^*} - \delta| < \varepsilon$. Then $|y_m - \sigma_{m^*}| = |(y_m - \delta) - (\sigma_{m^*} - \delta)| \leq |y_m - \delta| + |\sigma_{m^*} - \delta| \leq 2\varepsilon$. Consequently, $\sigma_{m^*} \to y_m$. The levels of δ_1 and δ_2 can be related to each other through consideration of area of confident convergence at rather large m^* and m. As was shown in References 1, 2, 167, σ_{m^*} begins to converge quickly after several iterations, and, therefore, we can take, for example, $m^* \approx 4$–5 as a boundary of the area of confident convergence. This determines the level of δ_1. The frequency y_m, as can be seen from Figure 3.27, begins to converge quickly to δ near the inner scale l_0 for $m \approx 17$–30 (level δ_2). Consequently, the levels δ_1 and δ_2, and the numbers m^*, m appear to be related by the approximate relations $\delta_2 \approx 2^{\delta_1}$, $m \approx 2^{m^*}$.

With allowance for these relations, from solution of the equation $\sigma_m(\mu) = y_n$, $n \approx 2^m$, we can find

$$\mu_m = c \cdot y_n^{-m} + \mu_\infty, \quad y_n = f_n/(n \cdot f_1) = v/(2\pi \cdot n \cdot f_1 \cdot R_n), c = \text{const}, n \approx 2^m. \quad (3.31)$$

The constant c here can be determined after selection of the type of the parameter μ, for example, in the form of the Reynolds number Re, Rayleigh number

Ra, or others. For approximate estimates, relation (3.31), as well as the equality $\sigma_m(\mu) = \delta$, is applicable at $m \geq 0$.

As an example, we demonstrate how equality (3.31) can be used, when the parameter μ is taken in the form of the Rayleigh number. We assume $\mu_m = \text{Ra}_m/$ const in Equation 3.31. A new constant, as can be seen from Equation 3.31, leads just to redetermination of the constant c.

We can follow the process of breakdown of the main vortex in the Benard cell down to the level when the existence of stationary periodic flows (vortices) becomes impossible. In this case, the number Ra decreases from some maximal value Ra_0 down to the critical value Ra_{cr}. In the process of breakdown, $\text{Ra}_m \ll \text{Ra}_{cr}$ at the rather large m, and, consequently, we can take $\text{Ra}_\infty = 0$ in the first approximation. Then, from Equation 3.31, we obtain $\text{Ra}_m = \text{Ra}_{m_0} y_{n_0}^{m_0} y_n^{-m}$, where m_0 is the value of m, at which Ra_{m_0} is known ($n_0 = 2^{m_0}$ since $n = 2^m$). If $m_0 = 0$, then, according to the Feigenbaum enumeration, Ra_{m_0} is the Rayleigh number for the main vortex in the Benard cell. This number can be determined approximately, if we take the layer thickness h equal to the diameter of the main vortex in Ra.

Table 3.2 presents the bifurcation diagram of breakdown of the main vortex in the Benard cell (dependence of the Rayleigh number Ra_m on the bifurcation number m) along with the harmonic numbers n corresponding to m and the vortex diameters $2R_n$ (in cm).

Table 3.2   Diagram of Bifurcations at Breakdown of the Main Vortex

Ra_m	$2R_n$ (cm)	m	n
$1.55 \cdot 10^9$	294	0	1
$5.15 \cdot 10^8$	49.1	1	2
$2.05 \cdot 10^8$	26.8	2	4
$2.76 \cdot 10^6$	4.5	3	8
$4.89 \cdot 10^5$	2.5	4	16
$1.94 \cdot 10^5$	1.5	5	32
$6.86 \cdot 10^4$	0.87	6	64
$2.65 \cdot 10^4$	0.50	7	128
$9.46 \cdot 10^3$	0.26	8	256
1474.1	0.12	9	512
444.7	0.06	10	1024
67.5	0.05	11	2048

As can be seen from Table 3.2, with an increase of m, the bifurcation numbers Ra_m decrease and intersect the level Ra_{cr} ($Ra_{cr} = 657$–1708 [1,2,161,162]) at $m = m_{cr} \approx 9$–10. The same value of m_{cr} can also be found from the number of recorded harmonics (maximal n in Figure 3.27 is 1849), $m_{cr} \approx E$ (log n/log 2) $= E$ $(10.85) = 10$, where $E(x)$ is an integer part of x.

Thus, in the pavilion, we observe the result of formation of the main vortex in the Benard cell and approximately 10 events of its discrete breakdown. The vortex diameters corresponding to the critical value m_{cr}, as can be seen from Table 3.2, fall within the range 0.6–1.2 mm. They coincide with the diameters of minimal vortices, which can exist in air [1,2].

Based on the analysis, the known stochastization scenarios can be divided into two groups. The first group incorporates the scenarios of formation (appearance, generation) of periodic flows from laminar flows. In the first turn, they are the Pomeau–Menneville scenario and the Rayleigh–Benard convection. The second group includes the scenarios of breakdown (destruction, degeneration) of formed vortex periodic flows. In this group, the main scenario is the Feigenbaum scenario. The Landau–Hopf and Ruelle–Takens scenarios (appearance of the limiting N-periodic flow) contain attributes of the both groups.

It follows from the results presented above, in the incipient turbulence the scenario of the both groups are confirmed. The formation of the periodic flow (main vortex in the Benard cell) and its discrete breakdown take place. These processes, in principle (in the simplest situations), are confirmed by the known solutions of nonlinear flow-dynamics equations. However, in the general case, these solutions are still unknown. It is clear that vortex solutions should exist, since they are observed in the experiment. It is also clear that the stability of vortex solutions is determined by nonlinear resonances (both between external forces and dissipative processes, and between harmonics with commensurable frequencies [169]). In this connection, the mechanism of formation and existence of hydrodynamic turbulence, in which the role of stochastization decreases significantly, becomes clearer. Then the turbulence considered usually as a purely random phenomenon should be significantly deterministic.

The turbulence determinacy appears to be much higher than expected from the results of the above analysis. Actually, if we consider the harmonics of the W_T spectrum from the data shown in Figure 3.26 upon the suppression of the threshold filter, then the same filter can separate the highest maxima (second-order harmonics) from them. It turns out that the local structure of arrangement of second-order harmonics (between neighboring first-order harmonics) is similar to the structure of the first-order harmonics in the entire W_T spectrum. The local self-similarity of the spectrum or, in other words, the fractality of the turbulence spectrum is observed.

Figure 3.29 shows the frequency dependence of the arguments of first-order and second-order harmonics (normalized frequency of the harmonics f_n/f_1).

This dependence is usually referred to as the devil's self-similarity staircase [169–171]. In this staircase, every internal interval between main stairs is similar to the entire staircase. In Figure 3.29, every dash (stair) corresponds to its own value of f_n/f_1 (for the data of Figure 3.29, all these values are integer numbers). Long dashes correspond to first-order harmonics, while short dashes are for the second-order harmonics.

Figure 3.30 shows the distances between arguments of neighboring harmonics $f_n - f_{n-1}$ ($n = 2,3,...$) as functions of frequency (as a function of frequencies f_n). The lower plot in Figure 3.30 shows the distances between arguments of the first-order harmonics for the entire W_T spectrum (see Figure 3.26). [As follows from Figure 3.24, in the W_T spectrum, it is sufficient to consider only the frequency range 0–10 Hz without the weak component, which is caused by the transfer of the motion energy into heat and is observed at frequencies higher than 10 Hz.] The central plot in Figure 3.30 gives the distances between arguments of the second-order harmonics for the longest range between first-order harmonics (0.19–0.64 Hz), while the upper plot is for the range 23.34–23.44 Hz.

As was seen from the devil's staircase, other ranges contain much smaller number of second-order harmonics (because of the limited experimental possibilities and, correspondingly, limited spectral resolution) and, therefore, are not convenient for analysis.

From comparison of the data shown in Figure 3.30, we can see that the local structure of arrangement of second-order harmonics is generally similar to the structure of arrangement of first-order harmonics in the entire W_T spectrum.

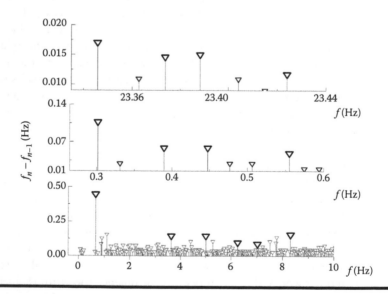

Figure 3.30 **Distances between neighboring harmonics** $f_n - f_{n-1}$ ($n = 2,3,...$): **between first-order harmonics for the entire spectrum (lower plot), between second-order harmonics for the ranges 0.19–0.64 Hz (central plot) and 23.34–23.44 Hz (upper plot).**

The devil's staircase actually turns out to be self-similar, while the spectrum itself is fractal. Since second-order maxima are much weaker in comparison with the first-order maxima, second-order harmonics can be called a fractal shadow of first-order harmonics. Thus, we can see that the process of formation and breakdown of turbulent vortices is deterministic even in fine spectral details.

3.3.5 Extension of the "Coherent Structure" Concept: Actual Turbulence

The conditions of chaos formation in typical dynamic systems, formulated in References 169–171, allow us to reveal versatile characteristic features observed at the formation of turbulence (in incipient turbulence). They include [169]: appearance of irregular long-lived spatial structures, whose form (character) is determined by dissipative factors, local instability and fractality of the phase space of such structures, and appearance of the central (at zero frequency) peak in the spectrum.

All these features were observed in the measurements as the spatial structure in the further study was taken the convective Benard cell. Its form depends on viscosity of the medium and geometry (shape) of the space it appears in (i.e., dissipative factors). The cell breaks down into smaller cells (vortices) in a cascade manner. The spectrum of a passive admixture in the cell (temperature) is fractal. Due to nonstationarity of a random process in the cell (bifurcation of stability alternation, see Figure 3.22), the average temperature was not constant. If we do not undertake additional measures to remove nonstationarity (average value of the random function can then be found by the time averaging not over the whole sample length, but only over the length of the characteristic scale of function variation). Then the centered random temperature contains the noncompensated constant component. Just this component lifts the low-frequency part of the Fourier spectrum, in particular, gives the central peak at the zero frequency. This phenomenon is observed for Fourier transforms of both the correlation function and the random nonstationary function itself (see definitions of these functions in Chapter 2).

It is convenient to join all these features by the common term "coherent structure," if we will extend this, already existing, concept and will include small-scale components into the coherent structure.

Monin and Yaglom (see References 1, 2) defined the coherent structure as a nonrandom nonlinear superposition of large-scale components of turbulence characterized by the high stability. It is likely the simplest definition of all available definitions of the coherent structure (see, for example, References 43, 44, 50–52). The earliest definitions [43,44] deal with the expansion of the instantaneous field of velocities $\mathbf{v}$ into the coherent large-scale $\mathbf{v}_{coh}$ and random incoherent turbulent $\mathbf{v}_T$ components, that is, $\mathbf{v} = \mathbf{v}_{coh} + \mathbf{v}_T$ (double expansion [43]). Later, more complex expansions (e.g., the triple expansion [50,51]) were used. According to these definitions, turbulence consists of nonrandom coherent (large-scale) and purely random (small-scale) motions. Random motions are superimposed on coherent ones and usually extend far beyond the coherent structure. Coherent large-scale vortices are believed to be just main independent energy carriers and usually are not included in the structure of turbulence.

Coherent structures have been actively studied during the recent decades (see, e.g., References 43–79). Near-wall small-scale turbulence, turbulent convection in the near-surface atmospheric layer in the presence of wind shear, "cloud streets" in the atmosphere, and "Langmuir circulation" in seas and lakes are objects of intense studies. In addition, periodic large vortices in jet engine wakes have been investigated. It was shown that the main energy carriers in turbulent flows are large-scale ordered vortices, which influence significantly the formation of all flow characteristics. It was found that large-scale turbulent motions are deterministic, that is, are not random. The studies apply different methods of turbulence visualization (usually, coloring of flow). However, the resolution of the used visualization methods is low. That is why, only large-scale components of turbulent flows can be clearly seen, as a rule. Small-scale inhomogeneities usually remain invisible.

In their research, Sreenivasan et al. [71,72], the assumptions that fractality manifests itself in turbulent flows were confirmed. The fractal dimension of some of them was measured. An assumption was made that turbulence is a set of slightly different fractals or an ensemble of semi-organized motions. The concept of multiplicative processes—multifractals—was used in application to turbulence. The statistical scale invariance of turbulent multifractals was used in the theoretical works by Novikov [79] and Chainais [78]. These works propose the models of generalized infinitely divisible cascades for phenomena of intermittence in hydrodynamic turbulence and introduce the concept of the similarity scale into the theory of infinitely divisible probability distributions, as well as consider the problems of self-similarity and asymptotic behavior of statistical characteristics.

As can be seen from the results presented above, the process of coherent breakdown of the main energy-carrying vortex into smaller ones does not terminate in the region of large-scale (low-frequency) vortices. It continues permanently into the small-scale (high-frequency) region up to the size of small vortices, which still can exist in air (0.6–1.2 mm, see the bifurcation diagram in Table 3.2). The frequencies of small-scale vortices are multiple to the frequency of the main vortex (see Figures 3.27 and 3.28). The oscillation phases in these vortices are rigidly related, and the vortices themselves are coherent (inphase). Therefore, the process of coherent breakdown cannot be limited to the region of large-scale vortices (as was done based on old definitions of the coherent structure).

According to the results presented above, we now can define the coherent structure as a wider concept including small-scale components as well.

The hydrodynamic coherent structure is a compact formation including a long-lived spatial vortex structure (cell) arising as a result of long action of thermodynamic gradients and products of its discrete coherent cascade breakdown.

The breaking-down spatial structure can be called a cell, giving rise to a coherent structure. The parent cell is the main energy-carrying vortex. The breakdown of the parent cell corresponds to breakdown of the main energy-carrying vortex. The frequency of this main vortex can be considered as one of the main attributes of both the parent cell and the coherent structure as a whole. The size of the coherent structure is blurred. Flows, external with respect to the parent cell, can transport its breakdown products to long distances, forming a long turbulent wake. The lifetime of this coherent structure is determined by the time of action of thermodynamic gradients. As a limiting case of the very weak local instability (when the parent structure is locally stable and does not break down), the coherent structure can consist of only one long-lived parent structure. In this case, the parent structure is some configuration of the laminar flow (usually, vortex). We have this situation at the observation of, for example, Benard cells in a thin layer of very viscous fluid.

Thus, the process of turbulence formation (generation) can be attributed to appearance of coherent structures. The compact formation observed in the described above experiments includes a long-lived main energy-carrying vortex and products of its simultaneous cascade discrete breakdown (coherent vortices with multiple frequencies). The frequency of the main vortex is determined by the size of the convective Benard cell. According to our definition, this formation is just the coherent structure.

The issue of formation of coherent structures in fluid, for example, in water seems to be interesting. An example of the smoothed turbulence spectrum in ocean was published in References 1, 2. This spectrum, analogously to the spectrum of coherent structure in air (see Figure 3.26), has frequency intervals with the 5/3-law decrease arranged stepwise. The stepwise character of this spectrum allows us to consider it as a spectrum of coherent structure in water.

Analogous features are also observed as fluid flows around some obstacles. This can be seen from the data of numerous observations [43,44,50–52,62,64,71,172,181] and available numerical solutions of the Navier–Stokes equations [47,48,56–58,63,65–70,182–184]. On the obstacle surface (near the rear side), a long-lived main vortex is formed (parent structure; there can be several such structures). Behind the obstacle (at some distance from it), well-defined large vortices appear. Their sizes do not exceed the size of the main vortex. The vortex shape is distorted by the external flow (the front surface is concave). With the distance, we observe few smaller vortices and then a continuous turbulent jet consisting of small-scale vortices.

Behind obstacles like a sphere, several parent structures can be formed. Then the presence of even small asymmetry of the obstacle surface distorts the axial symmetry of the flow-around pattern and leads to the competition between parent cells. In this case, the zone behind the obstacle, in which parent structures are formed, extends in length, and the parent cells are usually arranged behind the obstacle as an extended chessboard-like structure. This structure is called a von Kármán vortex street. The length of such a vortex street depends on the obstacle shape, flow velocity, fluid viscosity, and other parameters. The breakdown of parent cells is observed at the end of the von Kármán vortex street, where the process of formation of parent vortices terminates and their stability is lost.

As is shown below (see Section 3.3.6), the actual atmospheric turbulence can be considered as a mixture of different coherent structures. The frequencies of the main vortices of these structures are not multiple (not commensurable). The sizes of parent cells can differ significantly from each other. Consequently, the mixture of different coherent structures is not coherent in the general case. Based on our data, if $\mathbf{v}_{\text{c.str, }i}$ is the velocity of motion arising due to existence of the i-th coherent structure, and $\mathbf{v}$ is the total velocity, then

$$\mathbf{v} = \sum_{i=1}^{N_c} \mathbf{v}_{\text{c.str, }i}.$$

The constant N_c determines the number of coherent structures existing in the observation area (usually, $N_c \gg 1$). Thus, the turbulence can be studied through observation and study of all coherent structures arising in the considered spatial area.

Consider some isolated area, in which the influence of the ambient space is weak or absent. Let it include N_c coherent structures. If the size of the parent cell of one of these structures (assume that with the number $i = 1$) is far larger than the sizes of parent cells of other structures, then the velocities $\mathbf{v}_{\text{c.str, }i}$ ($i = 1, \dots, N_c$) can be divided into large-scale and small-scale components

$$\mathbf{v}_{\text{c.str,1}} = \mathbf{v}_{\text{c.str,1}}^{\text{large}} + \mathbf{v}_{\text{c.str,1}}^{\text{small}}, \quad i = 1,$$

$$\mathbf{v}_{\text{c.str,}i} = (\mathbf{v}_{\text{c.str,}i})^{\text{small}}, \quad i = 2, \dots, N_c.$$

This separation is quite conditional and determined by limiting possibilities of the used visualization methods. Then the total velocity of the medium can be represented as

$$\mathbf{v} = \sum_{i=1}^{N_c} \mathbf{v}_{\text{c.str, }i} = \mathbf{v}_{\text{c.str, 1}}^{\text{large}} + \mathbf{v}_{\text{c.str, 1}}^{\text{small}} + \sum_{i=2}^{N_c} (\mathbf{v}_{\text{c.str, }i})^{\text{small}} = \mathbf{v}_{\text{coh}} + \mathbf{v}_{\text{T}},$$

where

$$\mathbf{v}_{\text{coh}} = \mathbf{v}_{\text{c.str, 1}}^{\text{large}}, \quad \mathbf{v}_{\text{T}} = \mathbf{v}_{\text{c.str, 1}}^{\text{small}} + \sum_{i=2}^{N_c} (\mathbf{v}_{\text{c.str, }i})^{\text{small}}.$$

This equation is the double expansion [43]. It is true for the conditions described above and was, in its time, the experimental basis for the earlier definitions of the coherent structure [43,44,51,52]. The considered spatial area can be called the area with decisive influence of one coherent structure or the area with coherent turbulence.

3.3.6 Coherent Structures in Open Air

3.3.6.1 Coherent and Incoherent Turbulence

As is known from meteorology, there are rather stable vortex formations (cells) of different scales in the atmosphere. Ferrel and Hadley cells are the largest, with a radius up to 5000 km. They can be considered as a sort of Benard cells in a thin spherical layer (in Earth's scales). There are also smaller cells (cyclones, anticyclones, storm cells, tornado, and others). The products of breakdown of these vortices having the pronounced deterministic character (corresponding to a coherent structure or non-Kolmogorov incipient turbulence) can be observed in open air.

As above, we compare the result of measurements carried out in the open air conditions and in the rooms' areas. Thus, Figure 3.31 presents the data of measurements in open air and the result of simulation of the wind transport of the frozen spatial pattern of flows in different LSVT rooms of the data recorded at seven neighboring points located along a straight line at the same height [80,81]. This straight line, in

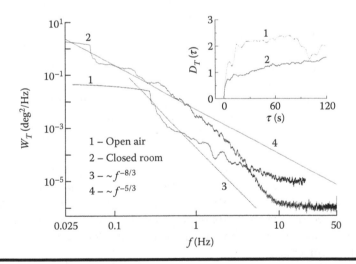

Figure 3.31 **Kolmogorov turbulence as a result of mixing of different coherent structures. W_T—smoothed spectra, D_T—structure functions of temperature fluctuations. (1) summer daytime measurements in mountains (at a height of 2032 m), (2) wind transport of the frozen flow pattern in different LSVT rooms through the same point.**

addition to points 5, 4, and 2 in the spectrograph pavilion, includes also four other points located in neighboring closed LSVT rooms. These rooms are isolated from the pavilion, and the frequencies of the main energy-carrying vortices in them and in the pavilion are different. Comparison of data presented in Figures 3.31, 3.24, and 3.23 shows that the results depicted in Figure 3.31 for open air are indicative of the predominant influence of one coherent structure ($W_T \sim f^{-8/3}$ in the inertial range). To the contrary, the transport of deterministic vortices formed in closed rooms through one point gives ultimately the turbulence close to Kolmogorov ($W_T \sim f^{-5/3}$).

Thus, we can conclude that the actual Kolmogorov atmospheric turbulence is a result of mixing of deterministic vortices from different coherent structures. Inequality (nonmultiplicity, incommensurability) of the frequencies of main vortices of different coherent structures leads to out-of-phase (incoherent) oscillation of vortex families, being products of vortex breakdown. That is why the turbulence arising at the mixing of coherent structures with incommensurate frequencies of the main vortices is naturally referred to as incoherent.

As a result of research missions in the 2000s under mountain and valley conditions, we have accumulated a large experimental database of near-surface measurements of the main turbulence parameters in different geographic regions and different meteorological situations. It follows from these data that extended areas, in which one coherent structure has the decisive influence, are often observed in open air. Moreover, characteristic attributes of coherent structure are present, in some or other degree, in the most of accumulated data. The incoherent Kolmogorov developed turbulence is usually observed only at wide areas of smooth and homogeneous underlying surface.

Coherent turbulence (characterizing an area with the decisive influence of one coherent structure) differs from incoherent one, first of all, by the faster decrease of the smoothed W_T spectrum in the inertial range ($\sim f^{-8/3}$) and by the smaller contribution of high-frequency components (small-scale vortices). Therefore, in the atmosphere in areas with the decisive influence of one (local) coherent structure, the spectrum in the inertial range has two pronounced parts of decrease: first, the rather fast decrease is observed (usually, $\sim f^{-8/3}$, sometimes even faster) and then the decrease becomes slower ($\sim f^{-5/3}$) as the frequency increases. The second Kolmogorov section characterizes the mixture of breakdown products of other largest parent structures present in the atmosphere. In some cases, the entire inertial range of the spectrum has two pronounced parts with the 8/3-law decrease located by steps. Then, we can speak about the presence of two local coherent structures in the measurement area.

Thus, the data of our measurements show that the actual atmospheric turbulence can be considered as a result of mixture of different coherent structures with incommensurate frequencies of the main vortices. The coherent structure in this case is understood in its extended meaning, including the Feigenbaum breakdown scenario. Every coherent structure contains a long-lived parent structure arising under effect of local thermodynamic gradients and products of its coherent cascade breakdown.

The parent structures can take different forms (from a solitary ordered structure like a Benard cell to systems of periodically spatially distributed hydrodynamic perturbations like systems of various shafts). The sizes of parent structures (cells) in the atmosphere can differ 10^8–10^9 times: from few centimeters (near-wall turbulence) to several thousands of kilometers (Ferrel and Hadley cells).

In the cases of high stability, parent structures not always break down. However, in the atmosphere due to the low viscosity of the medium (and, correspondingly, high values of the Rayleigh number), the lifetime of nondecomposing cells is not long. These cells are usually observed at the relatively short stage of their generation (at the stage of origination and formation). Sometimes large nondecomposing cells with the sizes larger than the outer scale of turbulence can be considered and studied as laminar flows. However, in any case, regardless of the size, nondecomposing cells should be considered as a part of the single process of turbulence formation and evolution.

In the area with the decisive influence of one large coherent structure, large-scale products of its breakdown are coherent (coherent turbulence). Small-scale products are mixed with products of breakdown of other small coherent structures present in this area, thus forming the incoherent mixture (incoherent Kolmogorov turbulence). If several mixing coherent structures have comparable sizes of parent structures (which is usually observed over regions with smooth underlying surface), then their mixture is incoherent (Kolmogorov) in the ranges of both small and large scales (up to scales comparable with the size of the parent structure).

The data of our measurements [80–82,86,87,91–93,95,98,102–106,128] and known data of other authors on the problem of turbulence formation (both experimental works on flow visualization [43,44,50–52,62,64,71,172,173,181] and theoretical works presenting analytical results [42,45,46,48,49,52–55,61,66,72–75,78,79,161–163,168–171] and numerical solutions of the Navier–Stokes equations [47,48,56–58,63,65–70,182–184]) show that the coherent structure in its extended meaning can be considered as a structure element in hydrodynamic turbulence. This structure element is a vortex formation arising as a result of transition of the energy of thermodynamic perturbation into the energy of motion of a continuous medium. A vortex can have various shapes and sizes, and under conditions of instability, it breaks down coherently and in a cascading manner into smaller vortices. From the mathematical point of view, the structure element of turbulence is a soliton-like solution of flow-dynamics equations. It can be either a solitary one-soliton solution or one soliton in a many-soliton solution.

This view of coherent structures and, in general, the problem of turbulence formation became possible due to the indoor study of the processes of formation and breakdown of a solitary ordered structure (solitary Benard cell arising in the air medium of a closed room) with the use of small-size sensors and methods of spectral analysis.

For the further investigation of the local structure of actual turbulence, it is necessary to distinguish successfully the products of breakdown of different coherent structures. It can be possible, because the products of breakdown of one particular coherent structure have their own, inherent of only this structure, harmonic composition.

3.3.6.2 Turbulent Characteristics in Coherent Turbulence

Since coherent turbulence observed in the atmosphere differs from the Kolmogorov one, it becomes necessary to refine the domain of applicability of the Kolmogorov–Obukhov law. In particular, it is needed to refine the values of the Kolmogorov C and Obukhov C_θ constants.

According to the Kolmogorov–Obukhov law, the structure function of fluctuations of the longitudinal velocity $D_{rr}(r)$ in the inertial range of scales r can be expressed through the structure characteristic C_V^2 of fluctuations of the longitudinal velocity: $D_{rr}(r) = C_V^2 \cdot r^{2/3}$ (see also Chapter 1). The structure function of temperature fluctuations $D_T(r)$ can be expressed through the structure characteristic C_T^2 of temperature fluctuations: $D_T(r) = C_T^2 \cdot r^{2/3}$. The structure characteristics C_V^2, C_T^2 determine the intensity of velocity and temperature fluctuations and are important parameters of the turbulent motion of the medium. In their turn, C_V^2 and C_T^2 depend on the average dissipation rate of kinetic energy ε, temperature dissipation N, and Kolmogorov C and Obukhov C_θ constants: $C_V^2 = C \cdot \varepsilon^{2/3}$, $C_T^2 = C_\theta \cdot \varepsilon^{-1/3} \cdot N$. It is seen from here that at known ε and N, velocity and temperature fluctuations are determined by the constants C and C_θ. The constants C and C_θ were measured in different media by different methods. The values $C = 1.9$ and $C_\theta = 3.0$ are recommended as the most probable estimates of C and C_θ in References 1, 2. These estimates are average over data of different authors. The deviation from the average is rather large [1,2], for example, the values of 0.9, 1.6, and 2.8 were observed for C; and for C_θ the observed values were 1.1, 1.4, 2.5, 2.7, 3.3, 3.5, 5.6, 5.8, 6.5, and 9.0. As was shown in References 1, 2, 4, the Kolmogorov constant C is connected with the asymmetry S of probability distribution of the longitudinal velocity difference as

$$C = (0.8/|S|)^{2/3}, \quad S = D_{uuu}/|D_{uu}|^{3/2}.$$

The third moment of the longitudinal difference of wind velocities

$$D_{uuu}(r) = \langle [u'(r'+r,t) - u'(r',t)]^3 \rangle$$

can be found from the temporal moment

$$D'_{uuu}(\tau) = \langle [u'(r',t+\tau) - u'(r',t)]^3 \rangle$$

and the condition of frozen turbulence [1,2,4]

$$D_{uuu}(\mathbf{v} \cdot \tau) = D'_{uuu}(\tau).$$

Here, $u'(r,t) = u(r,t) - \langle u(r,t) \rangle$, where $u(r,t)$ is the random value of the longitudinal velocity (usually, the projection of the random velocity vector onto the direction of the average velocity vector) at the point r at time t. The constant C_θ is connected with the asymmetry S' of the probability distribution of temperature difference as [1,2,4]

$$C_\theta = -(4/3)/(C^{1/2} \cdot S'), \quad S' = D_{uTT}/(D_{uu}^{1/2} \cdot D_T).$$

The third spatial moment of the difference

$$D_{uTT}(r) = \langle [u'(r'+r,t) - u'(r',t)] \cdot [T'(r'+r,t) - T'(r',t)]^2 \rangle$$

can also be found from the temporal moment

$$D'_{uTT}(\tau) = \langle [u'(r',t+\tau) - u'(r',t)] \cdot [T'(r',t+\tau) - T'(r',t)]^2 \rangle$$

and the condition of frozen turbulence

$$D_{uTT}(\mathbf{v} \cdot \tau) = D'_{uTT}(\tau).$$

Here, $T'(r,t) = T(r,t) - \langle T(r,t) \rangle$, $u'(r,t) = u(r,t) - \langle u(r,t) \rangle$ are random centered temperature and longitudinal velocity.

Figures 3.32 and 3.33 illustrate the process of determination of the Kolmogorov C and Obukhov C_θ constants for the case of non-Kolmogorov turbulence in the atmosphere. The frequency spectrum depicted in Figure 3.32 has the long inertial range, in which $W_T \sim f^{-8/3}$.

The analysis of the data of more than 30 measurement points as revealed the following. If turbulence is close to Kolmogorov (smooth underlying surface, temporal spectra in the inertial range $W_u(f) \sim f^{-5/3}$, and others), then the values of the Kolmogorov and Obukhov constants can be taken $C = 1.9$ with an error of 1%–12% and $C_\theta = 3.0$ with an error not exceeding 30%. If turbulence deviates from the Kolmogorov one and is close to coherent ($W_u(f) \sim f^{-8/3}$ and others), then the Kolmogorov constant C falls within the range 0.9–3.7, while the Obukhov constant C_θ is in the range 1.3–5.1. In this case, the error of $C = 1.9$ is 19%–93%, and the error of $C_\theta = 3.0$ can achieve 70%.

As the experience shows, the measurements of turbulent characteristics of the atmosphere (usually conducted with the use of the Kolmogorov–Obukhov

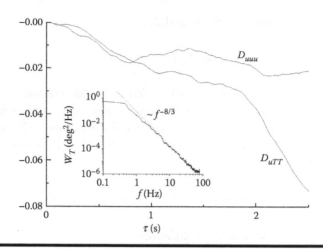

Figure 3.32 **Third moments of the longitudinal difference of velocity and temperature** $D_{uuu}(\text{m}^3/\text{s}^3)$, $D_{uTT}(\text{deg}^2\text{m}/\text{s})$. **Smoothed spectrum (bottom left inset). Coherent turbulence, summer daytime measurements in mountains at a height of 680 m, July 2, 2007.**

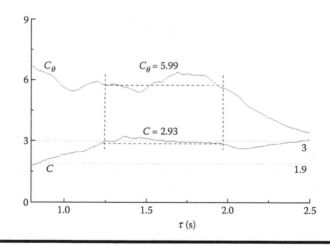

Figure 3.33 **Kolmogorov** C **and Obukhov** C_θ **constants. Coherent turbulence measured in mountains at a height of 680 m, July 2 of 2007. Vertical lines show the boundaries of the inertial range, and dashed lines are the average values of the constants** C **and** C_θ.

law) are accompanied by significant errors, as a rule. The data of our measurements indicate the main cause for appearance of these errors. Variations of the Kolmogorov and Obukhov constants within 100% (depending on the observation point) lead to practically the same errors in determination of the characteristics C_T^2, C_V^2, and C_n^2.

Another source of errors is the structure function of temperature fluctuations $D_T(r)$ itself, if we determine the value of C_T^2 using the Kolmogorov–Obukhov law $D_T(r) = C_T^2 \cdot r^{2/3}$. As known [4], the function $D_T(r)$ can be represented in the form of

$$D_T(r) = 8\pi \int\limits_0^\infty d\,\text{æ} \cdot [1 - \sin(\text{æ} \cdot r)/(\text{æ} \cdot r)] \cdot \Phi_T(\text{æ}) \cdot \text{æ}^2.$$

After the substitution of von Kármán spectrum (3.30) written in the exponential form,

$$\Phi_T(\text{æ}) = A_0 \cdot C_T^2 \cdot (6.6\text{æ}_{0e})^{2(\nu - 1/3)} \cdot \text{æ}^{-2(\nu + 3/2)} \cdot \exp\left(-\text{æ}^2/\text{æ}_m^2\right) [1 - \exp\left(-\text{æ}^2/\text{æ}_{0e}^2\right)],$$

$$\text{æ}_{0e} = 2\pi/L_0, \nu = 5/6, 1/3.$$

$$(3.32)$$

Spectrum (3.32) was obtained by us in References 82, 128 with allowance for the approximate relation between the outer scales for $\nu = 5/6$ and $1/3$ (according to Table 3.1, they differ, on average, by the coefficient 2.3) and the relation at $\nu = 1/3$ between the used von Kármán outer scale L_0^K and exponential L_0 (usually, $L_0 = 0.54\,L_0^K$, see References 175–177). Exponential spectrum (3.32) deviates from von Kármán spectrum (3.30) only in the energy range, where $\text{æ}^2/\text{æ}_0^2 \ll 1$. However, at $\nu = 1/3$, it gives practically the same results as (3.30) [175]. At the same time, it significantly simplifies the calculations. After calculation of the corresponding ranges, we obtain the asymptotic representations for the structure function $D_T(r)$ in coherent turbulence ($\nu = 5/6$):

$$D_T(r) = 53.45 C_T^2 \cdot L_0^{K-1} \cdot l_0^{-1/3} \cdot r^2, \quad r \ll l_0,$$

$$D_T(r) = C_T^2 \cdot r^{2/3} \cdot f(r/L_0^K), \quad f(x) = 49.87(x - 1.34x^{4/3}), \quad l_0 \ll r \ll L_0^K,$$

$$D_T(r) = 3.56 C_T^2 \cdot (L_0^K)^{2/3}[1 - 0.63(r/L_0^K)^{-1/3}], \quad L_0^K \ll r.$$

As can be seen, the function $D_T(r)$ deviates from the Kolmogorov–Obukhov 2/3 law. At small r ($r \ll l_0$), $D_T(r) \sim r$, as in the Kolmogorov turbulence, but the coefficient in the asymptotic becomes dependent on the outer scale L_0^K. In the inertial range ($l_0 \ll r \ll L_0^K$), there is an extended initial part, in which $D_T(r) \sim r^{5/3}$. The parts with $\sim r^2$ and $\sim r^{5/3}$ intersect at $r \approx 0.8\,l_0$ (remind that the used inner scale l_0 of coherent turbulence is an order of magnitude larger than the Kolmogorov one, see Table 3.1). The function $f(x)$ has a maximum. Therefore, in the inertial range of coherent turbulence $3.0 l_0 \leq r \leq 0.4 L_0^K$, where this function can be considered approximately as a constant [$f(x) \approx f_{max} = 2.17$], the approximately 2/3

Kolmogorov part is observed, in which $D_T(r) \approx C_T^2 \cdot f_{max} \cdot r^{2/3}$. It follows from here that in the measurements of C_T^2 from the 2/3-asymptotic of $D_T(r)$ we can expect that the values of C_T^2 will be more than twice overestimated.

3.3.6.3 Attenuation of Amplitude and Phase Fluctuations of Optical Wave in Coherent Turbulence

With the use of the model of 3D spectrum (3.32), we can obtain equations for the variance of fluctuations of the log amplitude $\sigma_\chi^2 = \langle \chi^2 \rangle$ (under conditions of applicability of the method of smooth perturbations [4]), the variance of displacements of the energy centroid of laser beam [176] $\sigma_c^2 = \langle \rho_c^2 \rangle$, and the variance of displacements of the image of optical sources [177] $\sigma_t^2 = \langle \rho_t^2 \rangle$. Consider the parameters b_χ, b_c, and b_t, being ratios of these variances for coherent and incoherent Kolmogorov turbulence

$$b_\chi = \sigma_\chi^2 |_{ch} / \sigma_\chi^2 |_{nch}, \quad b_c = \sigma_c^2 |_{ch} / \sigma_c^2 |_{nch}, \quad b_t = \sigma_t^2 |_{ch} / \sigma_t^2 |_{nch}.$$

Let a_t be the radius of optical receiver, $a_e(x)$ be the effective radius of optical beam at the path with the length x ($a_e(x) \geq a_e(0)$); λ be wavelength, $R_F = (\lambda x)^{1/2}$ be the radius of the first Fresnel zone. The calculations show that in the coherent turbulence ($\nu = 5/6$) the variances σ_c^2, σ_t^2, and σ_χ^2 can be found (accurate to constant factors) from known equations for them in Kolmogorov turbulence ($\nu = 1/3$) through replacements $a_e(x) \rightarrow L_0$; $a_t \rightarrow L_0$; $l_0 \rightarrow l_0^{4/7} L_0^{3/7}$ at $l_0 \gg R_F$ and $x \rightarrow x (R_F/L_0)^{6/11}$ at $l_0 \ll R_F$; respectively.

Actually, consider, for example, displacements of images of astronomical sources. For the case of incoherent Kolmogorov turbulence, the variance of angular displacements (jitter) of images σ_α^2 ($\sigma_\alpha = \sigma_t/F_t$, F_t is the focal length of the receiving telescope) can be expressed through the integral value I of the structure characteristic of refractive index C_n^2 (integral intensity of incoherent turbulence) in the known way [4]:

$$\sigma_\alpha^2 = 4.51 a_t^{-1/3} \cdot \sec\theta \cdot I, \quad I = \int\limits_{h_0}^{\infty} dh \cdot C_n^2(h),$$

where a_t and h_0 are, respectively, the radius of entrance aperture of the telescope and the height of the aperture center above the underlying surface, θ is the zenith angle of an observed astronomical object (measured at the place of receiver location from the zenith direction), $C_n^2(h)$ is the structure characteristic of fluctuations of the refractive index of air dependent on the height h above the underlying surface (vertical profile of C_n^2). For every value of the angle θ, the value of I

determines the integral intensity of atmospheric Kolmogorov turbulence at optical paths of the given tilt angle. This equation allows reconstruction of the integral intensity of Kolmogorov turbulence I from measured values of the variance of jitter σ_α^2.

As can be seen from Equation 3.32, in coherent turbulence ($\nu = 5/6$), the spectrum of atmospheric turbulence differs from the case of Kolmogorov turbulence ($\nu = 1/3$). Therefore, the equation for the variance σ_α^2 changes. After calculation of the corresponding integrals, we obtain the approximate estimative representation for the variance σ_α^2 in coherent turbulence [126,127,140]:

$$\sigma_\alpha^2 = 4 \cdot 8.06 L_0^{-1/3} \cdot \sec\theta \cdot I_c, \quad I_c = \int_{h_0}^{\infty} dh \cdot C_n^2(h),$$

where $L_0 = L_0(h_0)$ is the outer (exponential) scale of turbulence at the height of the center of the receiving aperture h_0, I_c is the integral intensity of coherent turbulence at optical paths with the given tilt angle. It can be seen from this equation that in coherent turbulence the variance of jitter σ_α^2 is independent of the receiver radius a_t in contrast to the case of Kolmogorov turbulence. In place of the receiver radius a_t, the equation for the variance of jitter includes the outer scale of turbulence L_0. Consequently, theoretical representation of the variance of jitter in coherent turbulence ($\nu = 5/6$) follows (accurate to a constant factor) from the corresponding equation in Kolmogorov turbulence ($\nu = 1/3$), as should be expected, by the replacement $a_t \rightarrow L_0$.

We believe, for comparison, that the coherent and incoherent turbulences have the same outer exponential scales L_0, inner scales l_0, and the turbulence intensity C_T^2 (or C_n^2). Then

$$b_\chi \approx l_0/L_0 \text{ at } l_0 \gg R_F, \quad b_\chi \approx R_F/L_0 \text{ at } l_0 \ll R_F; \quad b_c \approx [a_e(x)/L_0]^{1/3}; \quad b_t \approx (a_t/L_0)^{1/3}.$$

Since the values of $a_e(x)$, a_t, l_0, and R_F are usually much smaller than the outer scale of turbulence L_0 [$a_e(x)$, a_t, l_0, $R_F \ll L_0$], it is seen from here that for typical optical paths and typical values of the source and receiver parameters these ratios are small: b_χ, b_c, $b_t \ll 1$. This means that, in comparison with incoherent Kolmogorov turbulence, in coherent turbulence, the significant attenuation of both the amplitude (under conditions of weak radiation intensity fluctuations [4]) and phase (refractive) fluctuations of optical radiation takes place.

In Reference 174, it was found experimentally that for large receiver size, the variance of displacements of astronomical images (see Figure 3.34; measurements at the top of a 2000-m high mountain; displacements of the image of moon disk edge; the maximal receiver radius $a_t = 22$ cm; afternoon transient turbulent conditions; northern wind through Sayan mountain ridge and the deep river valley) is

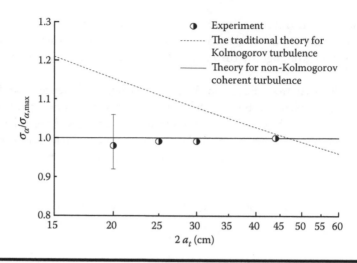

Figure 3.34 **Normalized root-mean-square deviation of jitter of astronomical images. Comparison of traditional incoherent theory with coherent theory and experiment [174].**

independent of the receiver radius a_t in contrast to the usual dependence $\sigma_t^2 \sim a_t^{-1/3}$ for the Kolmogorov spectrum (dashed line in Figure 3.34).

This result can now be explained by the predominant action of one large coherent structure during the measurements carried out in Reference 174.

In 2010–2015, it was conducted the experimental observations of the effect of attenuation of phase (refractive) fluctuations of optical radiation at the coherent turbulence [114,116,126,140–142]. For this purpose, optical measurements were carried out analogous to those reported in Reference 174. The measurements were conducted at the Sayan Sun Observatory of the Institute of Solar-Terrestrial Physics SB RAS (Russia) with the automated horizontal sun telescope. Simultaneously with optical measurements, the state of the near-surface atmosphere was monitored with the mobile ultrasonic meteorological system.

As an example, we present the data of the experiment of 2011 (Figure 3.35). The main parameters of the experiment were the following: sun zenith angle $\theta \approx 55°$; structure characteristic of refractive index fluctuations at a height of 4.5 m from the underlying surface $C_n^2 = 4.2 \cdot 10^{-15}$ cm$^{-2/3}$; the angular radius of the astronomical source (solar disk edge) corresponded to the limiting angular resolution of the used receiver of 0.1 arc s.

The measured results have shown that when large coherent structures (spectrum of temperature fluctuations W_T in the inertial range is proportional to $f^{-8/3}$) are present in the atmosphere over the Sayan Sun Observatory (Russia), the results of measurements coincide with predictions of the coherent theory (horizontal line in Figure 3.35). If in the atmosphere, there are no large coherent structures (incoherent turbulence, $W_T \sim f^{-5/3}$), our results coincide with data of the traditional

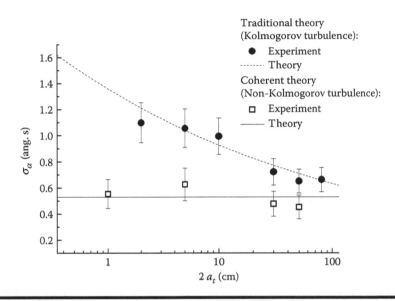

Figure 3.35 **Root-mean-square deviation σ_α of jitter of astronomical image of the solar disk edge as a function of diameter of the telescope entrance aperture $2a_t$. Sayan Sun Observatory (Russia). Summer measurements of August 4, 2011. Experimental points correspond to spectra: either $W_T \sim f^{-8/3}$ (open squares, coherent turbulence) or $W_T \sim f^{-5/3}$ (closed circles, Kolmogorov turbulence).**

incoherent theory (slant line in Figure 3.35). As can be seen from Figure 3.35, the standard deviation of image jitter of the solar disk edge in coherent turbulence for the same aperture is smaller (more than two times for small receivers) than for the Kolmogorov turbulence. Consequently, in the presence of large coherent structures (area of coherent turbulence) in the atmosphere, refractive fluctuations of optical radiation decrease significantly. This means that the effect of attenuation of phase (refractive) fluctuations of optical radiation in coherent turbulence is confirmed experimentally.

Thus, in References 114, 117–119, 123–126, 128, 129, 131, 133, 134, 139, 140, 142, the effect of attenuation of radiation fluctuations in coherent turbulence was found theoretically and experimentally. The effect appears in the presence of large coherent structures (areas of coherent turbulence) in the atmosphere, consists in attenuation of phase (refractive) fluctuations of optical radiation in comparison with Kolmogorov turbulence, and is caused by the faster decrease of the spectrum of coherent turbulence and the smaller contribution of small-scale components.

The effect of attenuation of refractive fluctuations decreases the jitter and improves the quality of astronomical images. Therefore, for installation of ground-based telescopes, we can recommend regions, over which during measurements there are large coherent structures [123–125,128,131,140–142]. At the same time, large coherent structures themselves can be found from characteristic jitter of astronomical images. It should be noted that, for every ground-based astronomical observatory, time intervals of the best images are individual, and their position in the 24-h interval and duration are determined by the regional meteorological situation.

The experiments of 2010–2015 carried out under different meteorological conditions extended our understanding of the influence of coherent turbulence on the propagation of optical radiation in a turbulent medium. The measurements show that over the territory of the Sayan Sun Observatory the coherent turbulence is observed for a long time (up to 20–120 min), usually, at the northern wind (from mountains of Sayan ridge). In summer, this wind is usually observed at night. Therefore, in this observatory, night astronomical observations are preferable. At the southern wind (from relatively smooth underlying surface, usually observed in daytime), the maximal average lifetime of areas of coherent turbulence ranges from 10 to 30 min. The lifetimes of turbulence of different types (time intervals of the presence of turbulence of one type, when it does not alternate to other type) observed in daytime optical measurements of different years are given in Table 3.3.

In general, we can say that long daytime astronomical observations in Sayan Sun Observatory are accompanied by often transitions from the Kolmogorov to the coherent turbulence (especially, in late fall at a strong wind). This transition corresponding to the alternation of turbulence type gives intermittence in the jitter of astronomical images, which shows itself in the frequent alternation of intervals of strong and weak image jitter. Since coherent turbulence leads to improvement of the quality of optical images [123–125,128,131,140–142], this effect can be considered as positive for short-exposure measurements (within tens of minutes), when the highest-quality images can be selected from a series of recorded images based on the meteorological data.

Table 3.3 Average Lifetimes of Turbulence of Different Types Observed in Daytime Optical Measurements

Date of Measurements	Wind Direction (deg)	Horizontal Wind Velocity (m/s)	Time of Continued Observation (min)	
			of Coherent Turbulence	of Kolmogorov Turbulence
June 19, 2010	190–230	4.4	6–14	8–33
August 4, 2011	20–40	1.9	7–12	8–15
July 14, 2012	40–70	1.7	18–39	7–29
September 25, 2013	180–200	6.0	10–26	6–7

3.3.7 Properties of Coherent Structures

As we can see, coherent structures are important elements for understanding the processes of turbulence formation (appearance) and further evolution of the turbulence structure. Therefore, here we list briefly the revealed properties of single coherent structures and the properties of the mixtures (sums) of different coherent structures.

3.3.7.1 Properties of Single-Coherent Structures

The properties of single coherent structures were discussed thoroughly in References 82, 86–90, 95, 117, 128. We studied characteristics of a single coherent structure experimentally (by small-size acoustic sensor) with the following theoretical analysis of experimental spectra of fluctuations of random temperature and velocity components.

1. As a result of action of thermodynamics gradients (temperature or pressure) at boundaries of some selected volume, a spatial vortex structure (cell, energy-carrying vortex) arises in a fluid medium. There can be one cell or many cells. The cells are results of transformation of energy perturbations at the volume boundaries into the motion of fluid medium. In our extended definition, one such (usually long-lived) cell together with products of discrete coherent cascade breakdown of this cell is referred to as a coherent structure [82,86–90,95,117,128].

2. The breaking-down spatial structure, being the main energy-carrying vortex, can be called the parent cell (structure). The frequency of the breaking-down coherent main vortex (parent cell) is the main attribute of a coherent structure [82,86–90,95,117,128].

3. The size of the coherent structure is blurred. Currents, external with respect to the main vortex, can transport breakdown products to significant distances, thus forming a long turbulent wake [117,128].

4. The lifetime of the coherent structure is determined by the time of action of thermodynamic gradients (temperature and pressure gradients) [36,37,86–89,95,117,118,128].

5. As a limiting case of high stability, the coherent structure can consist of one long-lived parent structure. Then the parent structure is some configuration of laminar flow (undecomposing topological soliton) [95,128].

6. In the convective coherent structure arising in a closed room, all the main scenarios of turbulence formation from laminar flows [82,86–90,95,128] (Landau–Hopf, Ruelle–Takens, Pomeau–Menneville, and Feigenbaum stochastization scenarios) are confirmed. The Rayleigh–Benard scenario of breakdown of a convective coherent structure (Rayleigh–Benard convection [1,2]) is confirmed as well. The Feigenbaum scenario can be

considered as the main scenario. Based on our results [82,86–90,95,112–115,128,130,133,134,137,138], we can state that all these scenarios reflect different aspects of the process of turbulence formation.

7. The breakdown of the main energy-carrying vortex of a convective coherent structure follows the Feigenbaum scenario. The main vortex in a cell breaks down into smaller ones as a result of the series of period-doubling bifurcations (in the atmosphere, about 10 bifurcations). The resultant turbulence is coherent and deterministic [82,86–90,95,128].

8. The spectrum of passive admixture (temperature) is the breaking-down cell is fractal (locally self-similar) [82,86–90,95,128].

9. Turbulence arising in a coherent structure, as was shown in References 82, 86–90, 95, 128, satisfies all the attributes characterizing the appearance of chaos in typical dynamic systems. These attributes usually include formation of irregular long-lived spatial structures, whose type (character) is determined by dissipative factors, local instability and fractality of the phase space of such structures, appearance of the central (at zero frequency) peak in the spectrum. As was found, the central peak arises in this spectrum due to nonstationarity of random processes in the coherent structure [86–89,95,128].

10. The known processes of transformation of laminar flows into turbulent ones (Rayleigh–Benard convection, flowing-around of obstacles by fluid, and others) can be considered as processes of formation of either single coherent structures or a sum of different coherent structures [82,86–90,95,113,117,120,128].

11. A coherent structure contains both large-scale and small-scale turbulence. The 1D spectrum of turbulence (velocity components and temperature) is characterized by the faster decrease in the inertial range (usually, the 8/3-law decrease, which in the high-frequency part of the inertial range transforms into even faster 12/3-law decrease) in comparison with the Kolmogorov 5/3-law decrease [82,86–90,95,128].

12. The outer scale of turbulence in a single coherent structure can be considered as a product of the first breakdown event of the coherently breaking-down main vortex [82,86–90,95,128].

3.3.7.2 Properties of Mixtures of Different Coherent Structures

The properties of mixtures (sums) of different coherent structures are considered in detail and studied in References 36, 37, 112–115, 117–126, 128–131, 133, 134, 137–140, 142, 143. As a result of many-year (more than 10 years) field measurements under various climatic conditions and in regions with mountain and plane surface, we have accumulated a large experimental database of near-surface characteristics of atmospheric turbulence. The measurements were carried out in the Baikal Astrophysical Observatory, Sayan Sun Observatory, in mountains of the Kolyvan ridge, in mountains of North Caucasus, and other regions of Russia. Coherent structures with main vortices of different size were usually observed in the atmosphere. Therefore, the atmosphere can be considered as a medium convenient for investigations of the characteristics of mixtures of different coherent structures. The accumulated experimental data were analyzed via the prism of applications of the methods of spectral analysis of random processes occurring in the irregular atmosphere. This allows to point out that the following:

1. In one coherent structure, the products of its breakdown form the family of vortices inphase (coherent) to the main vortex. In the atmosphere, usually there are different coherent structures, in which the frequencies of main vortices are different (nonmultiple, incommensurate). At the mixing of such different coherent structures, the elements of one family are out-of-phase (incoherent) to elements of other family. Therefore, turbulence arising at mixing of coherent structures with main vortices of different size is naturally referred to as incoherent [80–82].

2. Turbulence observed in one coherent structure (coherent turbulence) differs from incoherent Kolmogorov turbulence by the faster decrease of the 1D spectrum of turbulence in the inertial range (usually, 8/3 power-law decrease instead of the 5/3 Kolmogorov's power-law decrease) and the smaller contribution of small-scale components [36,37,117–122].

3. Any turbulence spectrum in the atmosphere (in a wide range of frequencies from small-scale turbulence to synoptic vortices) can be represented as a sum of spectra of different coherent structures with the same turbulence intensity, but having different sizes of main vortices (different outer scales of turbulence) [130,137,138,143]. Therefore, turbulence in open air can be considered as a sum of different coherent structures having different sizes [112–115]. Consequently, the actual atmospheric turbulence can be believed to be an incoherent mixture (sum) of different coherent structures with incommensurate frequencies of the main energy-carrying vortices.

4. At the same turbulence intensity, the curve corresponding to the inertial range of the 1D spectrum of Kolmogorov turbulence is the upper envelope of the sum of 1D spectra of different coherent structures having different sizes of the main energy-carrying vortices (different outer scales) [112–115].

5. If the difference between the sizes of main vortices of different coherent structures in the considered spatial area of the medium is small (no more than two to eight times), then the sum of spectra of different coherent structures practically does not differ from the Kolmogorov dependence. If this difference is large (more than 20 to 30 times), then the sum of spectra has a deep dip, which "reveals" the spectrum of one largest structure with the 8/3 power-law decrease. In this case, turbulence in this region is called coherent [112–115].

6. If coherent structures have different, but close sizes (different by no more than two to eight times) and are located relatively close to each other ("well mixed") in the spatial area of the medium under consideration, then the local isotropy of turbulence described by the Kolmogorov spectrum is observed.

If one of the coherent structures is significantly larger than the others (or the structures are far from each other), then the anisotropy of turbulence described by the turbulence spectrum in one coherent structure (spectrum of coherent turbulence) is observed [112–115].

7. In the case of flow around obstacles, due to the permanent generation of large cells and transport of their breakdown products by the external current, the flows immediately after the obstacle are depleted with small vortices. Therefore, immediately after an obstacle, the spectrum of fluctuations corresponds to the coherent turbulence. With an increase of the distance from an obstacle, vortices (being breakdown products) from turbulent wakes of breaking-down coherent structures mix with the ambient turbulent atmosphere, and the turbulence gradually transforms from coherent to Kolmogorov's one [113,117,120].

8. The systems of parent cell structures (system of solitons being main energy-carrying vortices) can take different forms: from a solitary ordered structure like a toroidal Benard cell or arbitrary axisymmetric vortex to systems of periodically arranged hydrodynamic perturbations like systems of various shafts and others [1,2,86–89,129].

9. The sizes of parent cells in the atmosphere can differ hundreds of millions times: from few centimeters (near-wall turbulence, thermics) to several thousands of kilometers (Ferrel and Hadley cells—cells of general circulation of the atmosphere) [1,2,86–89,129].

10. Cyclones and anticyclones (axisymmetric vortices with different directions of revolution having a rather complex internal structure and sizes from 100 to 2000 km) are interesting examples of long-lived system of large parent cells in the atmosphere. The usually are not generated by one, but form stable quartets. A quartet is stable just because of the different directions of revolution, since air moves in the same direction in the area of contact of a cyclone and anticyclone and the pressure decreases. The external pressure ties up them to each other. A stable structure—four-soliton solution of flow-dynamic equations—was formed. Destruction of a cyclone or anticyclone occurs for several days and leads to considerable intensification of turbulence in the region. Airport flight services usually alert pilots about these events [115].

11. Coherent turbulence is the main cause of significant deviations of the Kolmogorov and Obukhov constants from their standard values. This leads to large (up to 100%) errors in the measurements of turbulence characteristics based on the Kolmogorov–Obukhov law [121].

12. Structure functions (fluctuations of temperature and velocity components) serve as other source of errors in measurements of turbulence characteristics. It was shown that, in comparison with the Kolmogorov turbulence, the structure function of temperature has the longer initial square interval (the inner scale of turbulence in the coherent structure is one-order larger than the Kolmogorov's one), the shorter inertial 2/3-range, and an

additional coefficient. Therefore, in the measurements from the 2/3-asymptotic of the structure function, we can expect that the intensity of the coherent turbulence is two times overestimated [122].

13. In the coherent turbulence, in comparison with the Kolmogorov's turbulence, the attenuation of phase fluctuations of optical radiation (due to depletion with small-scale inhomogeneities) occurs. Therefore, to improve the quality of astronomical images, we can recommend installing ground-based telescopes in the regions, over which the areas of coherent turbulence take place [123–126].

14. The depletion of coherent turbulence with small-scale inhomogeneities (in comparison with Kolmogorov's turbulence) can also lead to about two-time decrease of the integral intensity of atmospheric coherent turbulence defined as a height integral of the vertical profile of turbulence intensity [127].

15. In high-mountain observatories, as wind transports the areas of coherent turbulence formed in the atmosphere, the non-Kolmogorov effect of intermittence of the jitter of astronomical images is observed. The effect consists in the periodic weakening and intensification of phase fluctuations of optical radiation and is caused by the presence of a large number of areas with coherent turbulence on the windward side (and their following wind transport) [129].

16. In high-mountain observatories, the non-Kolmogorov effect of turbulence intermittence (alternation of the two types of turbulence: coherent and Kolmogorov) is observed. This effect characterizes the local structure of turbulence over particular region. In the continuous turbulent field over a region, there are areas of both isotropic Kolmogorov's turbulence, in which the products of breakdown of coherent structures similar in size are well mixed, and adjacent regions of anisotropic coherent turbulence with insufficient mixing. These areas are usually spaced from each other (or inserted in each other) and are transported by wind as a whole structure. By continuous recording at a fixed near-surface point, we observed the intermittence (alternation) of different types of turbulence [114,117–119,126,128,129,131,133,134,139,140,142].

References 95, 128 present the reviews of worldwide publications on the study of properties of coherent structures. The comparison of structure properties established by our group of researchers, listed above, with other results demonstrates that the obtained data extend considerably the understanding of coherent structures with respect to others. At the same time, our data are confirmed by the results of theoretical investigations of other authors, including the data of numerical solutions of the Navier–Stokes equations (see Section 3.3.8). They are also confirmed by experimental results obtained by meteorological and optical methods. In general, our data are in a precise agreement with the data of other well-known approaches concerning various aspects of the problem of turbulence and generalize them.

As it follows from the mentioned above, properties of both single coherent structures and mixtures of different coherent structures:

> *Coherent structure (in the extended definition), despite its complex internal constitution, can be considered as a main basic structural element or an elementary particle of the turbulence.*

This conclusion explains also the internal structure (internal constitution) of turbulence. It can be considered as a main result of the current studies of the local structure of turbulence discussed in detail in References 112–115. The possibility of seeing the coherent structures (visualizing them by flow lines in numerical solutions of the Navier–Stokes equations) is demonstrated in the Section 3.3.8.

3.3.8 Numerical Solutions of Navier–Stokes Equations: Visualization of Turbulence

A coherent structure corresponds to a solitary soliton solution of flow-dynamics equations (Navier–Stokes equations):

$$\partial u/\partial t + (u \cdot \nabla)u = \nu\Delta u - \rho^{-1}\nabla p + f,$$
$$\nabla \cdot u = 0, \tag{3.33}$$

where u is velocity vector, t is time, p is pressure, ν is kinematic viscosity, ρ is density, and f are accelerations due to external forces.

The solution of these equations allows us to model coherent structures. This makes it possible further investigation of the properties of turbulence, which is formed by different coherent structure (including processes of turbulence formation and evolution). These processes can be visualized by flow lines of medium motion in solution of Navier–Stokes equations.

The analytical solution of Navier–Stokes equations is a complicated problem because of their nonlinearity. Therefore, numerical methods are usually used for solution of such problems. In connection with the recent advent of high-performance computers by democratic prices, the numerical solution of flow-dynamic equations with a good accuracy within reasonable time became possible.

Now it is possible to solve the problems of flow dynamics numerically with the aid of free specialized software called "the Gerris Flow Solver" [185,186] as a free software for solution of partial differential equations describing problems of flow dynamics. This open-source software was developed by Stéphane Popinet [185,186,189,190]. The efficiency and needed accuracy of the software [185] were tested and confirmed at a rather wide class of 100 typical test problems [186–190], the solution of which gives good results.

To describe the convection in fluids and gases, Navier–Stokes equations are usually written in Boussinesq approximation. The software [185] solves system (3.33) in the following form:

$$\partial u/\partial t + (u \cdot \nabla)u = \nabla \cdot (\nu D) - \rho_0^{-1}\nabla p + \rho_0^{-1}\sigma \cdot \kappa \cdot \delta_S n + \beta Tg,$$
$$\partial T/\partial t + u \cdot \nabla T = \chi \Delta T, \qquad\qquad (3.34)$$
$$\nabla \cdot u = 0,$$

where D is deformation tensor $D_{ij} = (\partial u_j/\partial x_i + \partial u_i/\partial x_j)$, σ is the surface tension coefficient, κ is curvature of the surface, δ_S is delta function of the surface, n is normal to the surface, g is acceleration due to gravity ($g = -ge_3$, e_3 is the unit vector of the vertical axis ox_3), T is deviation of absolute temperature from equilibrium temperature T_0, ρ_0 is density at equilibrium temperature T_0, χ is thermal diffusivity, and β is the thermal expansion coefficient (usually $\beta = 1/T_0$).

Our contribution to numerical solutions of system of Equation 3.34 consists in the imaginative complementation of the software [185] with tools for visualization and spectral analysis. As a result, several computer animations have been created from the data of numerical solution of the Navier–Stokes equations.

As an example, we present the experimental data and the corresponding numerical solution of the boundary-value problem from Navier–Stokes equations (3.34).

In 2012 [144], experimental studies of the daytime astroclimate were carried out in the specialized (dome) room of the Big Alt-azimuthal Telescope (BAT, Special Astrophysical Observatory of the Russian Academy of Sciences, Figure 3.36a). The diameter of the BAT receiving mirror is 6.05 m. The results of measurements have shown that two large contra-rotating vortices with vertical axes and maximal diameters of about 16 m are observed in the BAT dome space (Figure 3.36b). At every measurement point, the spectrum of temperature fluctuations in the inertial range is characterized by the 8/3 power-law decrease (with the following

(a)

(b)

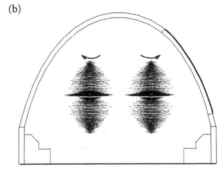

Figure 3.36 (a) Big alt-azimuthal telescope (BAT). (b) Experimentally recorded approximate pattern of air motion in the BAT dome space: two air contra-rotating vortices with vertically oriented axes. Side view.

faster decrease), which corresponds to coherent turbulence. The spectra become Kolmogorov (5/3-law decrease) only at the measurements directly in the open slit of the telescope.

For numerical simulation, we have formulated the following boundary-value problem corresponding to conditions of the experiment.

The Dirichlet boundary conditions are the following: zero velocities at the boundaries: $U = V = W = 0$ (U is the x-component of velocity; V is the y-component of velocity; W is the z-component), for the parameters: temperature $T_{min} = 293$ K; $T_{max} = 300$ K; $T_{dome\,surface} = T(\theta)$ is temperature distribution; T_{base} is not specified ($T\sim$); $\theta = 60°$ is sun zenith angle.

Initial conditions are: $T_0 = 290$ K; pressure $P_0 = 94$ kPa. In the dome, there is the air-like medium. The dome model (with diameter of 45.2 м) is empty, that is, without telescope and other equipment.

The dome side facing the sun (Figure 3.37a) is heated up to the maximal temperature T_{max} at the point under direct sunlight; with the distance from the point of temperature maximum, the temperature decreases gradually down to T_{min}; the temperature of other unlighted sides is T_{min} (Figure 3.37b).

As a result of solution of the formulated boundary-value problem, we have obtained the pattern of air motion inside the dome model in parameters of the vector velocity field and scalar temperature and pressure fields (Figure 3.38).

From the comparison of the data in Figures 3.36b and 3.38, it can be seen that the pattern of air motion obtained as a result of numerical simulation of coherent structures (Figure 3.38) practically coincides with the independently experimentally observed pattern of average air motion in the BAT dome space (Figure 3.36b).

Thus, the results of numerical simulation (for conditions of the experiment) confirm the presence of two contra-rotating air vortices with vertically oriented axes in the BAT dome space.

Figure 3.39 shows the calculated and experimental temporal frequency spectra of temperature fluctuations W_T in the BAT dome (at point P02, Figure 3.37c).

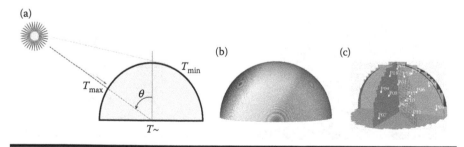

Figure 3.37 (a) Model (diagram) of the telescope dome with the side heated by the sun, right view; (b) Dome surface heated by the sun, right view; (c) Arrangement of measurement points, top front view.

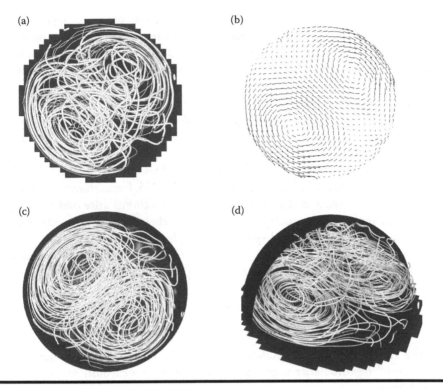

Figure 3.38 **Pattern of air motion inside the dome model (simulation): (a) top view, (b) horizontal cross section of the velocity field (top view), (c) bottom view, (d) general side view. Solid lines are flow lines.**

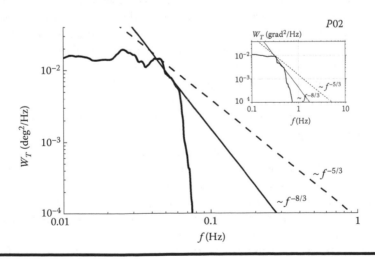

Figure 3.39 **Temporal frequency spectrum of temperature fluctuations** W_T**. Simulation. Experimental** W_T **spectrum (right top inset).**

It can be seen from Figure 3.39 that the theoretical spectrum appears to be practically identical to the experimental spectrum recorded. There is a small 8/3 power-law decrease part (in the inertial range) with the further, as expected, faster decrease at high frequencies. Both the theoretical and the experimental spectra correspond to the coherent turbulence.

The processes of turbulence formation and evolution were visualized in References 144–157 through drawing of the flow lines (motion of the medium, more often, the air) in numerical solutions of the Navier–Stokes equations for different boundary-value problems.

The structure of turbulent air motions in closed volumes over a heated surface was studied numerically in References 147–150, 152–155. It was shown that solitary toroidal vortices (coherent structures or topological solitons) arise over the irregularly heated surface. The number of vortices and their internal structure depend on the shape and size of heated spots. For complex shapes of the heating pattern (thermal mottling), toroidal vortices become markedly deformed. Vortices can both be elongated along the surface and have spiral flow lines.

In the process of evolution, the vortices mix markedly. This yields the Kolmogorov's (incoherent) turbulence.

The results presented in References 147–150, 152–155 allow to follow the evolution of the turbulence structure formed over homogeneously and irregularly heated surfaces.

Thus, at the initial phase of turbulence formation over the heated surface, a family of small (in comparison with the size of the considered volume) convective toroidal (mushroom-like) vortices, usually referred to as thermics, arise. Thermics are formed over both homogeneously and irregularly heated surfaces (for example, over individual heated spots on the surface). The rate of formation of thermics and their sizes depend on viscosity of the medium, shape of the volume, and the degree of heating of its surfaces. In the worldwide scientific literature, the phase of formation of a family of thermics is often called the near-wall turbulence.

In the absence of spatial limits, thermics rise up, increasing their volume. The presence of limits (closed volume) leads to the situation that small vortices (thermics), distorting their shapes, join in large vortices-cells, which form a certain order in the volume (which is characteristic of Benard cells). At this stage, first, the stationary pattern of motion (topological precursors) arises for the short time, and then it is alternated by the period of chaotization. In the following, the transition from chaos to stationary motions, being stable (or slowly breaking-down) vortices (coherent structures or topological solitons) occur. The cascade breakdown of these vortices shows itself as the coherent turbulence.

As follows from above discussions, the transition from small thermics (through the period of chaotization) to large stable vortices is inverse with respect to the usually observed cascade coherent breakdown of large vortices into smaller ones.

Consequently, the formation of near-wall turbulence (thermics) and its further evolution can serve an example of the inverse cascade process of energy transfer by the spectrum of motion scales.

Numerical calculations confirm our experimental conclusion [140] that the mixing of coherent structures with different close sizes (and with close frequencies of main vortices) yields the Kolmogorov turbulence. In addition, rather extended inertial ranges of the spectrum with Kolmogorov 5/3-law decrease are observed in media with high viscosity. Experimental data discussed earlier in the rooms of astronomical telescopes and in open air confirm our numerical calculations. The spectra of single coherent structures, observed experimentally in special campaigns of 2005–2007 and 2010–2016, are also confirmed.

Thus, the obtained results of numerical solution of Navier–Stokes equations in closed volumes allow us to state that the processes of turbulence formation and evolution, including formation of coherent turbulence, can be described based on solution of boundary-value problems of fluid dynamics (Navier–Stokes equations).

Numerical solutions of Navier–Stokes equations explain the mechanism of formation and existence of hydrodynamic turbulence, in which the role of stochastization decreases considerably.

In general, the accomplished analysis of all the accumulated data of measurement of the turbulence parameters both in open air and in closed volumes (as well as the data of numerical solution of the Navier–Stokes equations, presented above) shows that:

The laminar and turbulent flows can be considered as different phases of the single process of turbulence formation and evolution, which is also the single process of formation and breakdown of hydrodynamic topological solitons.

Within the framework of this main conclusion it can be stated that:

Non-Kolmogorov coherent turbulence and Kolmogorov incoherent turbulence can be considered as varieties (particular cases) of the single process of formation and breakdown of hydrodynamic topological solitons. These varieties are characterized by different intensities and different spatiotemporal resolution of energy sources of turbulence (thermodynamic gradients) at boundaries of continuous fluid medium.

As was already noted, we have revealed the properties of coherent structures through processing of accumulated experimental data by the methods of spectral analysis of random processes with the following formulation of logical conclusions. However, these methods do not allow one to see coherent structures and their mixtures. The possibility of seeing the coherent structures by visualizing them by flow lines in numerical solutions of the Navier–Stokes equations we demonstrated above in this section.

3.3.9 Main Results

Thus, Section 3.3 has presented the data of many-year experimental and theoretical studies of the local structure of turbulence. These data should be taken into account for the accurate prediction of characteristics of optical radiation in the atmosphere. The main results can be formulated as follows:

1. The processes of formation and breakdown of a Benard cell in air, studied experimentally, showed that temperature gradients cause the formation of the Benard cell. The breakdown of the Benard cell was found to be following the Feigenbaum scenario. The main vortex in the cell breaks down into smaller ones as a result of the series of period-doubling bifurcations. The resultant turbulence appears to be coherent and deterministic. The fractality (local self-similarity) of turbulence spectrum was revealed.

2. Turbulence arising as a result of breakdown of the Benard cell satisfies all attributes characterizing the appearance of chaos in typical dynamic systems. These attributes usually include appearance of irregular long-lived spatial structures, whose character is determined by dissipative factors, by local instability and fractality of the phase space of such structures; the appearance of the central peak in the spectrum (at zero frequency), and so forth.

3. It is convenient to joint these attributes of chaos formation in typical dynamic systems by the single term "coherent structure," if we extend this, already existing, concept and include small-scale components of cell breakdown into the composition of a coherent structure. We defined the coherent structure as a compact formation including long-lived spatial hydrodynamic vortex structure (cell) arising as a result of long action of thermodynamic gradients and the products of its discrete coherent cascade breakdown.

4. The obtained experimental and theoretical results show that the known processes of transition of laminar flows into turbulent ones (Rayleigh–Benard convection, flowing around obstacles, and others) can be believed coherent structures (or sums of these structures).

5. The results of investigation of the turbulence evolution over irregularly heated surfaces show that under conditions of spatial limits (in closed volumes) the transition (through chaotization) from small short-lived to rather large long-lived coherent structures (transition from small thermics to large stable vortices) occurs. This transition is inverse with respect to the usually observed cascade coherent breakdown of large vortices into smaller ones, mentioned in Chapter 1. Consequently, the formation of near-wall turbulence (thermics) and its further evolution can serve an example of the inverse cascade of the process of energy transfer by the spectrum of motion scales.

6. The actual atmospheric turbulence can be considered as an incoherent mixture of different coherent structures with incommensurate frequencies of main energy-carrying vortices. The mixing of coherent structures with different

close sizes (and close frequencies of the main vortices) yields the Kolmogorov turbulence. If one of the coherent structures is significantly larger than the others, then the turbulence corresponding mostly to one coherent structure is observed. This turbulence is called coherent, and the area, it is formed in, is called the area of coherent turbulence.

7. Coherent turbulence is the main cause of significant deviations of the Kolmogorov and Obukhov constants from their standard values. This leads to large errors in the measurements of turbulence characteristics based on the Kolmogorov–Obukhov law.

8. In coherent turbulence, in comparison with the Kolmogorov turbulence, the attenuation of phase fluctuations of optical radiation occurs. Therefore, to improve the quality of astronomical images, we recommend installing ground-based astronomical telescopes in the regions, over which the areas of coherent turbulence are observed during the measurements.

9. In high-mountain observatories, the non-Kolmogorov effect of turbulence occurs (alternation of two types of turbulence: coherent and Kolmogorov ones). The effect is characterized by the local structure of turbulence over particular region; it is caused by the wind transport of separate adjacent spatial areas containing either coherent or Kolmogorov turbulence (as a whole); it is observed at the continuous recording at a fixed near-surface point.

10. The coherent structure (in its extended definition), despite its complex internal constitution, can be considered as the main basic structural element or an elementary particle, the turbulence consists in. From the mathematical point of view, the structural element of turbulence is the soliton solution of flow-dynamic equations. It is either a solitary 3D topological soliton solution or one soliton in a many-soliton solution.

11. The analysis of the accumulated measured data on turbulence parameters and data of numerical solution of the Navier–Stokes equations shows that laminar and turbulent flows can be considered as different phases of the single process of turbulence formation and evolution, which is the single process of formation and breakdown of hydrodynamic topological solitons.

12. The non-Kolmogorov coherent turbulence and the Kolmogorov incoherent turbulence can be considered as varieties (particular cases) of the single process of formation and breakdown of hydrodynamic topological solitons. These varieties are characterized by different intensities and different spatiotemporal distributions of the energy sources of turbulence (thermodynamic gradients) at boundaries of a continuous fluid medium.

References

1. Monin, A. S. and Yaglom, M. A., *Statistical Fluid Mechanics: Mechanics of Turbulence*, vol. 1. Cambridge, MA: MIT Press, 1971.

2. Monin, A. S., and Yaglom, A. M., *Statistical Fluid Mechanics: Mechanics of Turbulence*, vol. 2. Cambridge, MA: MIT Press, 1975.

3. Monin, A. S., and Obukhov, A. M., Main regularities of turbulent mixing in the near-surface atmospheric layer, *Trudy Geofiz. Inst. AN SSSR.*, vol. 24, No. 151, pp. 163–187, 1954; *Dokl. Akad. Nauk SSSR*, vol. 93, No. 2, pp. 223–226, 1953.

4. Tatarskii, V. I., *Wave Propagation in the Turbulent Atmosphere*. Moscow: Science, 1967, 548 p.

5. Landau, L. D., and Livshits, E. M., *Fluid Dynamics*. Moscow: Science, 1988, 736 p.

6. Monin, A. S., *Hydrodynamics of the Atmosphere, Ocean, and Earth Interior*. St. Petersburg: Gidrometeoizdat, 1999, 524 p.

7. Zilitinkevich, S. S., *Dynamics of the Atmospheric Boundary Layer*. Leningrad: Gidrometeoizdat, 1970, 290 p.

8. Gurvich, A. S., Vertical profiles of wind velocity and temperature in the near-surface atmospheric layer, *Izv. AN SSSR. Phys. Atmos. Okeana*, vol. 1, No. 1, pp. 55–64, 1965.

9. Nosov, V. V., Emaleev, O. N., Lukin, V. P., and Nosov, E. V., Semi-empirical hypotheses of turbulence theory in the anisotropic boundary layer, *Atmospheric and Oceanic Optics*, vol. 18, No. 10, pp. 756–773, 2005.

10. Kozhevnikov, V. N., *Perturbation of the Atmosphere at Flowing around Mountains*. Moscow: Nauchn. Mir, 1999, 160 p.

11. Nosov, V. V., Emaleev, O. N., Lukin, V. P., and Nosov, E. V., Semiempirical hypotheses of the turbulence theory in anisotropic boundary layer, *Proc. SPIE*, vol. 5743, pp. 110–131, 2004.

12. Nosov, V. V., Lukin, V. P., Nosov, E. V., and Torgaev, A. V., Discrete-uninterrupted averaging in Taylor ergodic theorem, *Proc. SPIE*, vol. 6160, pp. 358–362, 2005.

13. Nosov, V. V., Grigorev, V. M., Kovadlo, P. G., Lukin, V. P., and Torgaev, A. V., Results of measurement of astroclimatic characteristics near Large Solar Vacuum Telescope, *Solar-Terrestrial Physics*, Issue 9, pp. 104–109, 2006.

14. Nosov, V. V., Grigorev, V. M., Kovadlo, P. G., Lukin, V. P., Papushev, P. G., Torgaev, A. V., Results of measurement of astroclimatic characteristics in the dome space of AZT-33 telescope of Sayan Sun Observatory of the Institute of Solar-Terrestrial Physics SB RAS, *Solar-Terrestrial Physics*, Issue 9, pp. 101–103, 2006.

15. Nosov, V. V., Lukin, V. P., Emaleev, O. N., and Nosov, E. V., Semiempirical hypothesis of turbulence theory in the atmospheric anisotropic boundary layer (for mountain region), in *Instrumentation, Measure, Metrologie (RS-I2M), 1631–4670*, vol. 6, No. 1–4, Paris, France: Lavoisier Pub., 2006, pp. 155–160.

16. Nosov, V. V., Grigorev, V. M., Kovadlo, P. G., Lukin, V. P., and Torgaev, A. V., Measurements of local astroclimate characteristics near the large solar vacuum telescope, *Proc. SPIE*, vol. 6522, pp. 65220T-1– 65220T-8, 2006. doi: 10.1117/12.723063.

17. Nosov, V. V., Lukin, V. P., Emaleev, O. N., and Nosov, E. V., Semiempirical hypotheses of the turbulence theory in the atmospheric anisotropic boundary layer, in *Vision for Infrared Astronomy*, Lavoisier service editorial, Paris, France: Hermes, 2006, pp. 219–223.

18. Lukin, V. P., Lavrinov, V. V., Botygina, N. N., Emaleev, O. N., and Nosov, V. V., Turbulence and wind velocity measurements under differential image motion meter, *Proc. SPIE*, vol. 6830, pp. 56–59, 2007.

19. Lukin, V. P., Lavrinov, V. V., Botygina, N. N., Emaleev, O. N., and Nosov, V. V., Differential turbulence and wind velocity meters, *Proc. SPIE*, vol. 6733, p. 6733ON, 2007.

20. Nosov, V. V., Emaleev, O. N., Lukin, V. P., and Nosov, E. V., Results of measurements of surface turbulence of the atmospheric air in the mountains of Baikal astrophysical observatory, in *X Joint Intern. Symp. Atm. and Ocean Optics. Atm. Phys.*, Publ. IAO SB RAS, ISTP SB RAS, Tomsk, 2003, pp. 70–71.

21. Nosov, V. V., Emaleev, O. N., Lukin, V. P., and Nosov, E. V., Surface values of outer and Kolmogorov inner scales of anisotropic turbulence, in *X Joint Internat. Symp. Atm. and Ocean Optics. Atm. Phys.*, Tomsk, 2003, pp. 71–72.

22. Nosov, V. V., Emaleev, O. N., Lukin, V. P., and Nosov, E. V., Experimental check of hypotheses of semiempirical theory of anisotropic turbulence, in *X Joint Internat. Symp. Atm. and Ocean Optics. Atm. Phys.*, Tomsk, 2003, pp. 76.

23. Nosov, V. V., Emaleev, O. N., Lukin, V. P., and Nosov, E. V., Dissipation rates of kinetic energy and temperature for anisotropic turbulence, in *XI Joint Internat. Symp. Atm. and Ocean Optics. Atm. Phys.*, Tomsk, 2004, pp. 89–90.

24. Nosov, V. V., Emaleev, O. N., Lukin, V. P., and Nosov, E. V., Outer scale of anisotropic turbulence, in *XI Joint Internat. Symp. Atm. and Ocean Optics. Atm. Phys.*, Tomsk, 2004, pp. 88–89.

25. Nosov, V. V., Turbulence in the anisotropic boundary layer, in *XII Joint Internat. Symp. Atm. and Ocean Optics. Atm. Phys.*, Tomsk, 2005, pp. 179–180.

26. Nosov, V. V., Lukin, V.P., Nosov, E.V., and Torgaev, A.V., Measurements of the anisotropic turbulence characteristics in the Mondy's observatory region, in *XIII Joint Intern. Symp. Atm. and Ocean Optics. Atm. Phys.*, Tomsk, 2006, B–49, pp. 88–89.

27. Nosov, V. V., Emaleev, O. N., Lukin, V. P., and Nosov, E. V., Turbulence in the anisotropic boundary layer, in *XIII Joint Intern. Symp. Atm. and Ocean Optics. Atm. Phys.*, Tomsk, 2006, P-12, pp. 41.

28. Nosov, V. V., Grigor'ev, V. M., Kovadlo, P. G., Lukin, V. P., and Torgaev, A. V., Local astroclimate nearby the Great Solar Vacuum Telescope, in *XIII Joint Intern. Symp. Atm. and Ocean Optics. Atm. Phys.*, Tomsk, 2006, pp. 90.

29. Nosov, V. V., and Lukin, V. P., Turbulence theory in the atmospheric anisotropic boundary layer, in *XV Joint Internat. Symp. Atm. and Ocean Optics. Atm. Phys.*, Tomsk, 2008, pp. 62.

30. Nosov, V. V., Emaleev, O. N., Lukin, V. P., and Nosov, E. V., The principle of interchange ability in the semiempirical theory of anisotropic turbulence, in *X Joint Internat. Symp. Atm. and Ocean Optics. Atm. Phys.*, Tomsk, 2003, pp. 72.

31. Nosov, V. V., Lukin, V. P., and Nosov, E. V., Optimization of the place for installation of ground-based astronomical telescopes, in *Proc.IV Int. Symposium Atmospheric and Ocean Optics*, Tomsk, 1997, pp. 57–58.

32. Nosov, V. V., Lukin, V. P., and Nosov, E. V., Recommendations on improving the efficiency of ground-based astronomical telescopes, in *Proc. IV Int. Symposium Atmospheric and Ocean Optics*, Tomsk, 1997, pp. 58–59.

33. Nosov, V. V., Lukin, V. P., and Nosov, E. V., On improving the image quality in ground-based astronomical telescopes, in *Proc. VI Int. Symposium Atmospheric and Ocean Optics*, Tomsk, 1999, pp. 76–77.

34. Nosov, V. V., Lukin, V. P., and Nosov, E. V., Influence of temperature inhomogeneities of the underlying surface on the vertical profile of atmospheric turbulence intensity, in *VI Int. Symposium Atmospheric and Ocean Optics*, IAO SB RAS, Tomsk, 1999, pp. 54.

35. Nosov, V. V., Lukin, V. P., and Nosov, E. V., Reconstruction of vertical profiles of atmospheric turbulence intensity from the inhomogeneous boundary field of

near-surface values, in *VI Int. Symposium Atmospheric and Ocean Optics*, IAO SB RAS, Tomsk, 1999, pp. 55.

36. Nosov, V. V., Lukin, V. P., Torgaev, A. V., Grigor'ev, V. M., and Kovadlo, P. G., Astroclimate parameters of the surface layer in the Sayan solar observatory, *Proc. SPIE*, vol. 7296, pp. 72960D-1–8, 2009.

37. Nosov, V. V., Lukin, V. P., Torgaev, A. V., Grigor'ev, V. M., and Kovadlo, P. G., Result of measurements of the astroclimate characteristics of astronomical telescopes in the mountain observatories, *Proc. SPIE*, vol. 7296, pp. 72960C-1–5, 2009.

38. Lukin, V. P., Grigor'ev, V. M., Antoshkin, L. V., Botygina, N. N., Emaleev, O. N., Konyaev, P. A., Kovadlo, P. G., Nosov, V. V., Skomorovskii, V. I., and Torgaev, A. V., Possibilities of using adaptive optics in solar telescopes, *Optika Atmosfery i Okeana*, vol. 22, No. 5, pp. 499–511, 2009.

39. Nosov, V. V., Grigor'ev, V. M., Kovadlo, P. G., Lukin, V. P., Nosov, E. V., and Torgaev, A. V., Turbulent scales of velocity and temperature in the atmospheric boundary layer, *Izv. Vuz., Phys.*, vol. 56, No. 8\3, pp. 331–333, 2013.

40. Nosov, V. V., Grigor'ev, V. M., Kovadlo, P. G., Lukin, V. P., Nosov, E. V., and Torgaev, A. V., Turbulent scales of the Monin–Obukhov similarity theory in anisotropic boundary layer, in *Proc. Int. Conf. in Memory of Academician A.M. Obukhov "Turbulence, Atmospheric Dynamics, and Climate," I. Turbulence*, Moscow, IPhA RAS, 2013, pp. 38–43.

41. Nosov, V. V., Grigorev, V. M., Kovadlo, P. G., Lukin, V. P., Nosov, E. V., and Torgaev, A. V., Turbulent Prandtl number, in *Proc. XVII Int. Symp. "Atmospheric and Ocean Optics. Atmospheric Physics"*, IAO SB RAS, Tomsk, D-66, 2011, pp. D235–D238.

42. Zhigulev, V. N., and Tumin, A. M., *Formation of Turbulence*. Novosibirsk: Science, 1987, 283 p.

43. Townsend, A. A., Measurements in the turbulent wake of a cylinder, *Proc. Roy. Soc., London, Ser.A*, vol. 190, pp. 551–561, 1947.

44. Townsend, A. A., *The Structure of Turbulent Shear Flow*, 1st ed. Cambridge: Cambridge University Press, 1956, 429 pp.

45. McNaughton, K. G., and Brunet, Y., Townsend's hypothesis, coherent structures and Monin–Obukhov similarity, *Boundary-Layer Meteorology*, vol. 102, No. 2, pp. 161–175, 2002.

46. McNaughton, K. G., Turbulence structure of the unstable atmospheric surface layer and transition to the outer layer, *Boundary-Layer Meteorology*, vol. 112, No. 2, pp. 199–221, 2004.

47. Kit, E., Krivonosova, O., Zhilenko, D., and Friedman, D., Reconstruction of large coherent structures from SPIV measurements in a forced turbulent mixing layer, *Experiments in Fluids*, vol. 39, No. 4, pp. 761–770, 2005.

48. Solomon, T. H., and Gollub, J. P., Chaotic particle transport in time-dependent Rayleigh–Benard convection, *Physical Review A*, vol. 38, No. 12, pp. 6280–6286, 1988.

49. Solomon, T. H., and Gollub, J. P., Thermal boundary layers and heat flux in turbulent convection: The role of recirculating flows, *Physical Review A*, vol. 43, No. 12, pp. 6683–6693, 1991.

50. Blackwelder, R. F., and Kovasznay, L. S. G., Time scale and correlation in a turbulent boundary layer, *Phys. Fluids*, vol. 15, pp. 1545–1554, 1972.

51. Blackwelder, R. F., Coherent structures associated with turbulent transport, in *Proc. 2nd Int. Sump. On Transport Phenomena in Turbulent Flows*, Tokyo, 1987, pp. 1–20.

52. Perry, A. E., Lim, T. T., Chong, M. S., and The, E. W., *The fabric of turbulence, AIAA Paper*, vol. 80, p. 1358, 1980.
53. McComb, W. D., *The Physics of Fluid Turbulence*. Oxford: Oxford University Press, 1991, 595 p.
54. Dodonov, I. G., Zharov, V. A., and Khlopkov, Yu. I., Localized coherent structures in the boundary layer, *Journal of Applied Mechanics and Technical Physics*, vol. 41, No. 6, pp. 1012–1019, 2000.
55. Sadani, L. K., and Kulkarni, J. R., A study of coherent structures in the atmospheric surface layer over short and tall grass, *Boundary-Layer Meteorology*, vol. 99, No. 2, pp. 317–334, 2001.
56. Zhang Zhaoshun, Cui Guixiang, and Xu Chunxiao, Modern turbulence and new challenges, *Acta Mechanica Sinica (English series)*, vol. 18, Issue 4, pp. 309–327, 2002.
57. Chen, J., and Hu, F., Coherent structures detected in atmospheric boundary-layer turbulence using wavelet transforms at Huaihe River Basin, *China, Boundary-Layer Meteorology*, vol. 107, No. 2, pp. 429–444, 2003.
58. Kim Si-Wan, and Park Soon-Ung, Coherent structures near the surface in a strongly sheared convective boundary layer generated by large-eddy simulation, *Boundary-Layer Meteorology*, vol. 106, No. 1, pp. 35–60, 2003.
59. Koprov, B. M., Koprov, V. M., Makarova, T. I., and Golitsyn, G. S., Coherent structures in the atmospheric surface layer under stable and unstable conditions, *Boundary-Layer Meteorology*, vol. 111, No. 1, pp. 19–32, 2004.
60. Feigenwinter, C., and Vogt, R., Detection and analysis of coherent structures in urban turbulence, *Theoretical and Applied Climatology*, vol. 81, No. 3–4, pp. 219–230, 2005.
61. Koprov, B. M., Koprov, V. M., Ponomarev, V. M., and Chkhetiani, O. G., Experimental studies of turbulent helicity and its spectrum in the atmospheric boundary layer, *Doklady Physics*, vol. 50, No. 8, pp. 419–422, 2005 (Translated from *Doklady Akademii Nauk*, vol. 403, No. 5, pp. 627–630, 2005.)
62. Liu Jian-hua, Jiang Nan, Wang Zhen-dong, and Shu Wei, Multi-scale coherent structures in turbulent boundary layer detected by locally averaged velocity structure functions, *Applied Mathematics and Mechanics (English Edition)*, vol. 26, No. 4, pp. 495–504, 2005.
63. Maslov, V., and Shafarevich, A., Rapidly oscillating asymptotic solutions of the Navier–Stokes equations, coherent structures, Fomenko invariants, Kolmogorov spectrum, and flicker noise, *Russian Journal of Mathematical Physics*, vol. 13, No. 4, pp. 414–424, 2006.
64. Pavageau, M., Loubiere, K., and Gupta, S., Automatic eduction and statistical analysis of coherent structures, *Experiments in Fluids*, vol. 41, No. 1, pp. 35–55, 2006.
65. Das, S. K., Tanahashi, M., Shoji, K., and Miyauchi, T., Statistical properties of coherent fine eddies in wall-bounded turbulent flows by direct numerical simulation, *Theoretical and Computational Fluid Dynamics*, vol. 20, No. 2, pp. 55–71, 2006.
66. Elperin, T., Kleeorin, N., Rogachevskii, I., and Zilitinkevich, S. S., Tangling turbulence and semi-organized structures in convective boundary layers, *Boundary-Layer Meteorology*, vol. 119, No. 3, pp. 449–472, 2006.
67. Pukhnachev, V. V., Symmetries in Navier–Stokes equations, *Usp. Mehaniki*, Issue 1, pp. 6–76, 2006.
68. Pylaev, A. M., Solution of problem on critical natural-convective motions in closed holes, in *Proceedings of ICHIT*, pp. 11–16, 2006.

69. Barthlott, C., Drobinski, P., Fesquet, C., Dubos, T., and Pietras, C., Long-term study of coherent structures in the atmospheric surface layer, *Boundary-Layer Meteorology*, vol. 125, No. 1, pp. 1–24, 2007.

70. Narasimha, R., Wavelet diagnostics for detection of coherent structures in instantaneous turbulent flow imagery. A review, *Sadhana*, vol. 32, Parts 1 and 2, pp. 29–42, 2007.

71. Sreenivasan, K. R., and Meneveau, C., The fractal facets of turbulence, *J. Fluid Mech.*, vol. 173, pp. 357–386, 1986.

72. Sreenivasan, K. R., and Meneveau, C., The multifractal nature of turbulent energy dissipation, *J. Fluid Mech.*, vol. 224, pp. 429–484, 1991.

73. Nelkin, M., What do we know about self-similarity in fluid turbulence, *J. Statist. Phys.*, vol. 54, No. 1/2, pp. 1–15, 1989.

74. Frish, U. et al., Does multifractal theory of turbulence have logarithms in the scaling relations, *J. Fluid Mech.*, vol. 542, pp. 97–103, 2005.

75. Arneodo, A. et al., Universal intermittent properties of particle trajectories in highly turbulent flows, *Phys. Rev. Lett.*, vol. 100, pp. 254–504, 2008.

76. Arneodo, A. et al., Structure functions in turbulence, in various flow configurations, at Reynolds number between 30 and 5000, using extended self-similarity, *Europhys. Lett.*, vol. 34, No. 6, pp. 411–416, 1996.

77. Muzy, J. F., Bacry, E., and Arneodo, A., Wavelets and multifractal formalism for singular signals Application to turbulence data, *Phys. Rev. Lett.*, vol. 67, No. 25, pp. 3515–3518, 1991.

78. Chainais, P., Multi-dimensional infinitely divisible cascades to model the statistics of natural images, in *Proc. of ICIP*, Genova, Italy, 2005.

79. Novikov, E. A., Infinitely divisible distributions in turbulence, *Phys. Rev. E*, vol. 50, No. 5, pp. R 3303, 1994.

80. Nosov, V. V., Grigorev, V. M., Kovadlo, P. G., Lukin, V. P., Nosov, E. V., and Torgaev, A. V., Astroclimate of specialized rooms of the large solar vacuum telescope. Part 1, *Atmospheric and Oceanic Optics*, vol. 20, No. 11, pp. 926–934, 2007.

81. Nosov, V. V., Grigor'ev, V. M., Kovadlo, P. G., Lukin, V. P., Nosov, E. V., and Torgaev, A. V., Astroclimate of specialized rooms at the Large Solar Vacuum Telescope. Part 2, *Atmospheric and Oceanic Optics*, vol. 21, No. 3, pp. 180–190, 2008.

82. Nosov, V. V., Grigor'ev, V. M., Kovadlo, P. G., Lukin, V. P., Nosov, E. V., and Torgaev, A. V., Coherent structures in the turbulent atmosphere. Experiment and theory, *Solar-Terrestrial Physics*, Issue 14, pp. 117–126, 2009.

83. Nosov, V. V., Grigor'ev, V. M., Kovadlo, P. G., Lukin, V. P., Papushev, P. G., and Torgaev, A. V., Measurements of characteristics of intradome astroclimate for the AZT-33 astronomic telescope (Sayan solar observatory), *Proc. SPIE*, vol. 6522, pp. 65220S-1–65220S-5, 2006. doi: 10.1117/12.723062.

84. Nosov, V. V., Grigor'ev, V. M., Kovadlo, P. G., Lukin, V. P., Nosov, E. V., and Torgaev, A. V., Practical recommendations on choosing sites for ground-based astronomical telescopes, *Solar-Terrestrial Physics*, vol. 18, pp. 86–97, 2011.

85. Nosov, V. V., Grigor'ev, V. M., Kovadlo, P. G., Lukin, V. P., Papushev, P. G., and Torgaev, A. V., Intradome astroclimate of telescope AZT- 33 in Mondy observatory, in *XIII Joint Int. Symp. Atm. and Ocean Optics. Atm. Phys.*, Tomsk, 2006, pp. 89.

86. Nosov, V. V., Grigor'ev, V. M., Kovadlo, P. G., Lukin, V. P., Nosov, E. V., and Torgaev, A. V., Astroclimate of specialized stations of the Large Solar Vacuum Telescope: Part I., *Proc. SPIE*, vol. 6936, pp. 69360P: 1–11, 2007.

87. Nosov, V. V., Grigor'ev, V. M., Kovadlo, P. G., Lukin, V. P., Nosov, E. V., and Torgaev, A. V., Astroclimate of specialized stations of the Large Solar Vacuum Telescope: Part II, *Proc. SPIE*, vol. 6936, pp. 69360Q: 1–12, 2008.
88. Nosov, V. V., Grigor'ev, V. M., Kovadlo, P. G., Lukin, V. P., Nosov, E. V., and Torgaev, A. V., Astroclimate of specialized rooms of the Large Solar Vacuum Telescope. Part 1, *Atmospheric and Oceanic Optics*, vol. 20, pp. 926–934, 2007.
89. Nosov, V. V., Grigor'ev, V. M., Kovadlo, P. G., Lukin, V. P., Nosov, E. V., and Torgaev, A. V., Astroclimate of specialized rooms of the Large Solar Vacuum Telescope. Part 2, *Atmospheric and Oceanic Optics*, vol. 21, pp. 180–190, 2008.
90. Nosov, V. V., Grigor'ev, V. M., Kovadlo, P. G., Lukin, V. P., Papushev, P. G., and Torgaev, A. V., Astroclimate inside the dome of AZT–14 telescope of Sayan Solar Observatory, *Proc. SPIE*, vol. 6936, pp. 69361R: 1–4, 2007.
91. Nosov, V. V., Lukin, V. P., Nosov, E. V., and Torgaev, A. V., Result measurements of A.N. Kolmogorov and A.M. Obukhov constants in the Kolmogorov–Obukhov law, *Proc. SPIE*, vol. 7296, [7296-09], pp. 70–77, 2008.
92. Nosov, V. V., Lukin, V. P., and Torgaev, A. V., Decrease of the light wave fluctuations in the coherent turbulence, *Proc. SPIE*, vol. 7296, [7296-10], pp. 77–82, 2008.
93. Nosov, V. V., Lukin, V. P., and Torgaev, A. V., Structure function of temperature fluctuations in coherent turbulence, *Proc. SPIE*, vol. 7296, [7296-13], pp. 94–97, 2008.
94. Nosov, V. V., Grigorev, V. M., Kovadlo, P. G., Lukin, V. P., Papushev, P. G., and Torgaev, A. V., Repeated testing of under dome astroclimate of AZT-33 telescope, *Proc. SPIE*, vol. 7296, [7296-07], pp. 48–53, 2008.
95. Nosov, V. V., Lukin, V. P., Nosov, E. V., Torgaev, A. V., Grigorev, V. M., and Kovadlo, P. G., Coherent structures in turbulent atmosphere, *Proc. SPIE*, vol. 7296, [7296-09], pp. 53–70, 2009.
96. Nosov, V. V., Grigor'ev, V. M., Kovadlo, P. G., Lukin, V. P., and Torgaev, A. V., Astroclimate of specialized stations of the large solar vacuum telescope, in *XIV Joint Internat. Symp. Atm. and Ocean Optics. Atm. Phys.*, Tomsk, 2007, pp. 105.
97. Nosov, V. V., Grigor'ev, V. M., Kovadlo, P. G., Lukin, V. P., Papushev, P. G., and Torgaev, A.V., Intradome astroclimate of telescope AZT-14 in Sayan solar observatory, in *XIV Joint Internat. Symp. Atm. and Ocean Optics. Atm. Phys.*, Tomsk, 2007, pp. 225.
98. Nosov, V. V., Grigor'ev, V. M., Kovadlo, P. G., Lukin, V. P., Nosov, E. V., and Torgaev, A. V., Benard cells and the stochastization scenario, in *XIV Joint Int. Symp. Atm. and Ocean Optics. Atm. Phys.*, Tomsk, 2007, pp. 76–77.
99. Nosov, V. V., Grigor'ev, V. M., Kovadlo, P. G., Lukin, V. P., and Torgaev, A. V., Frequency spectra of temperature fluctuations in the incipient turbulence, in *XIV Joint Int. Symp. Atm. and Ocean Optics. Atm. Phys.*, Tomsk, 2007, pp. 103–104.
100. Nosov, V. V., Grigor'ev, V. M., Kovadlo, P. G., Lukin, V. P., and Torgaev, A. V., Intensity of incipient turbulence, in *XIV Joint Int. Symp. Atm. and Ocean Optics. Atm. Phys.*, Tomsk, 2007, pp. 77.
101. Nosov, V. V., Grigor'ev, V. M., Kovadlo, P. G., Lukin, V. P., and Torgaev, A. V., Outer and internal scales of incipient convective turbulence, in *XIV Joint Int. Symp. Atm. and Ocean Optics. Atm. Phys.*, Tomsk, 2007, pp. 104–105.
102. Nosov, V. V., Lukin, V. P., Nosov, E. V., and Torgaev, A. V., Measurements of A.N. Kolmogorov constant in Kolmogorov–Obukhov law, in *XV Joint Int. Symp. Atm. and Ocean Optics. Atm. Phys.*, Tomsk, 2008, pp. 72–73.

103. Nosov, V. V., Lukin, V. P., Nosov, E. V., and Torgaev, A. V., Measurements of A.M. Obukhov constant in Kolmogorov–Obukhov law, in *XV Joint Int. Symp. Atm. and Ocean Optics. Atm. Phys.*, Tomsk, 2008, pp. 73–74.

104. Nosov, V. V., Lukin, V. P., and Torgaev, A. V., A temperature fluctuations structural function in coherent turbulence, in *XV Joint Internat. Symp. Atm. and Ocean Optics. Atm. Phys.*, Tomsk, 2008, pp. 67–68.

105. Nosov, V. V., Lukin, V. P., and Torgaev, A. V., Decrease of the light wave fluctuations in coherent turbulence, in *XV Joint Int. Symp. Atm. and Ocean Optics. Atm. Phys.*, Tomsk, 2008, pp. 68–69.

106. Nosov, V. V., Grigorev, V. M., Kovadlo, P. G., Lukin, V. P., and Torgaev, A. V., Astroclimatic measurements in HST—telescope, in *XV Joint Int. Symp. Atm. and Ocean Optics. Atm. Phys.*, Tomsk, 2008, pp. 81.

107. Nosov, V. V., Grigor'ev, V. M., Kovadlo, P. G., Lukin, V. P., and Torgaev, A. V., Measurements of the astroclimate characteristics nearby LSVT entrance mirror, in *XV Joint Int. Symp. Atm. and Ocean Optics. Atm. Phys.*, Tomsk, 2008, pp. 82.

108. Nosov, V. V., Grigor'ev, V. M., Kovadlo, P. G., Lukin, V. P., Papushev, P. G., and Torgaev, A. V., Repeated testing of the under dome astroclimate of telescope AZT-33, in *XV Joint Int. Symp. Atm. and Ocean Optics. Atm. Phys.*, Tomsk, 2008, pp. 80–81.

109. Nosov, V. V., Grigorev, V. M., Kovadlo, P. G., Lukin, V. P., Papushev, P. G., and Torgaev, A. V., Repeated testing of under dome astroclimate of AZT-33 telescope, *Proc. SPIE*, vol. 7296, [7296-08], pp. 729608-1–5, 2008.

110. Nosov, V. V., Grigorev, V. M., Kovadlo, P. G., Lukin, V. P., and Torgaev, A. V., A surface layer astroclimatic characteristics in the Sayan solar observatory, in *XV Joint Intern. Symp. "Atm. and Ocean Optics. Atm. Phys."*, Tomsk, 2008, BP-05, pp. 82–83.

111. Nosov, V. V., Lukin, V. P., Nosov, E. V., and Torgaev, A. V., Some generalizations of the Taylor ergodic theorem, in *XII Joint Int. Symp. Atm. and Ocean Optics. Atm. Phys.*, Tomsk, 2005, D-45, pp. 202.

112. Nosov, V. V., Grigorev, V. M., Kovadlo, P. G., Lukin, V. P., Nosov, E. V., and Torgaev, A. V., Coherent structures are elementary components of the atmospheric turbulence, *Izv. Vuz., Phys.*, vol. 55, pp. 236–238, 2012.

113. Nosov, V. V., Lukin, V. P., Nosov, E. V., Torgaev, A. V., Grigorev, V. M., and Kovadlo, P. G., The Solitonic Hydrodynamical Turbulence, in *Proc. VI Int. Conf. "Solitons, Collapses and Turbulence: Achievements Developments and Perspectives"*, Novosibirsk, 2012, pp. 108–109.

114. Nosov, V. V., Grigor'ev, V. M., Kovadlo, P. G., Lukin, V. P., Nosov, E. V., and Torgaev, A. V., The problem of coherent turbulence, *Vestnik MSTU "Stankin"*, vol. 24, No. 1, pp. 103–107, 2013.

115. Nosov, V. V., Grigor'ev, V. M., Kovadlo, P. G., Lukin, V. P., Nosov, E. V., and Torgaev, A. V., Coherent components of the turbulence, in *The Int. Conf. dedicated to the memory of academician "Turbulence, Atmosphere and Climate Dynamics"*, Obukhov, A. M. Ed. Moscow: GEOS, IPA RAS, 2013, pp. 43–47.

116. Banakh, V. A., Belov, V. V., Zemlyanov, A. A., Krekov, G. M., Lukin V. P., Matvienko, G. G., Nosov, V. V., Sukhanov, A. Ya., and Falits, A. V., Optical Waves Propagation in the Inhomogeneous, in *Random, Non-Linear Media*, Zemlyanov, A. A. Ed. Tomsk: Publishing House of the V.E. Zuev Institute of Atmospheric Optics SB RAS, 2012, p. 404.

117. Nosov, V. V., Lukin, V. P., Torgaev, A. V., and Kovadlo, P. G., Atmospheric coherent turbulence, *Atmospheric and Oceanic Optics*, vol. 26, pp. 201–206, 2013.

118. Nosov, V. V., Grigor'ev, V. M., Kovadlo, P. G., Lukin, V. P., Nosov, E. V., and Torgaev, A. V., The coherent turbulence near the receiving aperture of astronomical telescope, *Izv. Vuz., Phys.*, vol. 55, pp. 212–214, 2012.

119. Nosov, V. V., Grigor'ev, V. M., Kovadlo, P. G., Lukin, V. P., Nosov, E. V., and Torgaev, A. V., Coherent turbulence on the territory of the Baikal Astrophysical Observatory, *Izv. Vuz., Phys.*, vol. 55, pp. 204–205, 2012.

120. Nosov, V. V., Lukin, V. P., and Torgaev, A. V., Coherent structures in the atmosphere, arising in the flow around obstacles, in *XVI Int. Symp. "Atmospheric and Ocean Optics. Atmospheric Physics"*, Tomsk: IAO SB RAS, 2009, pp. 645–648.

121. Nosov, V. V., Lukin, V. P., Nosov, E. V., and Torgaev, A. V., Results of measurements of A. N. Kolmogorov and A. M. Obukhov constants in the Kolmogorov–Obukhov law, *Proc. SPIE*, vol. 7296, pp. 72960A-1–8, 2009.

122. Nosov, V. V., Lukin, V. P., and Torgaev, A. V., Structure function of temperature fluctuations in coherent turbulence, *Proc. SPIE*, vol. 7296, pp. 72960E-1–7, 2009.

123. Nosov, V. V., Lukin, V. P., and Torgaev, A. V., Decrease of the light wave fluctuations in the coherent turbulence, *Proc. SPIE*, vol. 7296, pp. 72960B-1–6, 2009.

124. Nosov, V. V., Grigor'ev, V. M., Kovadlo, P. G., Lukin, V. P., Nosov, E. V., and Torgaev, A. V., Recommendations for the selection of sites for the ground-based astronomical telescopes, *Atmospheric and Oceanic Optics*, vol. 23, pp. 1099–1110, 2010.

125. Nosov, V. V., Grigor'ev, V. M., Kovadlo, P. G., Lukin, V. P., Nosov, E. V., and Torgaev, A. V., Optimal arrangement of the ground-based shortwave receivers for atmospheric telecommunication systems, *Radio and Telecommunication Systems*, vol. 3, pp. 76–82, 2011.

126. Nosov, V. V., Grigor'ev, V. M., Kovadlo, P. G., Lukin, V. P., Nosov, E. V., and Torgaev, A. V., Fluctuations of astronomical images in the coherent turbulence, *Izv. Vuz., Phys.*, vol. 55, pp. 223–225, 2012.

127. Nosov, V. V., Grigor'ev, V. M., Kovadlo, P. G., Lukin, V. P., Nosov, E. V., and Torgaev, A. V., The integral intensity of atmospheric turbulence according to highland optical measurements, *in XVII Int. Symp. Atmospheric and Ocean Optics. Atmospheric Physics*, Tomsk: IAO SB RAS, 2011, pp. B113–B116.

128. Nosov, V. V., Lukin, V. P., Nosov, E. V., Torgaev, A. V., Grigor'ev, V. M., and Kovadlo, P. G., Coherent structures in the turbulent atmosphere, in *Mathematical Models of Non-linear Phenomena, Processes and Systems: From Molecular Scale to Planetary Atmosphere*, Chap. 20. Nadycto, A. B. et al., Eds. New York: Nova Science Publishers, 2013, pp. 297–330.

129. Nosov, V. V., Grigor'ev, V. M., Kovadlo, P. G., Lukin, V. P., Nosov, E. V., and Torgaev, A. V., Intermittency of the astronomical images jitter in the high-mountain observations, *Proc. SPIE*, vol. 9292, pp. 92920V1–4, 2014.

130. Nosov, V. V., Grigor'ev, V. M., Kovadlo, P. G., Lukin, V. P., Nosov, E. V., and Torgaev, A. V., Coherent components of the synoptic spectra of atmospheric turbulence, *Izv. Vuz., Phys.*, vol. 58, pp. 206–209, 2016.

131. Lukin, V. P., Bol'basova, L. A., and Nosov, V. V., Comparison of Kolmogorov's and coherent turbulence, *Applied Optics*, vol. 53, pp. B231–236, 2014.

132. Lukin, V. P., Nosov, V. V., and Torgaev, A. V., Features of optical image jitter in a random medium with a finite outer scale, *Applied Optics*, vol. 53, B196–204, 2014.

133. Lukin, V. P. Nosov, V. V., Kovadlo, P. G., Nosov, E. V., and Torgaev, A. V., Non-Kolmogorov's and Kolmogorov's solitonic hydrodynamical turbulence, in *Imaging and Applied Optics Congress. Propagation through and Characterization of Distributed*

Volume Turbulence (pcDVT), OSA, USA, Seattle, Washington, 2014, Paper PM4E.2, ISBN: 978-1-55752-308-2, Control Number: 2041404.

134. Lukin, V. P., Nosov, V. V., Kovadlo, P. G., Nosov, E. V., and Torgaev, A. V., Intermittency of jitter of the astronomical images is non-Kolmogorov turbulence effect, in *Imaging and Applied Optics Congress. Propagation through and Characterization of Distributed Volume Turbulence (pcDVT)*, OSA, USA, Seattle, Washington, 2014, Paper PM4E.3, *ISBN: 978-1-55752-308-2*, Control Number: 2041428.

135. Nosov, V. V., and Lukin, V. P., Measurement technique of turbulence characteristics from jitter of astronomical images onboard an aircraft: Part 1. Main ergodic theorems, *Atmospheric and Oceanic Optics*, vol. 27, No. 1, pp. 75–87, 2014.

136. Nosov, V. V., and Lukin, V. P., Measurement technique of turbulence characteristics from jitter of astronomical images onboard an aircraft: Part 2. Accounting for photoreceiver response time, *Atmospheric and Oceanic Optics*, vol. 27, No. 1, pp. 88–99, 2014.

137. Nosov, V. V., Lukin, V. P., Nosov, E. V., and Torgaev, A. V., Approximations of the synoptic spectra of atmospheric turbulence by sums of spectra of coherent structures, *Proc. SPIE*, vol. 9910, pp. 99101Y1–6, 2016.

138. Lukin, V. P., Nosov, V. V., Nosov, E. V., and Torgaev, A. V., Approximations of the synoptic spectra of atmospheric turbulence by sums of spectra of coherent structures, in *Conference "SPIE Astronomical Telescopes + Instrumentation"*, Edinburgh, United Kingdom, Paper 9910-75, 2016.

139. Nosov, V. V., Lukin, V. P., Nosov, E. V., and Torgaev, A. V., Causes of non-Kolmogorov turbulence in the atmosphere, *Applied Optics*, vol. 55, pp. B163–168, 2016.

140. Nosov, V. V., Lukin V. P., Kovadlo P. G., Nosov E. V., and Torgaev A. V., *Optical Properties of the Turbulence in the Atmospheric Mountain Boundary Layer*. Novosibirsk: Publishing House of the Siberian Branch of the Russian Academy of Sciences, 2016, p. 153.

141. Nosov, V. V., Lukin, V. P., Nosov, E. V., and Torgaev, A. V., Intermittency of the astronomical images jitter in the high-mountain observations, *Proc. SPIE, 20th International Symposium on Atmospheric and Ocean Optics: Atmospheric Physics (AOO14)*, Novosibirsk, vol. 9292, pp. 92920V-1-4, 2014.

142. Nosov, V. V., Grigor'ev, V. M., Kovadlo, P. G., Lukin, V. P., Nosov, E. V., and Torgaev, A. V., The effect of turbulence intermittency in the mountainous observations, *Izv. Vuz., Phys.*, vol. 58, No. 8/3, pp. 210–213, 2015.

143. Nosov, V. V., Lukin, V. P., Nosov, E. V., and Torgaev, A. V., Approximations of the synoptic spectra of atmospheric turbulence by sums of spectra of coherent structures, *Proc. SPIE*, vol. 9680, pp. 9680 OQ, 2015.

144. Nosov, V. V., Lukin, V. P., Nosov, E. V., and Torgaev, A. V., Simulation of coherent structures (topological solitons) inside closed rooms by solving numerically hydrodynamic equations, *Atmospheric and Oceanic Optics*, vol. 28, pp. 120–133, 2015.

145. Nosov, V. V., Lukin, V. P., Nosov, E. V., and Torgaev, A. V., Simulation of coherent structures (topological solitons) indoors by numerical solving of hydrodynamics equations, *Proc. SPIE*, vol. 9292, pp. 92920U-1–14, 2014.

146. Nosov, V. V., Lukin, V. P., Nosov, E. V., and Torgaev, A. V., Structure of air motion along optical paths inside specialized rooms of astronomical telescopes. Numerical simulation, *Atmospheric and Oceanic Optics*, vol. 28, No. 7, pp. 614–621, 2015.

147. Nosov, V. V., Lukin, V. P., Nosov, E. V., and Torgaev, A. V., Structure of turbulence at specialized optical paths in astronomical telescopes, *Izv. Vuz., Phys.*, Issue 11, pp. 905–910, 2016.

148. Nosov, V. V., Lukin, V. P., Nosov, E. V., and Torgaev, A. V., Structure of turbulence over inhomogeneously heated surfaces. Numerical simulation, *Izv. Vuz., Phys.*, vol. 58, No. 8/3, pp. 187–190, 2015.

149. Nosov, V. V., Lukin, V. P., Nosov, E. V., and Torgaev, A. V., Turbulence structure over heated surfaces. Numerical solutions, *Atmospheric and Oceanic Optics*, vol. 29, No. 1, pp. 23–30, 2016.

150. Nosov, V. V., Lukin, V. P., Nosov, E. V., and Torgaev, A. V., Turbulence structure over heated surfaces: Numerical solutions, *Atmospheric and Oceanic Optics*, vol. 29, No. 3, pp. 234–243, 2016.

151. Nosov, V. V., Lukin, V. P., Nosov, E. V., and Torgaev, A. V., Structure of air turbulent motion inside Primary mirror shaft at Siberian lidar station of IAO SB RAS. Experiment and simulation, *Atmospheric and Oceanic Optics*, vol. 29, No. 11, pp. 905–910, 2016.

152. Nosov, V. V., Lukin, V. P., Nosov, E. V., and Torgaev, A. V., Turbulence structure over inhomogeneous heated surface, *Proc. SPIE*, vol. 9680, pp. 96800R-1–4, 2015.

153. Nosov, V. V., Lukin, V. P., Nosov, E. V., and Torgaev, A. V., Formation of turbulence over irregularly heated surfaces. Numerical solutions, in *Proc. Int. Conf. "Urgent Problems of Computational and Applied Mathematics– 2015"*, Novosibirsk: Akademizdat, 2015, pp. 67.

154. Nosov, V. V., Lukin, V. P., Nosov, E. V., and Torgaev, A. V., Structure of turbulence over irregularly heated surfaces, in *Proc. XXI Int. Symp. Atmospheric and Ocean Optics. Atmospheric Physics*, Tomsk: IAO SB RAS, 2015-1 CD-ROM, pp. 106–109.

155. Nosov, V. V., Lukin, V. P., Nosov, E. V., and Tonrgaev, A. V., Turbulence and heat exchange inside the dome room of lidar station. Experiment and simulation, *Journal of Physics: Conference Series*, pp. 754–758, 2016.

156. Nosov, V. V., Lukin, V. P., Nosov, E. V., and Torgaev, A. V., Structure of air motion along optical paths inside specialized rooms of astronomical telescopes. Numerical simulation, *Atmospheric and Oceanic Optics*, vol. 28, pp. 614–621, 2015.

157. Nosov, V. V., Lukin, V. P., Nosov, E. V., and Torgaev, A. V., Structure of turbulent air motion inside Primary Mirror shaft at Siberian Lidar station of IAO SB RAS. Experiment and simulation, *Proc. SPIE*, vol. 10035, pp. 100351Z-1–100351Z-6, 2016. doi:10.1117/12.2249265.

158. Van Dyke, M., *An Album of Fluid Motion*. The Parabolic Press, 1982.

159. Azbukin, A. A. et al., Automated ultrasonic meteorological complex AMK-03, *Meteorologiya i Gidrologiya*, 2006, pp. 89–97.

160. Tikhonov, A. N., and Arsenin, V. Ya., *Method for Solution of Ill-Posed Problems*. Moscow: Science, 1979, p. 285.

161. Getling, A. V., *Rayleigh-Benard Convection: Structures and Dynamics*. Singapore, New Jersey, London, and Hong Kong: World Scientific, 1998, 248 pp.

162. Gershuni, G. Z., and Zhuhovitskii, E. M., *Convective Stability of Incompressible Fluid*. Moscow: Science, 1972, 696 pp.

163. Pomeau, Y., and Manneville, P., Intermittent transition to turbulence dissipative dynamical system, *Comm. Math. Phys.*, vol. 74, No. 2, pp. 189–197, 1980.

164. Jenkins, G. M., and Watts, D. G., *Spectral Analysis and Its Applications*. San Francisco, CA: Holden Day, 1968, p. 525.

165. Bendat, J. S., and Piersol, A. G., *Random Data: Analysis and Measurement Procedures*. New York: John Wiley & Sons, 1971, p. 407.

166. Press, W. H., Teukolsky, S. A., Vetterling, W. T., and Flannery, B. P., *Numerical Recipes in C*, 2nd ed. Cambridge: Cambridge Univer. Press, 2002, p. 994.

167. Feigenbaum, M. J., Quantitative universality for a class of nonlinear transformations, *J. Stat. Phys.*, vol. 19, No. 1, pp. 25–32, 1978.
168. Ruelle, D., and Takens, F., On the nature of turbulence, *Comm. Math. Phys.*, vol. 20, No. 2, pp. 167–192, 1971; Ruelle, D., Strange attractors, *Math. Intellengen*cer, vol. 2, No. 3, pp. 126–137, 1980.
169. Zaslavskii, G. M., and Sagdeev, R. Z., *Introduction to Nonlinear Physics: From Pendulum to Turbulence and Chaos.* Moscow: Science, 1988, p. 368.
170. Zaslavskii, G. M., *Stochasticity of Dynamic Systems.* Moscow: Science, 1984, p. 272.
171. Schuster, H. G., *Deterministic Chaos: An Introdcution.* Weinheim: Physik-Verlag, 1984, p. 240.
172. Van Atta, C. W., Effect of coherent structures on structure functions of temperature in the atmospheric boundary layer, *Arch. Mech.*, vol. 29, No. 1, pp. 161–171, 1977.
173. Danaila, I., Dusek, J., and Anselmet, F., Coherent structures in a round, spatially evolving, unforced, homogeneous jet at low Reynolds numbers, *Phys. Fluids*, vol. 9, No. 11, pp. 3323–3342, 1997.
174. Darchiya, Sh. P., Ivanov, V. I., and Kovadlo, P. G., Results of astroclimatic studies in SibIZMIR SB AS USSR in 1971-1976, *Nov. Tekhn. Astronom., Leningrad: Science,* Issue 6, pp. 168–176, 1979.
175. Mironov V. L., and Nosov V. V., On influence of the outer scale of atmospheric turbulence on the spatial correlation of random displacements of optical beams, *Izv. Vuz., Radiophys.*, vol. 17, No. 2, pp. 274–281, 1974.
176. Kon, A. I., Mironov, V. L., and Nosov, V. V., Fluctuations centroids of optical beams in the turbulent atmosphere, *Izv. Vuz., Radiophys.*, vol. 17, No. 10, pp. 1501–1511, 1974.
177. Mironov, V. L., Nosov, V. V., and Chen, B. N., Jitter of optical images of laser sources in the turbulent atmosphere, *Izv. Vuz., Radiophys.*, vol. 23, No. 4, pp. 461–470, 1980.
178. Gurvich, A. S., Kon, A. I., Mironov, V. L., and Khmelevtsov, S. S., *Laser Radiation in the Turbulent Atmosphere.* Moscow: Science, 1976, p. 277.
179. Nosov, V. V., Lukin, V. P., and Nosov, E. V., Influence of the underlying terrain on the jitter of astronomic images, *Atmospheric and Oceanic Optics*, vol. 17, No. 4, pp. 321–328, 2004.
180. Nosov, V. V., Lukin, V. P., and Nosov, E. V., Effect of underlying terrain on jitter of astronomic images, *Proc. SPIE*, vol. 5396, pp. 132–141, 2003.
181. Kline, J., Reynolds, W. C., Schraub, F. A., and Runstadler, P. W., The structure of turbulent boundary layers, *J. Fluid Mech.*, vol. 30, pp. 741–773, 1967.
182. Temam, R., *Navier–Stokes Equations and Nonlinear Functional Analysis.* Philadelphia, PA: The Society for Industrial and Applied Mathematics, 1995, p. 142.
183. Drobinski, P., Carlotti, P., Redelsberger, J.-L., Banta, R. M., Masson, V., and Newsom, R. K., Numerical and experimental investigation of the neutral atmospheric surface layer, *J. Atmos. Sci*, vol. 64, pp. 137–156, 2007.
184. Foster, R.C., Vianey, F., Drobinski, P., and Carlotti, P., Near-surface coherent structures and the vertical momentum flux in a large-eddy simulation of the neutrally-stratified boundary layer, *Boundary-Layer Meteorology*, vol. 120, pp. 229–255, 2006.
185. Popinet, S., The Gerris Flow Solver. A free, open source, general-purpose fluid mechanics code (2001-2013), 2013, http://gfs.sf.net
186. Popinet, S., Gerris: A tree-based adaptive solver for the incompressible Euler equations in complex geometries, *J. Comput. Phys.*, vol. 190, pp. 572–600, 2003.

187. Popinet, S., Smith, M., and Stevens, C., Experimental and numerical study of the turbulence characteristics of air flow around a research vessel, *J. Ocean Atmos. Technol.*, vol. 21, pp. 1574–1589, 2004.
188. Popinet, S., 100 Gerris Tests. V. 1.3.2, http://gerris.dalembert.upmc.fr/gerris/tests/tests/index.html
189. Popinet, S., Gerris: Bibliography. http://gfs.sf.net/wiki/index.php/Bibliography
190. Popinet, S., List of recent publications. http://gfs.sf.net/wiki/index.php/User:Popinet

Chapter 4

Nonlinear Propagation of Laser Radiation in the Atmosphere

Alexander Zemlyanov

Contents

In the historical aspect, the main contribution to the foundation of nonlinear optics was done by V. E. Zuev from former USSR with his colleagues and followers (see Reference 1 and the bibliography therein). By now, a major cycle of basic investigations in the optics of high-power laser radiation in the atmosphere has been accomplished [2–4]. In these works, mechanisms of optical nonlinearity of liquid-droplet media have been studied, and efficient calculation methods and semiempirical models have been developed for the prediction of losses in energy and direction of high-power laser beams at long atmospheric paths. Physical principles for new methods of diagnosis of dispersive media have been formulated.

This research field remained actual till nowadays, growing at a new stage due to advent of high-power femtosecond lasers systems. The application of these systems in the atmosphere opens new ways of the usage of optical technologies for studying natural phenomena occurring in the inhomogeneous and irregular atmosphere.

As a rule, nonlinear optical effects deteriorate the conditions of atmospheric propagation of laser beams, and, therefore, they should be taken into account when estimating the efficiency of laser energy transfer to long distances. Thus, in the case of continuous-wave laser radiation, thermal aberrations of laser beams in air have been observed [3]. In fogs, however, the transparency of medium increases for infrared (IR) radiation and long pulses attenuate due to evaporation of droplets [2]. As high-power picosecond and femtosecond laser pulses propagate in air, the phenomenon of filamentation takes place, which is connected with self-focusing of laser beam into filaments, formation of plasma, and generation of broadband radiation or the so-called supercontinuum [5]. The longer laser pulses propagate in aerosol medium, the stronger optical breakdown of air occurs near aerosol particles (see also Chapter 1 and bibliography in Reference 4). This is accompanied by additional attenuation of the radiation.

Many nonlinear optical effects have served as a basis for new methods of diagnostics of the atmosphere and water medium. The optical breakdown of aerosol media has attracted the attention as a source of emission spectrum of the component material of suspended particles [6–9]. The two-photon absorption in aerosols containing organic molecules causes their fluorescence, which serves an indicator of these molecules [10,11]. With the use of broadband supercontinuum radiation, the diagnostics of gas and aerosol media by the method of multiwave sensing has become possible [12].

In this chapter, we present the high-priority theoretical results obtained from the study of the interaction of laser radiation of various spectral composition and pulses of different durations with atmospheric constituents based on the framework of nonlinear atmospheric optics.

4.1 Generation of Stimulated Raman Scattering of Radiation in a Spherical Microparticle

The study of generation of stimulated Raman scattering (SRS) of radiation in microparticles under exposure to high-intensity laser radiation is important, first of all, as there exist problems of Raman spectroscopy with respect to droplets concerning sensitivity and informative level [13]. On the other hand, the literature actively discusses issues of generation of laser radiation at whispering gallery modes in micro-resonators, that is, when a microparticle serves as a microlaser [14,15]. The reviews of papers on these issues can be found, for example, in References 4, 13, 16.

The nature of stimulated radiation from a microspherical resonator is connected with generation of radiation being in resonance with Eigen modes of this particle. Therefore, both experimental and theoretical aspects of these two research fields have much in common. It should be noted that most publications on these issues are experimental. They consider physical principles and obtain quantitative data about such processes as SRS, stimulated Brillouin scattering (SBS), third harmonic generation, stimulated fluorescence, and lasing in microparticles. For the completion of these studies, it is necessary to have the consecutive quantitate description of the generation process, which would allow interpretation and correct explanation of experimental facts, as well as prediction of desirable results in new experiments.

In this chapter, we consider the theoretical description of the process of SRS of radiation in a transparent microparticle based on the method of expansion of the solution in series in eigen functions of the stationary linear scattering problem. The SRS effect is analyzed in a particle at the time of its origination and under stationary scattering conditions.

4.1.1 Basic Equations

We assume that only two waves take part in the process of nonlinear scattering: the pump wave and the Raman scattering wave (Stokes wave) with the frequencies ω_L and ω_S, respectively, which are related through the equation of phase synchronism $\omega_S = \omega_L - \Omega_R$ (Ω_R is the frequency of molecular oscillations). The wave equations under these conditions have the following form:

$$\text{rot rot}\, \mathbf{E}_S(\mathbf{r};t) + \frac{\varepsilon_a}{c^2}\frac{\partial^2 \mathbf{E}_S(\mathbf{r};t)}{\partial t^2} + \frac{4\pi\sigma}{c^2}\frac{\partial \mathbf{E}_S(\mathbf{r};t)}{\partial t} = -\frac{4\pi}{c^2}\frac{\partial^2 \mathbf{P}_N^S(\mathbf{r};t)}{\partial t^2}, \quad (4.1)$$

$$\text{rot rot } \mathbf{E}_L(\mathbf{r};t) + \frac{\varepsilon_a}{c^2}\frac{\partial^2 \mathbf{E}_L(\mathbf{r};t)}{\partial t^2} + \frac{4\pi\sigma}{c^2}\frac{\partial \mathbf{E}_L(\mathbf{r};t)}{\partial t} = -\frac{4\pi}{c^2}\frac{\partial^2 \mathbf{P}_N^L(\mathbf{r};t)}{\partial t^2}, \quad (4.2)$$

where $\mathbf{E}_L$ and $\mathbf{E}_S$ are the real electric vectors of the pump and Stokes waves, respectively; ε_a and σ are the permittivity and specific conductivity of particulate matter; c is the speed of light in vacuum; $\mathbf{P}_N^L$, and $\mathbf{P}_N^S$ are the real vectors of nonlinear polarization of the medium at the frequencies ω_L and ω_S. The medium is assumed to be nonmagnetic and isotropic. Dispersion effects are ignored. Field Equations 4.1 and 4.2 are complemented with the corresponding boundary conditions [16] consisting in continuity of the tangential spherical field components (θ- and φ-components) at the transition through the particles surface:

$$(\mathbf{E}_L)_{\theta,\varphi} = (\mathbf{E}_L^{sc})_{\theta,\varphi} + (\mathbf{E}_L^{i})_{\theta,\varphi}; \quad (\mathbf{H}_L)_{\theta,\varphi} = (\mathbf{H}_L^{sc})_{\theta,\varphi} + (\mathbf{H}_L^{i})_{\theta,\varphi},$$

$$(\mathbf{E}_S)_{\theta,\varphi} = (\mathbf{E}_S^{sc})_{\theta,\varphi}; \quad (\mathbf{H}_S)_{\theta,\varphi} = (\mathbf{H}_S^{sc})_{\theta,\varphi},$$

where the superscripts sc and i correspond to the scattered and incident fields, respectively.

Now passing from real field vectors to their complex representation, we get

$$2\mathbf{E}(\mathbf{r};t) = \tilde{\mathbf{E}}(\mathbf{r};t)e^{i\omega t} + \tilde{\mathbf{E}}^*(\mathbf{r};t)e^{-i\omega t};$$

$$2\mathbf{P}_N(\mathbf{r};t) = \tilde{\mathbf{P}}_N(\mathbf{r};t)e^{i\omega t} + \tilde{\mathbf{P}}_N^*(\mathbf{r};t)e^{-i\omega t},$$

where $\tilde{\mathbf{E}}, \tilde{\mathbf{P}}_N$ are the slowly varying functions of time. We now represent the fields of the interacting waves, as series of eigen functions of the resonator particle $\mathbf{E}_{np}^{TE,TH}(\mathbf{r}), \mathbf{H}_{np}^{TE,TH}(\mathbf{r})$, describing the spatial profile of the fields of oscillation modes of *TE*- and *TH*-polarizations with Eigen frequencies $\omega_{np}^{TE,TH}$:

$$\mathbf{E}_{L,S}(\mathbf{r};t) = \sum_{n=1}^{\infty}\sum_{p=1}^{\infty}\left[A_{np}^{L,S}(t)\mathbf{E}_{np}^{TE}(\mathbf{r}) - iB_{np}^{L,S}(t)\mathbf{E}_{np}^{TH}(\mathbf{r})\right];$$

$$\mathbf{H}_{L,S}(\mathbf{r};t) = \sqrt{\varepsilon_a}\sum_{n=1}^{\infty}\sum_{p=1}^{\infty}\left[iA_{np}^{L,S}(t)\mathbf{H}_{np}^{TE}(\mathbf{r}) + B_{np}^{L,S}(t)\mathbf{H}_{np}^{TH}(\mathbf{r})\right], \quad (4.3)$$

where the coefficients $A_{np}^{L,S}(t)$, $B_{np}^{L,S}(t)$ reflect the temporal behavior of the fields. The functions $\mathbf{E}_{np}^{TE,TH}(\mathbf{r}), \mathbf{H}_{np}^{TE,TH}(\mathbf{r})$, forming the orthogonal systems within a sphere, fulfill the homogeneous Maxwell equations and can be expressed through vector spherical harmonics $\mathbf{M}_{np}(r,\theta,\varphi), \mathbf{N}_{np}(r,\theta,\varphi)$ [6].

Substituting Equation 4.3 into Equations 4.1 and 4.2, we, after some transformations (see Reference 17), are led to the system of differential equations for expansion coefficients of the main and Stokes waves. Below we consider the waves with the *TE*-polarization. The corresponding equations have the form

$$\frac{d^2}{dt^2} A_{np}^{L,S}(t) + 2\Gamma_{np}^{L,S} \frac{d}{dt} A_{np}^{L,S}(t) + \omega_{np}^2 A_{np}^{L,S}(t) = J_{np}^{L,S}(t), \tag{4.4}$$

where the "stimulating" forces at the right-side of Equation 4.4 are expressed as

$$J_{np}^L(t) = F_{np}^i(t) + \frac{4\pi}{\varepsilon_a} \int_{V_a} \mathbf{E}_{np}^* \frac{\partial^2 \mathbf{P}_N^L}{\partial t^2} \, d\mathbf{r},$$

$$J_{np}^S(t) = -\frac{4\pi}{\varepsilon_a} \int_{V_a} \mathbf{E}_{np}^* \frac{\partial^2 \mathbf{P}_N^S}{\partial t^2} \, d\mathbf{r}. \tag{4.5}$$

Here, $\Gamma_{np}^{L,S} = (\omega_{np}/(2Q_{np}(\omega_{L,S})))$ is the mode attenuation factor; Q_{np} is the total Q-factor of the resonator particle, which takes into account the total loss of the mode caused by absorption and emission of the optical wave [16]; V_a is the particle volume. The term $F_{np}^i(t)$ is connected with the inflow of electromagnetic energy to the particle at the expense of the incident radiation and can be found from the linear problem of elastic scattering [18].

Let the spherical particle be exposed to an incident plane optical wave of the form

$$\mathbf{E}^i(\mathbf{r};t) = E_0 \mathbf{p}_e \tilde{f}(t) \exp\{i(\omega_L t - k_L z)\},$$

where E_0 is the real amplitude; $\mathbf{p}_e$ is the wave polarization vector; $k_L = \omega_L/c$, and $\tilde{f}(t)$ is the function of time (temporal profile of radiation). In what follows, we restrict our consideration to the particle exposed to radiation pulses with the duration so that the effects of delay of optical fields at the scattering can be neglected. Under these conditions, we obtain for $F_{np}^i(t)$

$$F_{np}^i(t) = -\frac{ic}{\varepsilon_a} \left[\int_{S_a} \left\{ \omega_{np} \left[\mathbf{E}^i \cdot \mathbf{H}_{np}^* \right] - i \frac{\partial}{\partial t} \left[\mathbf{H}^i \cdot \mathbf{E}_{np}^* \right] \right\} \cdot \mathbf{n}_r ds \right] = E_0 f(t) K_{np}^n. \tag{4.6}$$

Here, $f(t) = \tilde{f}(t) \exp\{i\omega_L t\}$; $\mathbf{E}^i, \mathbf{H}^i$ are the vectors of electric and magnetic fields in the incident wave; $\mathbf{n}_r$ is the external normal to the particle surface bounded by the surface S_a. In this equation, the coefficient K_{np}^n takes into account the degree of excitation of the mode of internal field (with the subscript np) by each mode of the external field (with the superscript n). At the circular polarization of the pump wave ($\mathbf{p}_e = \mathbf{e}_x + i\mathbf{e}_y$, where $\mathbf{e}_x$, $\mathbf{e}_y$ are the orths) for the considered *TE*-modes of the internal field, it is, in particular,

$$K_{np}^n = \frac{ic^2 R_n}{\varepsilon_a k_0 z_{np} V_a}\left[\psi_n(k_L a_0)\psi_n'^*(n_a k_{np} a_0) - \frac{1}{n_a}\frac{\omega_L}{\omega_{np}}\psi_n'(k_L a_0)\psi_n^*(n_a k_{np} a_0)\right], \quad (4.7)$$

where $k_{np} = \omega_{np}/c$; z_{np} is the normalization coefficient for the eigen functions

$$z_{np}^{-2} = \int\limits_0^{2\pi} d\varphi \int\limits_0^\pi \sin\theta d\theta \int\limits_0^{a_0} r^2 dr |\mathbf{M}_{np}|^2;$$

ψ_n are Riccati–Bessel spherical functions; $R_n = i^n((2n+1)/(n(n+1)))$. Primes denote derivatives with respect to the full argument of the function.

It should be noted that the equations for the coefficients of *TH*-modes $B_{np}(t)$ are fully analogous to system (4.4) with the only difference in the equation for the coefficients K_{np}^n.

4.1.2 Quasistationary Approximation

Differential Equation 4.4 is solved in the approximation of slowly varying amplitudes, that is, when it is assumed that $A_{np}^S(t) = \tilde{A}_{np}^S(t)e^{i\omega_S t}$, where $\tilde{A}_{np}^S(t)$ is the slowly varying amplitude, or, what is the same, in the quasi-stationary approximation. In the isotropic medium for waves with frequencies ω_L and ω_S, the vectors of nonlinear polarization responsible for the Raman scattering can be represented as follows:

$$\mathbf{P}_N^L(\mathbf{r},t) = \chi_R^{(3)}(\omega_L)\left(\mathbf{E}_S(\mathbf{r},t)\cdot\mathbf{E}_S^*(\mathbf{r},t)\right)\mathbf{E}_L(\mathbf{r},t),$$

$$\mathbf{P}_N^S(\mathbf{r},t) = \chi_R^{(3)}(\omega_S)\left(\mathbf{E}_L(\mathbf{r},t)\cdot\mathbf{E}_L^*(\mathbf{r},t)\right)\mathbf{E}_S(\mathbf{r},t) + \mathbf{P}_N^{sp}(\mathbf{r},t), \quad (4.8)$$

where $\chi_R^{(3)}$ is the nonlinear (Raman) third-order dielectric susceptibility of the medium, and $\operatorname{Im}\chi_R^{(3)}(\omega_L) = -\operatorname{Im}\chi_R^{(3)}(\omega_S)$; $\mathbf{P}_N^{sp}(\mathbf{r},t)$ is nonlinear polarization responsible for spontaneous Raman scattering.

At the Raman resonance, the Raman susceptibility becomes purely imaginary

$$\chi_R^{(3)}(\omega_S) = -i\frac{N_0 T_2}{16m\Omega_R}\left(\frac{\partial\alpha}{\partial q_k}\right)^2,$$

and it is usually related with the stationary SRS amplification coefficient g_s:

$$g_s = -\frac{32\pi^2 \omega_S}{\varepsilon_a c^2} \operatorname{Im}\left(\chi_R^{(3)}\right). \tag{4.9}$$

Here, q_k is the coordinate of displacement of nuclei in a molecule; α is polarizability of the medium; m is the reduced mass of the molecule; T_2 is the time of transverse relaxation; n_m is the population difference of the Raman-active transition, which is believed to be constant in our approximation; N_0 is the concentration of medium molecules.

Owing to the stochastic nature of spontaneous scattering to find the form of the last term in Equation 4.8, it is necessary to solve the equation of harmonic oscillator under the action of a random force for the complex function q_k (see, e.g., Reference 19):

$$\frac{\partial^2 q_k}{\partial t^2} + 2\Gamma_k \frac{\partial q_k}{\partial t} + \Omega_R^2 q_k = f_E(\mathbf{r};t) + f_{sp}(\mathbf{r};t), \tag{4.10}$$

where $2q_k = q_k + q_k^*$; $f_E = (1/2m)(\partial\alpha/\partial q_k)n_m(\mathbf{E}_L \cdot \mathbf{E}_S^*)$ is the stimulating force, that is, a source of stimulated radiation; $\Gamma_k = 1/T_2$; $f_{sp}(\mathbf{r};t)$ is the random distributed force. As for the last one, it is assumed to be delta-correlated in space and time:

$$\left\langle f_{sp}(\mathbf{r};t) f_{sp}(\mathbf{r}';t') \right\rangle = F_0^2 \delta(t-t')\delta(\mathbf{r}-\mathbf{r}'),$$

F_0 is the root-mean-square amplitude of random perturbations, which can be determined through the cross section of Raman scattering of matter.

At the Raman resonance, the solution of Equation 4.10 has the form

$$q_k(t) = \frac{e^{-\Gamma_k t}}{\Omega_R} \int\limits_0^t \sin\left(\widehat{\Omega}_R(t-t')\right) e^{\Gamma_k t'} \left(f_E(t') + f_{sp}(t')\right) dt' = q_k^E(t) + q_k^{sp}(t),$$

where $\widehat{\Omega}_R = \Omega_R\sqrt{1 - \left(\Gamma_k^2/\Omega_R^2\right)}$. For the spontaneous component of Stokes polarization, we obtain

$$P_N^{sp}(\mathbf{r},t) = N_0 \frac{\partial\alpha}{\partial q_k}\left(q_k^{sp}\right)^* \mathbf{E}_L(\mathbf{r};t) + \text{к.c.}$$

$$\approx -\frac{N_0}{\Omega_R}\frac{\partial\alpha}{\partial q_k} \tilde{\mathbf{E}}_L(\mathbf{r};t) e^{i\omega_{St}-\Gamma_k t} \int\limits_0^t \sin\left(\widehat{\Omega}_R(t-t')\right) e^{\Gamma_k t'} f_{sp}^*(\mathbf{r};t') dt' + \text{к.c.} \tag{4.11}$$

Here, $\tilde{\mathbf{E}}_L$ is the slowly varying function of time.

Equation 4.5 can be transformed for the source of the Stokes wave J_{np}^S. For this purpose, we again use the expansion of the fields in Eigen modes of the particle and take into account Equation 4.8. As a result, we obtain

$$
J_{np}^S(t) = -\frac{4\pi \chi_R^{(3)}(\omega_S)}{\varepsilon_a} \sum_{n'} \sum_{p'} \frac{d^2}{dt^2} \left(\left| A_{n'p'}^L(t) \right|^2 \sum_m \sum_q A_{mq}^S(t) \right)
$$

$$
\times \int_{V_a} \left(\mathbf{E}_{n'p'} \cdot \mathbf{E}_{n'p'}^* \right) \cdot \left(\mathbf{E}_{mq} \cdot \mathbf{E}_{np}^* \right) d\mathbf{r} + F_{np}^{sp}(t),
$$

$$
F_{np}^{sp}(t) = -\frac{4\pi}{\varepsilon_a} \int_{V_0} \mathbf{E}_{np}^* \frac{\partial^2 \mathbf{P}_N^{sp}}{\partial t^2} d\mathbf{r}
$$

is the source of spontaneous radiation at the frequency ω_S. By definition, the eigenmodes, the field is expanded in, are believed noninteracting with each other within the volume of the microresonator particle [20]. The energy exchange between modes is possible only in the presence of local inhomogeneities of the permittivity or through the mode interaction on the particle surface at particle deformations. Consequently, in the obtained equation, we can omit the summation over modes of the Stokes field. With allowance for Equation 4.11, we obtain

$$
J_{np}^S(t) = -i\frac{c^2 g_S}{8\pi \omega_S} \sum_{n'} \sum_{p'} \frac{d^2}{dt^2} \left(\left| A_{n'p'}^L(t) \right|^2 A_{np}^S(t) \right) S_{n'p'}^{np} + F_{np}^{sp}(t), \qquad (4.12)
$$

where $S_{n'p'}^{np}$ denotes the integral of spatial overlapping of optical modes of the main and Stoke's waves

$$
S_{n'p'}^{np} = \int_{V_0} \left[\left(\mathbf{E}_{n'p'} \cdot \mathbf{E}_{n'p'}^* \right) \cdot \left(\mathbf{E}_{np} \cdot \mathbf{E}_{np}^* \right) \right] d\mathbf{r}, \qquad (4.13)
$$

and the summation is performed only over the modes of the pump field.

Coming back to Equation 4.12, we can note that every mode of the pump field contributes to evolution of the selected mode of Raman scattering wave proportionally to the integral parameter of overlapping of these modes.

Within the framework of the quasistationary approximation, the initial equation takes the following form:

$$\left(2i\omega_S + 2\Gamma_{np}^S + \frac{2i}{\omega_S}\left[\frac{dG_{np}^S(t)}{dt} + i\omega_S G_{np}^S(t)\right]\right)\frac{d\tilde{A}_{np}^S(t)}{dt}$$

$$+ \omega_S^2\left(\Delta_{nps}^2 + \frac{2i\Gamma_{np}^S}{\omega_S} - \frac{2}{\omega_S^2}\frac{dG_{np}^S(t)}{dt} + \frac{i}{\omega_S}G_{np}^S(t)\right)\tilde{A}_{np}^S(t) = \tilde{F}_{np}^{sp}(t), \quad (4.14)$$

where $\Delta_{nps} = (\omega_{np} - \omega_S)/\omega_S$ is the relative frequency detuning of the Stokes field mode;

$$G_{np}^S(t) = \frac{c^2 g_S}{8\pi}\sum_{n'}\sum_{p'}\left|A_{n'p'}^L(t)\right|^2 S_{n'p'}^{np}.$$

The amplitude coefficient of the Stokes mode $A_{np}^S(t)$ can be expressed through the integral relationship. Provided that the stimulated scattering develops at long-living resonance modes $(\omega_{L,S}, \Gamma_{np}^S, G_{np}^S)$, this relationship can be written in the form

$$A_{np}^S(t) = \tilde{A}_{nps}^0(t)e^{D_{np}^S(t)}e^{i\hat{\omega}_{np}^S t + i\varphi_{np}^S(t)}. \quad (4.15)$$

In Equation 4.15, the following designations are introduced:

$$\hat{\omega}_{np}^S = \omega_S\left(1 - \Delta_{nps}^2/2\right)$$

is the frequency of generation of the Stokes wave field mode;

$$\varphi_{np}^S(t) = \frac{1}{2\omega_S}\int_0^t\left(G_{np}^S(t') - \Gamma_{np}^S\right)^2 dt';$$

$$D_{np}^S(t) = \frac{1}{2}\left(1 - \Delta_{nps}^2\right)\int_0^t\left(G_{np}^S(t') - 2\Gamma_{np}^S\right)dt'$$

is the function taking into account the mode amplification and attenuation;

$$\tilde{A}_{nps}^0(t) = \frac{1}{2i\omega_S}\int_0^t e^{-D_{np}^S(t')}\tilde{F}_{np}^{sp}(t')dt'$$

is the amplitude factor characterizing spontaneous Raman scattering.

The equation for the stimulated scattering wave intensity averaged over the particle volume

$$\bar{I}_S(t) = \frac{1}{V_a}\int\limits_{V_a} I_S(\mathbf{r};t)d\mathbf{r} = \frac{cn_a}{8\pi V_a}\sum_n\sum_p\left|A_{np}^S(t)\right|^2$$

with allowance for Equations 4.9–4.15 under the condition of resonance excitation ($\Delta_{npS} = 0$) can be written in the following form (modal indices are omitted):

$$\bar{I}_S(t) = \bar{I}_{sp}(t)e^{2D^S(t)}. \tag{4.16}$$

Here

$$\bar{I}_{sp}(t) = 2\left[\frac{8\pi N_0 F_0}{\Gamma_k n_a \Omega_R}\frac{\partial\alpha}{\partial q_k}\right]^2\int\limits_0^t e^{-2D^S(t')}\sum_{n'}\sum_{p'}\left|A_{n'p'}^L(t')\right|^2 S_{n'p'}^{np}dt'$$

characterizes the intensity of spontaneous Raman scattering. With an increase of the factor D^S, the function $\bar{I}_{sp}$ has a tendency to saturation.

For the pump wave, the integral formulation of the problem is the following:

$$A_{np}^L(t) = \tilde{A}_{npL}^0(t)e^{-D_{np}^L(t)}e^{i\widehat{\omega}_{np}^L t + i\varphi_{np}^L(t)},$$

where

$$\widehat{\omega}_{np}^L = \omega_L\left(1 - \frac{\Delta_{npL}^2}{2}\right); \quad \Delta_{npL} = \frac{\omega_{np} - \omega_L}{\omega_L};$$

$$\varphi_{np}^L(t) = -\frac{1}{2\omega_L}\int\limits_0^t\left(G_{np}^L(t') + \Gamma_{np}^L\right)^2 dt';$$

$$D_{np}^L(t) = \frac{1}{2}\left(1 - \Delta_{npL}^2\right)\int\limits_0^t\left(G_{np}^L(t') + 2\Gamma_{np}^L\right)dt';$$

$$\tilde{A}_{npL}^0(t) = \frac{1}{2i\omega_L}\int\limits_0^t e^{D_{np}^L(t')}\tilde{F}_{np}^i(t')dt';$$

$$G_{np}^L(t) = \frac{c^2 g_s \omega_L}{8\pi \omega_S} \sum_n \sum_p \left|A_{np}^S(t)\right|^2 S_{n'p'}^{np}.$$

4.1.3 Thresholds of SRS Generation

To find the SRS generation threshold, it is necessary to express the volume-averaged pump intensity inside the particle $\overline{I}_L$ though the intensity of incident radiation $I_0 = (c/8\pi)E_0^2$. Two formulations of this problem are possible: (1) determination of the threshold for initiation of SRS generation; (2) determination of the threshold for generation of stimulated radiation with the given intensity exceeding the intensity of spontaneous scattering. Consider first the former one. It is obvious that here we can use the approximation of the given pump field.

The function G_{np}^S can be represented in the following form:

$$G_{np}^S(t) = \frac{cg_s}{n_a} \overline{B}_c^{np} \overline{I}_L(t), \tag{4.17}$$

where

$$\overline{I}_L(t) = \frac{1}{V_a} \int_{V_a} I_L(\mathbf{r};t)d\mathbf{r} = \frac{cn_a}{8\pi V_a} \sum_{n'} \sum_{p'} \left|A_{n'p'}^L(t)\right|^2; \tag{4.18}$$

$$\overline{B}_c^{np}(\omega_L;\omega_{np}) = \frac{cn_a}{8\pi \overline{I}_L} \sum_{n'} \sum_{p'} \left|A_{n'p'}^L\right|^2 S_{n'p'}^{np} \tag{4.19}$$

is the normalized coefficient of spatial overlapping of interacting fields inside the particle. The coefficient $\overline{B}_c^{np}$ weakly depends on time at the initial stage and at the stationary stage of SRS. At the initial stage of the process, it can be calculated separately in the linear approximation, that is, within the framework of the Mie theory. In this case, $\overline{B}_c^{np}$ is determined mostly by the morphology of the particle and its optical properties. In the case that the field of the Stokes wave can be considered as unimodal ($\omega_{np} = \omega_S$), it follows from Equation 4.19 with allowance for Equation 4.13 that

$$\overline{B}_c^{np}(\omega_L;\omega_S) = V_a \left[\int_{V_a} \left(\mathbf{E}_L \cdot \mathbf{E}_L^*\right)d\mathbf{r}\right]^{-1} \int_{V_a} \left(\mathbf{E}_{np} \cdot \mathbf{E}_{np}^*\right) \sum_{n'} \sum_{p'} \left|A_{n'p'}^L\right|^2 \left(\mathbf{E}_{n'p'} \cdot \mathbf{E}_{n'p'}^*\right)d\mathbf{r}.$$

Upon convolution of the sum in the right-hand side of this equation and use of the identity following from the normalization conditions of eigen functions

$$\left|A_{np}^S\right|^2 = \int_{V_a} \left(\mathbf{E}_S \cdot \mathbf{E}_S^*\right) d\mathbf{r},$$

Equation 4.19 for the coefficient of field overlapping takes the following form (modal indices are omitted):

$$\bar{B}_c(\omega_L; \omega_S) = V_a \left[\int_{V_a} \left(\mathbf{E}_L \cdot \mathbf{E}_L^*\right) d\mathbf{r} \cdot \int_{V_a} \left(\mathbf{E}_S \cdot \mathbf{E}_S^*\right) d\mathbf{r}\right]^{-1} \int_{V_a} \left(\mathbf{E}_L \cdot \mathbf{E}_L^*\right)\left(\mathbf{E}_S \cdot \mathbf{E}_S^*\right) d\mathbf{r}. \quad (4.20)$$

Within the framework of approximation of the given field, the relation $\bar{I}_L/I_0$ is constant in time and determined through the following integral equation:

$$\frac{\bar{I}_L}{I_0} = \bar{B}_L = \frac{1}{V_a E_0^2} \int_{V_a} \mathbf{E}_L(\mathbf{r})\mathbf{E}_L^*(\mathbf{r}) d\mathbf{r}. \quad (4.21)$$

It should be noted that in the most cases the factor $\bar{B}_L$ equals unity and differs significantly from it only at the resonance excitation of the particle by the resonance field.

Then, finally, the energy density of the incident optical wave w_0, at which the stimulated radiation is generated in the particle, is determined by the following equation:

$$w_0 > w_0^{th} = \frac{n_a \omega_S t}{c g_e Q_S \bar{B}_L}. \quad (4.22)$$

Equation 4.22 includes the effective coefficient of SRS amplification in microresonator

$$g_e = g_S \bar{B}_c,$$

which reflects the difference in the rate of generation of the Stokes wave in the particle in comparison with an extended medium. This leads to the significant decrease of the process thresholds and, in some cases, allows the continuous-wave radiation to be used for pumping of a microresonator [21,22].

Figure 4.1 shows the dependence of the ratio g_e/g_S on the effective Q-factor of Eigen modes of the Stokes field Q_S. The calculation was made for water droplets

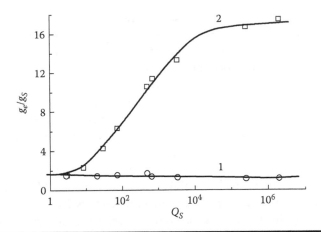

Figure 4.1 Ratio g_e/g_S as a function of the Q-factor of resonance modes Q_S of the Stokes field at the excitation of SRS of single (1) and double (2) field resonances in droplets. For illustration, the dots are connected by splines.

of different radius ($n_a = 1.33$; $\omega_L = 0.53$ μm; $\omega_S = 0.65$ μm) in two situations of nonlinear interaction of the waves: resonance of only the Stokes field (single resonance) and resonance of the both waves (double resonance).

It follows from Figure 4.1 that the ratio g/g_S at the nonresonance excitation of SRS is close to unity. The significant growth in the efficiency of nonlinear interaction is observed only in the case of double resonance of the fields. This circumstance was noticed, for the first time, in experimental paper [23] when studying thresholds of SRS excitation in drops of water solution of glycerol. Detailed theoretical investigations of the coefficient of spatial overlapping of the fields at different versions of SRS and SBS excitation in spherical particles can be found, for example, in References 16, 24.

If the incident radiation is rather long in time, then it is possible to pass from the energy density to the intensity in this equation. Thus, the condition of SRS generation at the continuous or quasicontinuous pumping takes the following form:

$$I_0 > I_0^{th} = \frac{n_a \omega_S}{c g_e Q_S \overline{B}_L} = \frac{n_a^2 V_a \omega_L \omega_S}{c^2 g_e \sigma_{ex}(a_0;\omega_L) Q_L Q_S}, \tag{4.23}$$

where $\sigma_{ex}(a_0;\omega_L)$ is the particle extinction cross section for the incident radiation. This equation was derived earlier in our papers [16,24] through consideration of the energy balance of the Stokes wave in a particle.

It should be repeated that the considered Raman wave excitation threshold corresponds, essentially, to fulfillment of the condition of positive feedback appearing in the resonator particle for the Stokes wave, when its total loss to the absorption and to the radiation escape through the particle surface becomes comparable with the amplification due to nonlinear interaction with the pump field. The intensity of the stimulated radiation under these conditions is low and, as can be seen from Equations 4.15 and 4.16, close to the intensity of spontaneous Raman scattering.

The problem on the excitation threshold of the SRS wave with the given intensity level is connected with determination of a certain level of the SRS gain coefficient G^S in the particle, which, in its turn, depends on the intensity of the pump wave. This solution can be obtained only numerically, and it will be reported in our following publications. Here, we restrict our consideration to the analysis of an important issue—the stationary state of SRS generation in the particle.

At $I_0 \geq I_0^{th}$, the stimulated radiation is generated in the particle. Normally, the generation has the nonstationary character, and at a rather long action of radiation the stationary SRS can be established. This stationary SRS generation in a microparticle was observed experimentally in References 21, 22.

The condition for the process of nonlinear scattering to become stationary is that the time derivative in the right-hand side of Equation 4.16 equals zero. This is achieved, when the volume-averaged intensity of the main wave inside the particle also achieves the stationary level

$$\bar{I}_L^{st} = 2n_a \Gamma^S / (cg_e). \tag{4.24}$$

Let us find the relation between the level of intensity of the incident radiation I_0 and the stationary value of the SRS intensity I_S^{st}. We can write the integral relationship for the field intensity of the pump wave inside the particle under conditions of double resonance, when the incident laser radiation is in resonance with one of the modes of the microresonator particle, and SRS is generated at another mode of this resonator (modal indices are omitted)

$$\bar{I}_L(t) = \bar{I}_L^0(t) e^{-2D^L(t)}, \tag{4.25}$$

where

$$\bar{I}_L^0(t) = \frac{cn_a}{8\pi} \tilde{A}_L^0(t) \left(\tilde{A}_L^0(t) \right)^*.$$

The sought relationship between the stationary intensities of the interacting waves in the particle and the pump intensity follows from Equations 4.24 and 4.25 under the condition of the quasicontinuous excitation:

$$\bar{I}_L^{st} = \frac{4\pi \left| K_{np}^n \right|^2 I_0}{\left(G_{st}^L + 2\Gamma^L \right)^2 cn_a \omega_L^2}. \tag{4.26}$$

Here, G_{st}^L is the stationary value of the factor G^L.

At the low intensity of the Stokes radiation, Equation 4.26 transforms into the above equation for the threshold intensity of the incident field leading to the SRS generation under the conditions of stationary pumping. For generation of the Stokes wave with higher intensity, the corresponding threshold value increases in $\left(1 + G_{st}^L/2\Gamma^L\right)^2$ times.

Thus, at the stationary generation of the Stokes radiation, additional energy losses of the incident optical field appear in the microresonator particle, which is equivalent to the decrease of the effective Q-factor of the resonator for the frequency of incident radiation. With this treatment of the processes, the equation for threshold intensity of the incident radiation leading to SRS in the particle keeps true for the case of excitation of the SRS wave with finite amplitude with the only difference that in Equation 4.23 Q_L should be replaced with $Q_L(1 + \eta)^{-2}$, where η is the pump depletion factor.

This factor can be determined from the numerical solution of the problem of stationary SRS. However, for tentative estimates, it is possible to use its approximate value, which can be found from the linear theory

$$\eta \approx \frac{c g_e \omega_L}{2\Gamma^L n_a \omega_S} \overline{I}_S^{st} = \frac{Q_L}{Q_S} \frac{\overline{I}_S^{st}}{\overline{I}_L^{st}}.$$

The level $\eta = 1$ is believed to be the condition, under which the pump depletion should be necessarily taken into account. Thus, the pump depletion takes place at

$$\overline{I}_S^{st} = \frac{Q_S}{Q_L} \overline{I}_L^{st}$$

in the case of stationary SRS.

4.2 Self-Action of Laser Beams in the Atmosphere

Nonlinear effects in gases changing the medium permittivity lead to self-action of laser beams. Insignificant changes of the wave phase due to a change in the refractive index in the elementary volume are accumulated at large distances into significant distortions of the wave phase and amplitude. At the beam self-action, its angular spectrum transforms, which leads to a change in the propagation trajectory, self-defocusing, and self-focusing of the radiation. The self-action of spatially modulated waves in the atmosphere causes thermal action of the laser radiation (heating, kinetic cooling) and effects of changes in the medium polarizability [25–31].

In the nonturbid atmosphere, the main factor decreasing the efficiency of laser energy transfer to long distances is the effect of thermal blooming having the lowest energy thresholds. In what follows, we restrict our consideration to only this effect.

There are numerous works devoted to the problem of thermal blooming of laser radiation [3,32–34]. These works systematize theoretical investigations and present experimental data on the thermal distortion of beams in model media. The current experimental study of this effect in the actual atmosphere is discussed in Reference 35.

It should be noted that laser experiments in the actual atmosphere are extremely expensive. In this connection, for the correct prediction of the atmospheric propagation of high-intensity laser beams, high requirements are imposed on the accuracy and reliability of analytical and numerical calculations of various parameters. This circumstance has led to development of numerous approaches and methods for theoretical study of the discussed effect and numerical algorithms for their implementation [3,33,34,36–44].

In the Institute of Atmospheric Optics (Tomsk, Russia), this problem was solved within the framework based on the comprehensive consideration of the atmospheric effect on the optical wave parameters. This formulation of the problem indicated that, along with consideration of the effect of self-action on the beam parameters, it is also necessary to take into account the influence of atmospheric turbulence distorting the beam coherence and making stochastization of the temperature field in the beam channel. It is also important to take into account the nonideal character of laser sources and some other features dictated by practical issues of application of high-power lasers in the actual atmosphere.

A characteristic feature of self-action of laser beams in the atmosphere is the mutual influence of different types of transformation of beam parameters (spatial, amplitude, frequency, temporal). This is caused by the participation of both linear (speckle structure of the beam due to scattering at turbulent and discrete inhomogeneities of the atmosphere) and nonlinear effects in this mutual influence of transformations. Thus, the amplitude nonlinear conversions of the beam lead to a change in the diffraction characteristics of the channel [39], while the stimulated Raman scattering influences the radiation divergence [45].

To study theoretically the combined influence of the atmosphere on the laser beam characteristics, we used two approaches. One of them is based on the field description of effects in the atmosphere. In this case, we speak about development and implementation of numerical methods for solution of the parabolic equation. Another approach assumes the development of the high-efficiency method for solution of similar diffraction problems, which is based on the method of splitting by physical factors in combination with the fast Fourier transform (FFT) method [36]. The results of investigation of the nonlinear propagation of beams with this method are reported in Reference 46.

Additionally to the approach regarding the propagation of high-power laser beams in the atmosphere based on the parabolic equation, the researchers have developed the original theory of the method of radiation transfer equation as a ray method of the wave theory. In the following, we generally characterize this method and illustrate the results of its application with particular examples.

In the nonlinear optics of the atmosphere, a lot of practically important problems require the study of self-action of wide-aperture laser beams under conditions of significant nonlinear distortions. This indicates that the main interaction in the nonlinear medium occurs in the zone of geometric shadow of the beam. The study of this problems based on the apparatus of quasi-optical equation appears to be quite difficult. Therefore, it is natural to refer to new approaches.

One of them in the methodology of the theory of wave processes is the ray approximation, which is treated in the theory of wave propagation in inhomogeneous media as the method of construction of shortwave asymptotic of the wave equation providing solution of diffraction problems based on the geometric optics equations [38].

The method of radiation brightness transfer equations allows one to extend significantly the domain of application of the ray methods and to study the problem of self-action at the wide range of process parameters.

For smoothly inhomogeneous low-attenuated media, the system of equations for the determination of intensity of pencil laser beams by the method of the radiation brightness transfer equation is formulated as follows:

$$\left[\frac{\partial}{\partial z} + \mathbf{n}\nabla_{\mathbf{R}} + \frac{1}{2}\nabla_{\mathbf{R}}\tilde{\varepsilon}(I)\nabla_{\mathbf{n}}\right]J(\mathbf{R},\mathbf{n},z,t) = 0; \tag{4.27}$$

$$J(\mathbf{R},\mathbf{n},z=0,t) = J_0(\mathbf{R},\mathbf{n},t); \tag{4.28}$$

$$I(\mathbf{R},z,t) = \int\int\limits_{-\infty}^{\infty} J(\mathbf{R},\mathbf{n},z,t)d^2n, \tag{4.29}$$

where J is the radiation brightness; $\mathbf{R}$ is the transverse vector of a point in the beam; z is the coordinate of propagation direction; $\mathbf{n}$ is the transverse vector of the tangent line to the trajectory of the geometric-optics ray; $\tilde{\varepsilon}$ is the perturbation of the medium permittivity caused by nonlinear effects (see also Chapter 2).

In every particular case, the system is supplemented with the equation determining the form of the dependence of permittivity perturbation $\tilde{\varepsilon}$ on the intensity $\tilde{\varepsilon} = \tilde{\varepsilon}(I)$. For integration of transfer Equation 4.27 with initial condition (4.28), the classical method is the method of characteristics. In this method, the intensity is connected with the brightness at the entrance aperture (4.28) by the integral form

$$I(\mathbf{R},z,t) = \int\int\limits_{-\infty}^{\infty} J_0(\mathbf{R}'(0),\mathbf{n}'(0),z,t)d^2n; \tag{4.30}$$

$$\frac{d\mathbf{R}'}{dz'} = \mathbf{n}'; \quad \frac{d\mathbf{n}'}{dz'} = \frac{1}{2}\nabla_R \tilde{\varepsilon}(\mathbf{R}, z', t) \tag{4.31}$$

with the initial conditions in the observation plane

$$\mathbf{R}'(z' = z) = \mathbf{R}', \quad \mathbf{n}'(z' = z) = \mathbf{n}', \quad z' = 0,\ldots,z. \tag{4.32}$$

The method of characteristics consists essentially in the fact that the integral for intensity determination (4.30) is written at the characteristics emitted from the spatial point $(\mathbf{R}, z)$ to the initial plane $z = 0$ in the direction $-\mathbf{n}'$ and intersecting the initial plane at the point $\mathbf{R}^0 = \mathbf{R}'(z' = 0)$ in the direction $\mathbf{n}^0 = \mathbf{n}'(z' = 0)$. The characteristics obey the geometric-optics equations.

The behavior of the ray characteristics demonstrates clearly peculiarities of the integrated solution. Ray trajectories become concentrated at the focus points and rarefied at the beam blooming. If in geometric optics the beam intensity is determined from the law of energy preservation in an elementary ray tube, the area of whose cross section is calculated with participation of one central ray, then in the method of transfer equation many rays are involved in the intensity calculation. This fact resolves the problem of caustic and allows the diffraction at the beam aperture to be taken into account.

The method of characteristics in different modifications was applied in References 37, 39, 41, 42, 44 to problems of propagation of coherent and partially coherent laser beams in the atmosphere for the wide range of the process parameters and paths of different lengths. Different modifications of the theory were used for investigation of self-action problems, whose solutions were *a priori* unknown, in particular, problems of self-induced caustic [37], transformation of coherent properties of radiation in a nonlinear medium [41], nonlinear refraction of wide-aperture beams [41], and fluctuation phenomena in a nonlinear medium [44]. Below we illustrate the use of the method of transfer equation for problems of propagation of high-power laser radiation in the atmosphere.

The results obtained in Reference 41 illustrate peculiarities of self-action of the partially coherent radiation at a vertical atmospheric path under conditions of kinetic cooling of the medium. Compare the coherent beam with the partially coherent beam having the same energy parameters and dimensions. For the coherent beam the effect of self-focusing is observed at the vertical path, while for the partially coherent beam, due to its higher initial diffraction divergence, the beam defocusing takes place (Figure 4.2).

This figure shows the dependence of the effective relative beam radius on the dimensionless distance for laser beams with different degrees of spatial coherence for the time $t = t_p$ and $\delta = t_{VT}/t_p = 0,5$ (t_p is the pulse duration; t_{VT} is the time of vibrational-translational relaxation of nitrogen molecules); 1 is the coherent radiation with the initial divergence $\Theta_0^c = 1.69 \times 10^{-6}$, the nonlinearity parameter

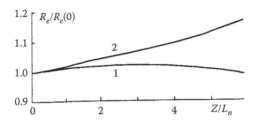

Figure 4.2 Relative effective beam radius *versus* the dimensionless distance. Atmospheric model—midlatitude summer.

$P = L_d^c/L_n = 123$; 2 is for the partially coherent radiation, $\Theta_0^{pc} = 1.69 \times 10^{-5}$, $P = L_d^{pc}/L_n = 12,3$.

Along with the mentioned approaches for studying the self-action of high-intensity laser radiation in homogeneous and inhomogeneous nonlinear refractive media, the methods were developed. They allow the *a priori* estimation of the influence of nonlinear effects for beams with different profiles and different mechanisms of radiation interaction with the medium [43] and, for some cases, provide exact solutions for effective (integral) parameters of the beam. Here, the criterion of energy transfer is chosen to be the effective beam intensity, whose initial value determines the degree of nonlinear distortions for beams of different profile

$$I_e(z) = P_0 \exp\left(-\int_0^z \alpha_g(z')dz'\right)\left[\pi\left(R_e^2(z) - R_c^2(z)\right)\right]^{-1}. \tag{4.33}$$

The parameters entering into the definition of the effective intensity, the effective beam radius R_e, and the vector of displacement of the beam centroid $\mathbf{R}_c$, obey the following equations [3,43]:

$$\frac{d^2\mathbf{R}_c}{dz^2} = \frac{1}{2P(z)}\int\int_{-\infty}^{\infty}\nabla_R\varepsilon(\mathbf{R},z,t)I(\mathbf{R},z,t)d^2R, \tag{4.34}$$

$$\frac{dR_e^2}{dz} = 2\frac{R_e^2}{F_{e1}}, \tag{4.35}$$

$$\frac{d}{dz}\frac{R_e^2}{F_{e1}} = \Theta_e^2 + \frac{1}{2P(z)}\int\int_{-\infty}^{\infty}\mathbf{R}\nabla_R\varepsilon(\mathbf{R},z,t)I(\mathbf{R},z,t)d^2R, \tag{4.36}$$

$$\frac{d\Theta_e^2}{dz} = \frac{k^{-1}}{2P(z)} \int\limits_{-\infty}^{\infty} \int \nabla_R \varepsilon(\mathbf{R}, z, t) \nabla_R \varphi(\mathbf{R}, z, t) I(\mathbf{R}, z, t) d^2 R, \qquad (4.37)$$

$$\frac{dP}{dz} = -\alpha_g P(z, t), \qquad (4.38)$$

where P is the beam power, φ is the wave phase, k is wavenumber, Θ_e is the effective width of the angular spectrum (directional pattern):

$$\Theta_e^2 = \frac{1}{k^2 \rho_{de}^2(z, t)} + \frac{R_e^2(z, t)}{F_e^2(z, t)}; \qquad (4.39)$$

The scale ρ_{de} characterizes the diffraction properties of the beam; F_e and F_{el} are the effective curvature radii of the beam phase front.

The parameters ρ_{de}, F_e, F_{el} have the following form:

$$\rho_{de} = \left[\frac{\int\int_{-\infty}^{\infty} k^{-2} (\nabla_R A)^2 d^2 R}{\int\int_{-\infty}^{\infty} A^2(\mathbf{R}, z, t) d^2 R} \right]^{-1/2},$$

where A is the wave amplitude;

$$F_e = \left[\frac{\int\int_{-\infty}^{\infty} (\nabla_R \varphi)^2 I(\mathbf{R}, z, t) d^2 R}{k^2 R_e^2 \int\int_{-\infty}^{\infty} I(\mathbf{R}, z, t) d^2 R} \right]^{-1/2},$$

$$F_{el} = \left[\frac{\int\int_{-\infty}^{\infty} R \nabla_R \varphi I(z, \mathbf{R}, t) d^2 R}{k R_e^2 \int\int_{-\infty}^{\infty} I(z, \mathbf{R}, t) d^2 R} \right]^{-1}.$$

The similarity was found in the behavior of effective parameters of collimated beams of different classes at the self-action under conditions of strong nonlinear distortions for different mechanisms of radiation interaction with the medium.

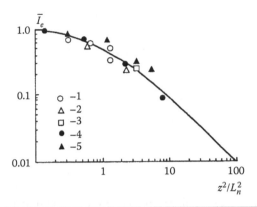

Figure 4.3 Relative effective beam intensity near the centroid as a function of the parameter z/L$_n$ for the nonlinear medium with stationary wind nonlinearity. Different dots correspond to beam with different initial profiles. (Zemlyanov, A. A., and Martynko, A. V., *Atmos. Ocean. Optics*, vol. 4, No. 11, pp. 834–838, 1991.)

This effect consists essentially in the fact that the form of dependence of the relative effective intensity $I_e(z)/I_e(z = 0)$ on the generalized distortion parameter $(z/L_n)^2$ is approximately identical for different beam classes (Figure 4.3). It can be explained by formation of the nonlinear layer near the radiator, in which the limiting directional pattern of the beam is formed. In this case, the nonlinear component of the limiting angular divergence of the beam is determined as $\Theta_{n\infty} = R_{e0}/L_n$.

The universal form of the longitudinal scale of nonlinearity was found

$$L_n = R_e(0) \left| \frac{1}{2P(0)} \int\limits_{-\infty}^{\infty}\int \mathbf{R}\nabla_R \varepsilon(\mathbf{R},0,t) I(\mathbf{R},0,t) d^2 R \right|^{-1/2}, \qquad (4.40)$$

the use of which for estimation solves automatically the well-known problem of underestimation of the thresholds of nonlinear effects in methods based on the "aberrationless" approximation or approximation of geometric optics.

Conditions for the formation of the limiting angular divergence in the initially homogeneous refractive medium have been studied. The solution was obtained for the effective parameters of the beam after the nonlinear layer

$$R_e^2(z) = R_e^{*2}\left[\left(1 + \frac{z-z^*}{F_{e1}^*}\right)^2 + \frac{(z-z^*)^2}{k^2 \rho_{de}^{*2} R_e^{*2}} + \frac{\theta_N^{*2}}{R_e^{*2}}(z-z^*)^2 + \beta^*(z-z^*)^2\right], \quad (4.41)$$

$$\theta_N^{*2} = \frac{1}{2P(z^*)}\int\limits_{-\infty}^{\infty}\int \mathbf{R}\nabla_R \varepsilon(\mathbf{R},z^*,t) I(\mathbf{R},z^*,t) d^2 R$$

is the squared nonlinear divergence of the beam;

$$\mathbf{R}_c(z) = \mathbf{R}_c^* + \Theta_c^*(z - z^*); \tag{4.42}$$

$$\Theta_c^* = \frac{1}{R_e^*} \int\limits_0^{z^*} \frac{dz}{P(z)} \int\limits_{-\infty}^{\infty}\int \nabla_{\mathbf{R}}\varepsilon(z)I(z,\mathbf{R})d^2R, \tag{4.43}$$

where $\beta = (F_e^*)^{-2} - (F_{el}^*)^{-2}$ is the factor of aberration distortions of the beam; $(F_{el}^*)^{-1} = (F)^{-1} + (F_n^*)^{-1}$, F_n^* is the nonlinear component of the effective wavefront curvature radius F_{el}; asterisked parameters are calculated at the boundary of the zone, where nonlinear effects manifest themselves (nonlinear layer). This zone can be found from the condition of saturation of the angular divergence

$$\Theta_\infty^2(z^*) = \Theta_e^2(0) + k^{-1} \int\limits_0^{z^*} \frac{dz}{P(z)} \int\limits_{-\infty}^{\infty}\int \nabla_{\mathbf{R}}\varepsilon(z)\nabla_{\mathbf{R}}\varphi(z)I(z,\mathbf{R})d^2R = \text{const.} \tag{4.44}$$

In the more general case, $F_e \leq F_{el}$ from definitions of the scales F_e and F_{el}. Consequently, the structure of the solution for the effective beam radius after the nonlinear layer is different for the aberration and aberrationless cases. The scale F_n can be both positive (self-defocusing) and negative (self-focusing). It is the most sensitive indicator of properties of a refractive medium. Therefore, it makes sense to determine the thresholds of nonlinear effects at an inhomogeneous path from the analysis of the scale F_n^*. The condition for significant self-defocusing (self-focusing) of the beam at the distance z is inequality $|F_n^*| \leq z$. If the beam is focused on the detection plane $F = z$, then the nonlinear effects show themselves against the background of the diffraction effects at $|F_n^*| \leq L_d$ or if the characteristic refraction angle $\tilde{\Theta}_n^* = R_{e0}/|F_n^*|$ exceeds the diffraction divergence of the beam: $\tilde{\Theta}_n \geq \Theta_0$.

For the case of weak nonlinear distortions at an inhomogeneous path, the equations for F_n^* and the nonlinear component of the limiting divergence $\Theta_{n\infty}$ can be written in the form $(\tilde{\varepsilon} = \tilde{\varepsilon}_{\max}(z)\overline{\varepsilon}(\mathbf{R}))$:

$$F_n^* \cong kR_{e0}^2 \left[\int\limits_0^{z^*} \tilde{\varepsilon}_{\max}(z')dz'\right]^{-1} \left(\int\limits_{-\infty}^{\infty}\int \mathbf{R}\nabla_{\mathbf{R}}\overline{\varepsilon}(\mathbf{R},0)I(\mathbf{R},0)d^2R\right)^{-1} = L_n^2/L_{eff};$$

$$\Theta_{n\infty} = R_0 L_{eff}/L_n^2; \tag{4.45}$$

$$L_{\mathit{eff}} = \int\limits_0^{z^*} \tilde{\varepsilon}_{\max}(z')dz' / \tilde{\varepsilon}_{\max}(0), \qquad (4.46)$$

where L_{eff} is the scale characterizing inhomogeneity of the atmospheric path parameters. If the condition $L_n \leq L_{\mathit{eff}}$ is fulfilled, the situation of strong nonlinear distortions of the beam takes place at the inhomogeneous path. It is obvious that there is an intermediate region, where $L_n \cong L_{\mathit{eff}}$. In this case, the beam parameters determining the effective intensity after the nonlinear layer can be obtained only numerically.

Figure 4.4 illustrates the dependence of the limiting angular divergence of the beam on the scale of nonlinearity. The data were obtained from the numerical solution of the problem of self-action of the pulsed radiation at a vertical atmospheric path by the method of brightness transfer equation.

This dependence has two asymptotics: strong $\Theta_{n\infty} = R_0/L_n$ and weak $\Theta_{n\infty} = R_0 L_{\mathit{eff}}/L_n^2$ nonlinear distortions. Since the solutions exactly corresponding to the case of strong distortions $L_n < L_{\mathit{eff}}$ were not obtained in Reference 41, the data of numerical calculations were extrapolated from the adjacent region to the region of strong distortions.

It turned out that the calculated dependence $\Theta_{n\infty}(L_n)$ can be well described by the approximation equation

$$\Theta_{n\infty} = R_0 L_{\mathit{eff}}/(L_n(L_n + L_{\mathit{eff}})).$$

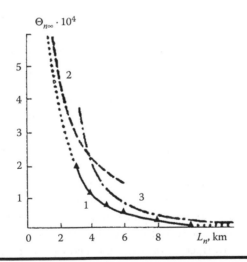

Figure 4.4 **Nonlinear component of the limiting beam divergence as a function of the nonlinearity length (self-action of the long pulse at the vertical atmospheric path): (1) calculation by the method of transfer equation (solid curve) and extrapolation dependence (dashed line), asymptotic of strong (2) and weak (3) nonlinear distortions, triangles are for the approximation dependence.**

The developed technique for estimating the efficiency of laser radiation energy transfer in nonlinear refractive media has allowed the qualitative and quantitative analysis of some practically important multiparameter problems of atmospheric nonlinear optics with the use of the equations for the integral beam parameters.

4.3 Influence of Diffraction on the Stimulated Raman Scattering of Laser Radiation in the Middle Atmosphere

The propagation of laser radiation in the middle atmosphere is accompanied by many nonlinear effects, among which the central place is occupied by the phenomenon of stimulated Raman scattering. This effect is caused by buildup of seed thermal vibrations of gas molecules exposed to the incident and scattered waves, which can lead (after excitation of some intensity threshold) to appearance of stimulated Raman scattering. The SRS effect in the atmosphere and its influence on the parameters of the propagating laser beam are studied theoretically and experimentally in numerous papers (see References 3, 23, and bibliography therein). It was found that at near-surface optical paths the SRS effect is masked by many other nonlinear effects, such as the thermal blooming, turbulent broadening of the beam, self-focusing due to the optical Kerr effect, and defocusing at the plasma of optical breakdown of air. As a laser source operates at long high-altitude paths under conditions of rarefied atmosphere, the role of SRS becomes predominant [47–49], and just this effect leads to the power loss of the main radiation owing to the radiation conversion into other frequency ranges [50] (Stokes and anti-Stokes components) and transformation of the spatial and energy spectrum of the beam [45,51,52].

At the same time, in addition to the nonlinear optical phenomena accompanying the laser energy transfer in the atmosphere, there are always linear effects, in particular, the radiation diffraction at the exit aperture of the laser source and absorption in atmospheric gases and aerosols. If the linear absorption can be often neglected in the case of an airborne laser (according to Reference 48, the mean coefficient of volume absorption of the middle atmosphere in the IR spectral range is $\sim 10^{-3}$ km^{-1}), the diffraction effects leading to a change in the transverse profile of the beam intensity should impact quite significantly the energy characteristics of the main radiation under the conditions of SRS. There are two different opinions on the effect of beam diffraction on the value and dynamics of energy exchange between waves at the stimulated scattering. On the one hand, it is believed that the diffraction broadening of the beam radius of the main radiation should decrease the efficiency of amplification of the Stokes components [45]. On the other hand, there are numerical calculations demonstrating the directly contrary role of diffraction as a process amplifying SRS [53].

In this section, using the oxygen–iodine laser radiation propagating along slant optical paths in the middle atmosphere (at altitudes exceeding 10 km), as an example, this issue is analyzed comprehensively. It is shown that the both opinions are true, and the result (amplification or reduction SRS due to beam diffraction) depends on the beam intensity and the optical path length.

4.3.1 Basic Equations of Forward SRS in the Atmosphere

The excitation of SRS at vibrational transitions ($\Delta \nu = \pm 1$, where ν is the vibrational quantum number) of molecular nitrogen and oxygen is considered. A radiation source installed on board an aircraft flying at an altitude of $h_0 = 10$ km operates along slant paths. The laser radiation is assumed to be monochromatic and spatially coherent, the pulse duration t_p satisfies the inequality $t_p \gg T_2$, where T_2 is the characteristic relaxation time of molecular vibrations, which corresponds to the stationary SRS conditions. In the parabolic approximation, the system of SRS equations for the slowly varying complex amplitudes $A_n(\mathbf{r}_\perp, z)$ of main radiation and the Stokes–anti-Stokes components with allowance for diffraction, linear absorption at the path, electronic Kerr effect, and parametric four-photon interaction effects has the following form [3,51]:

$$\left(\frac{\partial}{\partial z} + \frac{1}{2ik_n} \nabla_\perp^2 + \frac{\alpha_{ab}(\omega_n)}{2} \right) A_n(\mathbf{r}_\perp, z)$$
$$= \frac{cn_g(\omega_n)}{16\pi} g_R(z) \sum_{m,l,p} A_m A_l A_p^* \gamma_{nmlp} \delta_{nmlp} e^{i\Delta_{nmlp} z} + i \frac{\omega_n n_g(\omega_n) n_2(\omega_n)}{8\pi} |A_n|^2 A_n,$$

(4.47)

$$\delta_{nmlp} = \begin{cases} +1, l - p = n - m = -1, n \neq p, \\ -1, l - p = n - m = +1, n \neq p, \\ 0, \text{ in another cases} \end{cases} \quad \Delta_{nmlp} = k_{m_z} + k_{l_z} - k_{n_z} - k_{p_z},$$

where $E_n(\mathbf{r}; t) = A_n(\mathbf{r}_\perp, z) \exp\{i(\omega_n t - k_{n_z} z)\}$ is the complex strength of electric field of the n-th component $E_n(\mathbf{r}_\perp, z)$; $\omega_n = \omega_0 + n\Omega_R$, $n = 0, \pm1, \pm2, \ldots$; Ω_R the frequency of the Raman transition; $k_n = n_g(\omega_n) \omega_n/c$ the absolute value of the wave vector of the component $\mathbf{k}_n$; k_{n_z} is the projection of the wave vector $\mathbf{k}_n$ onto the z-axis. Characteristics α_{ab} and n_g are the linear absorption coefficient and the refractive index of the gas medium, respectively; g_R is the stationary Raman gain coefficient connected with the imaginary part of the cubic susceptibility of the medium; $\gamma_{nmlp} = \omega_0 \sqrt{\omega_n} / \sqrt{\omega_m \omega_l \omega_p}$ are the coefficients responsible for the condition of conservation of the photon number at molecular transitions (Manley–Rowe relations); n_2 the nonlinear part of the refractive index associated with the Kerr effect;

c is the speed of light in vacuum; z and $\mathbf{r}_\perp$ are, respectively, the longitudinal and transverse moving coordinates of radiation propagation. The pulse duration of the main radiation is assumed to be short, and the effects of group delay of interacting waves and the distortion of their envelope due to the time dispersion can be neglected.

To determine the phase factors Δ_{nmlp} at the parametric terms in Equation 4.47, it is necessary to take into account the conditions of spatial synchronism of interacting waves, which depend on the dispersion conditions of the medium. Within the framework of the collinear geometry of parametric interaction and the normal dispersion of air described by the Cauchy formula [3]:

$$n_g(\omega) - 1 = (a + b\omega^2)\rho_g, \qquad (4.48)$$

where $a = 2.23 \times 10^{-4}$; $b = 4.73 \cdot 10^{-37}$ Hz^{-2}; ρ_g is the air density in kg/m^3, and the frequency is in Hz, it can be easily shown that frequency detuning Δ_{nmlp} can be determined as follows:

$$\Delta_{nmlp} = \frac{3b}{c}\Omega_R^2[np(2\omega_0 + (n+p)\Omega_R) - ml(2\omega_0 + (m+l)\Omega_R)]. \qquad (4.49)$$

The effect of purely Raman scattering, when the interacting waves are always spatially synchronized along the path ($\Delta_{nmlp} = 0$), corresponds to the situation with $n = l$, $m = p = n \pm 1$. In this case, the cascade excitation of the Stokes component occurs. These components propagate simultaneously and collinearly with the initial radiation [48], and generally have the wider angular divergence than the radiation at the main frequency owing to diffraction. All the other admissible combinations of the indices n, m, l, p describe the four-wave parametric generation of combination frequencies. The angular structure of this radiation usually looks like a system of concentric rings with a center at the axis of the main beam and diverges much more strongly in comparison with axial SRS [54].

For quantitative comparison of the contributions of combination and parametric terms in Equation 4.47, the following parameter is usually used [55]:

$$\eta = |\Delta_{100-1}|/(\pi g_R I_0),$$

where $I_0 = cn_g|A_0(z=0)|^2/8\pi$ is the initial peak intensity of the power beam. In fact, this parameter compares the characteristic length $L_R = 1/g_R I_0$, at which the weak first Stokes component is amplified in e times, and the length $L_P = |\Delta_{100-1}|/\pi$, at which the phase factor of the first anti-Stokes component alternates the sign. At the length L_R, and under conditions $\eta < 1$, the interacting waves are always phase-matched and, consequently, the role of the parametric effects in SRS is important. The condition $\eta \geq 1$, contrary, points to the fast change in the phase of parametric terms and to the predominant effect of the Raman interaction.

Using Equations 4.48 and 4.49, we can estimate the parameter η for the radiation of three laser sources operating in different spectral ranges (according to data of Reference 48):

1. Oxygen–iodine laser (OIL) with $\lambda_0 = 1.315\ \mu m$, $\lambda_1 = 1.653\ \mu m$, $\Omega_R = 2231\ cm^{-1}$, $g_R = 3.5 \cdot 10^{-11}\ m/W$ (SRS at N_2);
2. CO laser with $\lambda_0 = 5.0\ \mu m$, $\lambda_{-1} = 22.52\ \mu m$, $\Omega_R = 1556\ cm^{-1}$, $g_R = 9.1 \cdot 10^{-11}\ m/W$ (SRS at O_2);
3. UV laser with $\lambda_0 = 0.308\ \mu m$, $\lambda_{-1} = 0.331\ \mu m$, $\Omega_R = 2231\ cm^{-1}$, $g_R = 7.6 \cdot 10^{-13}\ m/W$ (SRS at N_2).

At the peak intensity of radiation at the path start $I_0 = 1\ kW/cm^2$ and $h_0 = 10\ km$, we have $\eta \approx 10^3$ (OIL), 50 (CO), and 2×10^5 (UV). As can be seen, in all the cases $\eta \gg 1$ and the parametric generation of the SRS can be neglected.

It should be noted that the air density entering into Equation 4.48 indicates the decrease of the air dispersion with an increase of the height above the ground level. Consequently, we can expect the increasing role of the parametric interaction at high altitudes, as was predicted in Reference 51. However, as is shown in the following subsection, the Raman gain coefficient g_R also changes with height, and its decrease also obeys the barometric formula.

The situation can change dramatically either at an increase of the radiation intensity or at excitation of SRS at rotational sublevels of atmospheric gases (RSRS). In the last case, for molecular nitrogen the frequency shift of the Stokes component is too small ($\Omega_R \sim 76\ cm^{-1}$) to decrease significantly the wave detuning $|\Delta_{100-1}|$ and to increase the length of coherent parametric interaction. In addition, the higher RSRS gain coefficient in comparison with vibrational SRS also favors the decrease of the parameter η. From the practical point of view, for energy conversion from the main radiation into the Stokes components, the alternation of the predominant type of interaction from Raman to parametric leads to the suppression of the exponential gain of the Stokes wave and to its growth by the more slowly, power law [56] up to the almost complete termination of the pump depletion due to SRS [55,57].

So, in view of the obtained estimations and the experimental results [3], we can ignore RSRS and the parametric relation between components and restrict our consideration to only purely Raman interaction.

The influence of the Kerr effect on the laser beam propagation through the atmosphere is determined by the value of the parameter n_2. In the IR region of laser radiation wavelengths, the typical values of the nonlinear addition n_2 for the atmospheric air are $\sim 3.2 \times 10^{-23}\ m^2/W$ [3]. This gives the effective length of Kerr nonlinearity $L_K = (k_0 n_2 I_0)^{-1} \geq 6.5 \times 10^4\ km$ at $I_0 \leq 10\ kW/cm^2$, which falls far beyond actual optical paths considered in this chapter. Consequently, SRS is the strongest nonlinear effect within the framework of the considered model of radiation propagation in the middle atmosphere.

4.3.2 Analysis of Atmospheric Conditions

Now we consider the vertical model of parameters in Equation 4.47, for which the stationary Raman gain coefficient is described by the equation

$$g_R(h) = \frac{8c^3\pi^2 N(h)}{\hbar\omega_0\omega_s^2\Gamma_R n_s^2 n_0}\frac{d\sigma}{d\Omega},\tag{4.50}$$

where $\omega_s°\omega_{-1}$; Γ_R is the half-width of the transition line; N is the concentration of molecules at a path point; $d\sigma/d\Omega$ is the differential cross section of spontaneous Raman scattering; $n_i = n_g(\omega_i)$. The functional height dependence of g_R originates from the gas concentration and the line width of spontaneous Raman scattering.

The vertical model of Γ_R is determined by the Voigt dependence, which can be found from the following approximate equation [58]:

$$\Gamma_R = (1+\chi)\frac{\Gamma_L}{2} + \sqrt{\frac{\Gamma_L^2(1+\chi)^2}{4} + \Gamma_d^2}.\tag{4.51}$$

Here, $\chi = 0{,}07$;

$$\Gamma_L(Z) = \Gamma_0\frac{p_g(h)}{p_{g0}}\left(\frac{T_0}{T(h)}\right)^{1/2}$$

is the Lorentz line profile; Γ_0 is the natural line width; p_g and T are the gas pressure and temperature; h is the height above the ground level; $\Gamma_d \approx 3{,}58\cdot10^{-7}\Gamma_0(T[\kappa]/m_g[\text{a.e.}])^{1/2}$ cm^{-1} is the Doppler half-width; m_g is the molecule mass.

The calculations by using Equation 4.51 show that up to $h \sim 10$–15 km the half-width of the spontaneous Raman scattering is determined by the Lorentz profile Γ_L. At heights $h \sim 19$–20 km, the contributions of the Lorentz and Doppler broadening become nearly equal, and for higher levels Γ_R is determined by the Doppler profile Γ_d and only slightly depends on height.

The vertical model of the gas concentration N is determined by the corresponding models of atmospheric pressure and temperature. It is known that for heights $h \geq 10$ km the variation of the gas concentration is determined by the barometric dependence

$$N(h) = N(h_0)\exp\left(-\frac{g(h-h_0)}{R_gT(h_0)}\right) = N(h_0)\exp\left(-\frac{(h-h_0)}{h^*}\right),\tag{4.52}$$

where g is the free fall acceleration; R_g is the specific gas constant; $h^* \approx 6.8$ km. Since Equation 4.50 includes the ratio N/Γ_R, we take the following model. Up to heights $h_0 = 10$ km, the gain coefficient g_R is constant: $g_R(h) = g_R(h = 0)$ and at $h > h_0$ it becomes to decrease by law (4.52). If the radiation propagates along a slant path at an angle θ to zenith ($\theta \geq 0$), then, with regard for the curvature of the Earth's surface and without regard for refraction, the dependence $g_R(z)$ can be modeled by the following equation:

$$g_R(z) = g_R(h_0)\exp\left[-\frac{\left(\sqrt{z^2 + (R_E + h_0)^2 + 2z\cos\theta(R_E + h_0)} - (R_E + h_0)\right)}{h^*}\right],$$

$$h \geq h_0, \tag{4.53}$$

where R_E is the mean Earth's radius.

4.3.3 Technique of Numerical Calculations

Let us pass directly to details of the numerical solution of system of Equations 4.47. For this purpose, we first introduce the dimensionless variables:

$$\xi = z/L_D; \quad x = (\mathbf{r}_\perp \cdot \mathbf{e}_x)/R_0; \quad y = (\mathbf{r}_\perp \cdot \mathbf{e}_y)/R_0; \quad U_n = A_n/A_0(z = 0),$$

where $L_D = k_0 R_0^2$ is the free diffraction length of the original beam; R_0 is the initial radius of the main beam; $\mathbf{e}_x$, $\mathbf{e}_y$ are the orths of the Cartesian coordinate system. In new variables, the quasioptics equations for the main radiation U_0, two Stokes components U_{-1}, U_{-2}, and the first anti-Stokes component U_1 take the form

$$\frac{\partial U_0(x,y,\xi)}{\partial \xi} = -\frac{i}{2}\nabla_\perp^2 U_0 - \frac{L_D}{2L_a}\bar{\alpha}_0(\xi)U_0 + \frac{L_D}{2L_R}\bar{g}_R(\xi)\mathbf{R}(U_0), \tag{4.54}$$

$$\frac{\partial U_{-1}(x,y,\xi)}{\partial \xi} = -\frac{i}{2}\frac{\omega_0}{\omega_{-1}}\nabla_\perp^2 U_{-1} - \frac{L_D}{2L_a}\mu_{-1}\bar{\alpha}_{-1}(\xi)U_{-1} + \frac{L_D}{2L_R}\bar{g}_R(\xi)\mathbf{R}(U_{-1}), \tag{4.55}$$

$$\frac{\partial U_{-2}(x,y,\xi)}{\partial \xi} = -\frac{i}{2}\frac{\omega_0}{\omega_{-2}}\nabla_\perp^2 U_0 - \frac{L_D}{2L_a}\mu_{-2}\bar{\alpha}_{-2}(\xi)U_{-2} + \frac{L_D}{2L_R}\bar{g}_R(\xi)\mathbf{R}(U_{-2}), \tag{4.56}$$

$$\frac{\partial U_1(x,y,\xi)}{\partial \xi} = -\frac{i}{2}\frac{\omega_0}{\omega_1}\nabla_\perp^2 U_1 - \frac{L_D}{2L_a}\mu_1\bar{\alpha}_1(\xi)U_1 + \frac{L_D}{2L_R}\bar{g}_R(\xi)\mathbf{R}(U_1), \tag{4.57}$$

where $\nabla_\perp^2 = \partial^2/\partial x^2 + \partial^2/\partial y^2$ is the transverse Laplacian; $L_a = (\alpha_g(\omega_0; \xi = 0))^{-1}$ is the characteristic length of linear absorption;

$$\mu_n = \left[\alpha_g(\omega_n)/\alpha_g(\omega_0)\right]\Big|_{\xi=0}; \quad \bar{\alpha}_n(\xi) = \alpha_g(\omega_n; \xi)/\alpha_g(\omega_n; \xi = 0);$$
$$\bar{g}_R(\xi) = g_R(\xi)/g_R(\xi = 0)$$

are the normalized linear absorption coefficient and Raman gain factor, respectively. For reference, the operators $\mathbf{R}(U_n)$, describing the Raman interaction, are given with allowance for the parametric terms

$$\mathbf{R}(U_0) = \left(\gamma_{0101}\,|U_1|^2 - \gamma_{0-10-1}\,|U_{-1}|^2\right)U_0 + \gamma_{01-2-1}\exp(i\Delta_{01-2-1}z)U_1 U_{-2}U_{-1}^*$$
$$- \gamma_{0-1-1-2}\exp(i\Delta_{0-1-1-2}z)U_{-1}^2 U_{-2}^*,$$

$$(4.58)$$

$$\mathbf{R}(U_{-1}) = \left(\gamma_{-10-10}\,|U_0|^2 - \gamma_{-1-2-1-2}\,|U_{-2}|^2\right)U_{-1} + \gamma_{-1001}\exp(i\Delta_{-1001}z)U_0^2 U_1^*$$
$$- \gamma_{-1-210}\exp(i\Delta_{-1-210}z)U_{-2}U_1 U_0^*,$$

$$(4.59)$$

$$\mathbf{R}(U_{-2}) = \gamma_{-2-1-2-1}\,|U_{-1}|^2\,U_{-2} + \gamma_{-2-1-10}\exp(i\Delta_{-2-1-10}z)U_{-1}^2 U_0^*$$
$$+ \gamma_{-2-101}\exp(i\Delta_{-2-101}z)U_{-1}U_0 U_1^*,$$

$$(4.60)$$

$$\mathbf{R}(U_1) = -\gamma_{1010}\,|U_0|^2\,U_1 - \gamma_{100-1}\exp(i\Delta_{100-1}z)U_0^2 U_{-1}^*$$
$$- \gamma_{10-1-2}\exp(i\Delta_{10-1-2}z)U_0 U_{-1}U_{-2}^*.$$

$$(4.61)$$

The phase detunings Δ_{nmlp} are described by Equation 4.49.

It is convenient to solve the system of differential equations of the form (4.54)–(4.57) by the method of splitting by physical factors, when the initial self-consistent nonlinear problem is artificially divided into two conditionally independent subproblems: the problem of nonlinear self-action of the collimated radiation, and the problem of free diffraction of the beam with the profile formed as a result of nonlinearity, at every step by the spatial variable.

We calculated the free diffraction with the use of the spectral approach (see Reference 59). In this case, the solution of the diffraction equation

$$\frac{\partial U_n(\mathbf{r}_\perp, \xi)}{\partial \xi} = D_n \nabla_\perp^2 U_n(\mathbf{r}_\perp, \xi); \quad D_n = -\frac{i\omega_0}{2\omega_n} \qquad (4.62)$$

is sought for in the range of spatial frequencies $(\mathbf{k}_n \mathbf{r}_\perp)$. At the step $\Delta\xi$, solution (4.62) has the form

$$U_n(\mathbf{r}_\perp, \xi + \Delta\xi) = \mathbf{F}^{-1}\left[\mathbf{F}[U_n(\mathbf{r}_\perp, \xi)]\exp\left\{-D_n |\mathbf{r}_\perp|^2 \Delta\xi\right\}\right], \tag{4.63}$$

where F is the two-dimensional complex Fourier transform. The nonlinear part of Equation 4.54 was solved with the use of fourth-order Runge–Kutta numerical scheme.

As the initial conditions for system (4.54)–(4.57) and (4.58)–(4.61), we took the collimated beams with different spatial profiles (conventionally model the large-scale spatial inhomogeneities of the transverse intensity profile), which were specified by the following dependence [3]:

$$U_n(x, y, \xi = 0) = U_n^0(q,s)e^{-\frac{(x^q + y^q)}{2R_0^2}}\left[1 - e^{-\frac{(x^q + y^q)}{R_0^2}(s^{-q}-1)}\right]^{1/2}, \tag{4.64}$$

where q is the parameter determining the beam shape; $s = R_{01}/R_0$ is the shadowing parameter; R_{01} is the radius of the shadowed part of the beam at the e^{-1} intensity level; U_n^0 is the initial radiation amplitude taken from the condition of identical total power P_{n0} of beams of different profiles at the entrance into the medium. For beams of type (4.64), we have

$$P_{n0} = \iint_\infty |U_n(0)|^2\, dx\, dy = \frac{4R_0^2}{q^2}\left|U_n^0\right|^2 \Gamma(1/q)^2 (1 - s^2); \tag{4.65}$$

$$R_e^2(z = 0) = \frac{1}{P(z = 0)}\iint_\infty (x^2 + y^2)|U_n(z = 0)|^2\, dx\, dy = 2R_0^2 \frac{\Gamma(3/q)}{\Gamma(1/q)}(1 + s^2) \tag{4.66}$$

is the initial effective beam radius. Below we present the results for three types of beams: Gaussian ($q = 2, s = 0$), ring ($q = 2, s = 0,6$), and super-Gaussian ($q = 10$, $s = 0$).

4.3.4 Discussions

The role played by SRS in the propagation of a power beam in the atmosphere is indicated by some parameters, namely, the length of nonlinear interaction at SRS L_R, the length of free diffraction of the initial beam L_D, and the path length L.

The effect of SRS on the intensity of the main radiation is usually characterized by the integral gain factor

$$G_R = I_0 \int_0^L g_R(z) dz = I_0 L \langle g_R \rangle.$$

Under conditions of well-developed effect, when the ratio of the intensity of the first Stokes component I_{-1} to the pump intensity achieves the level $I_{-1}/I_0 \sim 1\%$, the gain factor amounts to $\sim 25 \div 30$, which corresponds to the initial level of the Stokes seed $\beta = (U_{-1}/U_0)\big|_{z=0} \approx 10^{-8}$. Correspondingly, the characteristic length L_R^*, at which the effect achieves the given level G_R^*, at small values of G_R, can be found as $L_R^* = G_R^*/(\langle g_R \rangle I_0)$. It is obvious that with an increase of the gain factor G_R, the effect of pump depletion should be taken into account, and L_R^* can be determined only numerically.

Another important factor is beam diffraction, which can affect significantly the efficiency of Raman interaction of the main radiation and the Stokes components.

Consider the simplified model of propagation of the power radiation at short paths, when the condition $L < L_D$ is valid and the excitation of only the first Stokes component is taken into account. Under these conditions, we can neglect the beam diffraction and obtain the analytical solution of system (4.54) and (4.55). Thus, the initial problem can be reduced to the following form:

$$\frac{\partial |U_0|^2 (\mathbf{r}_\perp, \xi)}{\partial \xi} = \left[-\frac{L_D}{L_a} \bar{\alpha}_0(\xi) - \frac{L_D}{L_R} \bar{g}_R(\xi) \frac{\omega_0}{\omega_{-1}} |U_{-1}|^2 \right] |U_0|^2,$$

$$\frac{\partial |U_{-1}|^2 (\mathbf{r}_\perp, \xi)}{\partial \xi} = \left[-\frac{L_D}{L_a} \mu_{-1} \bar{\alpha}_{-1}(\xi) + \frac{L_D}{L_R} \bar{g}_R(\xi) |U_0|^2 \right] |U_{-1}|^2.$$

If the absorption coefficient is constant all over the path ($\alpha_n = \text{const}$), the solution of Equation 4.61 is known to be a modified analog of SRS for plane waves [60]:

$$|U_0(\mathbf{r}_\perp, \xi)|^2 = \exp\left\{ -\frac{L_D}{L_a} \bar{\alpha}_0(\xi) \right\} |U_0(\mathbf{r}_\perp, 0)|^2 - \frac{\omega_0}{\omega_{-1}} |U_{-1}(\mathbf{r}_\perp, \xi)|^2,$$

$$|U_{-1}(\mathbf{r}_\perp, \xi)|^2 = |U_0(\mathbf{r}_\perp, 0)|^2 \frac{\beta^2 \exp\{G(\xi)\}}{1 + \beta^2 \gamma_{-1} \exp\{G(\xi)\}}.$$

Here,

$$G(\xi) = \exp\left(-\frac{L_D}{L_a}\bar{\alpha}_0\xi\right)|U_0(\mathbf{r}_\perp,0)|^2\,\frac{L_D}{L_R}\int_0^\xi \bar{g}_R d\xi' - \frac{L_D}{L_a}\mu_{-1}\bar{\alpha}_{-1}\xi; \quad (4.67)$$

$$\gamma_{-1} = \omega_0/\omega_{-1}.$$

The structure of Equation 4.62 allows us to introduce the combined parameter G^*, characterizing the degree of SRS manifestation at the optical path

$$G^* = T_0(L)I_0 \int_0^L g_R dz + \ln T_{-1}(L), \quad (4.68)$$

where $T_n(L) = \exp\left\{-\int_0^L \alpha_n(z)dz\right\}$ is the linear transmittance of the optical path for radiation with the frequency ω_n; $I_0 = P_0/(\pi R_e^2)$ is peak pump intensity.

Figure 4.5 shows the coefficient of conversion of the main radiation power P_0 into power of Raman components P_R:

$$\eta_R = P_R(L)/(P_0(0) + P_R(0)) \approx P_R(L)/P_0(0)$$

as a function of this parameter for different initial beam profiles ($\lambda_0 = 1.315\ \mu\text{m}$, $\alpha_n = 0$). Since the super-Gaussian beam has the smallest effective radius ($R_e(s\text{-}gauss) = 0.79$), while the ring and Gaussian beams are somewhat wider

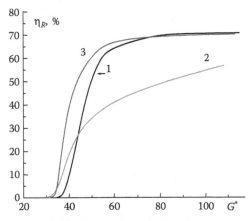

Figure 4.5 Coefficient of SRS conversion of radiation power η_R as a function of the parameter G^* for beams with different initial profiles: Gaussian (1), ring (2), super-Gaussian (3). Calculation without linear absorption and diffraction ($\lambda_0 = 1.315\ \mu\text{m}$).

(R_e(ring) = 1.2, R_e(gauss) = 1), the beam with the plateau-like profile has the higher peak intensity, the initial beam power being the same.

Consequently, SRS in this case progresses faster, which can be seen from Figure 4.5. Correspondingly, the conversion coefficient η_R at fixed G^* has the smallest values for beams with shadowing. The maximum achievable level of SRS conversion can be determined from the Manley–Rowe relations and, for the given conditions, is $\eta_R^{max} \leq 71\%$. The threshold value of the parameter G^*, at which the SRS becomes noticeable, is $G_1^* \approx 30 \div 35$.

What happens if the condition $L < L_D$ is no longer valid and it becomes necessary to take into account the wave diffraction in addition to the nonlinear interaction of waves? This question is answered in Figures 4.6–4.8. The calculation

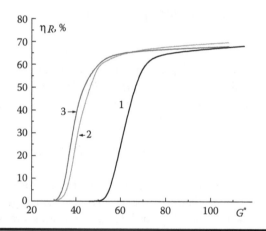

Figure 4.6 **The same as in Figure 4.5 but for** $\bar{L} = 1.05$.

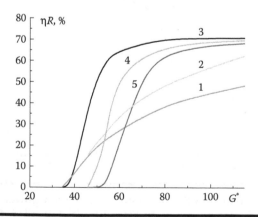

Figure 4.7 **Power conversion coefficient of the Gaussian beam with neglected diffraction (1), and at** $\bar{L} = 0,02$ **(2); 0,17 (3); 0,72 (4); 1,05 (5).**

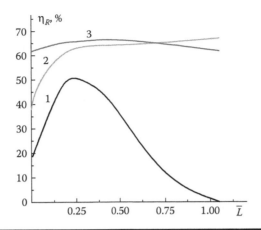

Figure 4.8 **Coefficient η_R as a function of the dimensionless path length at $G^* = 50$ for the Gaussian (1), ring (2), and super-Gaussian (3) beam profiles.**

parameters here are the same as in Figure 4.5, but Equations 4.61 are supplemented with the transverse Laplacian.

Figure 4.6 shows the dependence of the conversion coefficient of the power beam on the combined SRS gain parameter for different initial beam profiles and the dimensionless path length $\bar{L} = L/L_D = 1.05$.

It can be seen that the general character of the dependence $\eta_R(G^*)$ remains the same, but the relative position of the curves for beams with different transverse profiles becomes different: the SRS threshold G_1^* and the boundary (on the abscissa) of the developed effect G_2^* shift, when $\eta_R \approx \eta_R^{max}$.

This behavior is detailed in Figure 4.7, which shows only the Gaussian beam at different values of $\bar{L}$. It follows from Figure 4.7 that while G_1^* increases monotonically with an increase of the path length, G_2^* behaves differently: first increases and then decreases with an increase of $\bar{L}$ and has a maximum at $\bar{L} \approx 0.25$. The similar dependence is also characteristic for beams of other initial profile (see Figure 4.8).

To understand this effect, it is necessary to consider the evolution of transverse beam intensity profiles under conditions of joint manifestation of SRS ad diffraction. First, as the interaction length increases, the SRS threshold is achieved at the places with maximal values of the transverse intensity distribution of the main wave (Figure 4.9b). The active Raman scattering starting here leads to the distortion of the initial profile of the pump beam and its local depletion. Then other beam zones begin to satisfy the threshold SRS condition, thus providing the continuous increase of the conversion parameter η_R (Figure 4.9c).

The diffraction of radiation leads to the diffusion blooming of the beam in the transverse direction and to decrease of the peak intensity. Consequently, the increasing role of diffraction (increase of the parameter $\bar{L}$), on the one hand, should

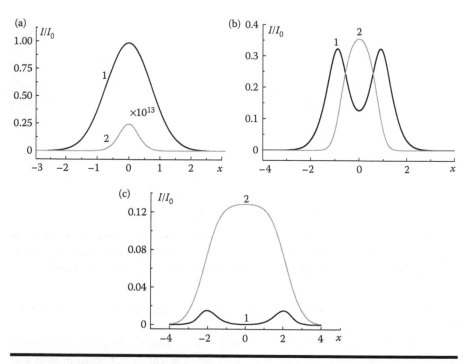

Figure 4.9 **Transverse intensity profile of the main (1) and the 1-st Stokes (2) beams propagating along the horizontal atmospheric path ($h_0 = 10$ km) at $\bar{L} = 0.05$ (a); 0.44 (b); 0.64 (c). Parameters of calculation: Gaussian beam with $I_0 = 1.6 \cdot 10^3$ W/cm², $R_0 = 20$ cm, $\lambda_0 = 1.315$ μm.**

manifest itself in the decrease of the coefficient η_R and the increase of the threshold G_1^*. In Figure 4.8, this situation is illustrated by the descending branches of the function $\eta_R(\bar{L})$.

This effect can be estimated through the solution of system of Equations 4.54 for two waves with regard for the diffraction of the main radiation and with the neglected depletion, as well as without regard for diffraction of the Stokes wave:

$$\frac{\partial U_0}{\partial \xi} = \left[D_0 \nabla_\perp^2 - \frac{L_D}{2L_a} \bar{\alpha}_0(\xi) \right] U_0, \tag{4.69}$$

$$\frac{\partial U_{-1}}{\partial \xi} = \left[-\frac{L_D}{2L_a} \mu_{-1} \bar{\alpha}_{-1}(\xi) + \frac{L_D}{2L_R} \bar{g}_R |U_0|^2 \right] U_{-1}. \tag{4.70}$$

The solution of Equation 4.69 for the Gaussian transverse profile can be expressed as

$$|U_0(\mathbf{r}_\perp;\xi)|^2 = \frac{|U_0(\mathbf{r}_\perp;0)|^2}{1+\xi^2} \exp\left(\frac{x^2+y^2}{1+\xi^2} - \frac{L_D}{L_a} \int_0^\xi \bar{\alpha}_0(\xi')d\xi' \right). \quad (4.71)$$

Upon the substitution of Equation 4.71 into Equation 4.70 at $x=y=0$ and path-constant Raman gain coefficient $\bar{g}_R$, we obtain the law of increase of the Stokes radiation intensity at the beam axis

$$|U_{-1}(0;\xi)|^2 = |U_{-1}(0;0)|^2 \exp\left(\frac{L_D \bar{g}_R |U_0(0;0)|^2}{L_R} T_0(\xi)\text{arctg}(\xi) + \ln T_{-1}(\xi) \right). \quad (4.72)$$

Equating the exponent in Equation 4.72 to G_i^* and assuming $T_0 = T_{-1} \approx 1$, we obtain the functional dependence of the threshold value of the SRS gain factor on the dimensionless path length $G_i^* \sim \text{arctg}(L/L_D)$, which qualitatively describes the data depicted in Figure 4.7.

On the other hand, diffraction by its nature tends to smooth the transverse profile of the beam, leading to diffusion of the wave amplitude from zones with high intensity to zones with the decreased intensity level. This is especially prominent in the zone, where the transverse amplitude gradient is large, that is, in the zone of active SRS manifestation. Thus, the pump depletion, for example, at the beam center due to intensification of the Stokes components is partially compensated by the energy inflow from peripheral zones. In this sense, the diffraction of the main beam favors SRS and intensifies it (ascending branches of the dependence $\eta_R(\bar{L})$), as was noted previously in Reference 53.

In the beams with the non-Gaussian initial profile, this effect is even more pronounced, because the transverse intensity gradient here is initially high. Numerical calculations of the intensity at the axis of collimated beams of different profiles along the path at their diffraction in the absence of the SRS effect show that, in contrast to the Gaussian profile, the intensity at the center of the ring and super-Gaussian beams first increases, as in the case of focusing, and then decreases due to the beam blooming.

Thus, the diffraction, redistributing intensity in the transverse profile of the beam, acts in two ways in the Raman active medium: it simultaneously intensifies and suppresses the SRS effect, which causes the ambiguous behavior of the beam power conversion coefficient for different values of the propagation path length. This effect is most pronounced for beams with the non-Gaussian initial spatial profile.

4.3.5 Effective Radius of Beams under SRS Conditions

The nonlinear self-action of radiation at atmospheric paths leads to a change in the beam cross size. In the case of thermal blooming [3] or propagation of laser radiation in a turbulent medium [59], the optical beam acquires additional divergence (in comparison with the initial diffraction one) proportional to the beam power, as well as to thermodynamic and structure parameters of the gas medium. This nonlinear divergence is connected with a change in the refractive index of the medium as a result of both thermal action of the radiation itself and local fluctuations of the refractive index due to the atmospheric turbulence. In any case, the phase of the optical wave changes nonlinearly, leading to a change in the transverse radiation intensity profile.

The physical mechanism of influence of the Raman scattering on the beam size is somewhat different, because SRS provides the amplitude effect rather than the phase effect. The modification of the transverse amplitude distribution of the main radiation as a result of the Raman amplification of the Stokes components leads to a change in the diffraction divergence of the beam at every path point and, as a consequence, to the significant increase of the effective beam radius.

Similar effect is also observed for the Raman radiation. Moreover, in this case, the parametric four-photon processes can additionally increase the divergence of the Stokes beam, because, owing to the requirement of wave synchronism and the presence of air permittivity dispersion, the Raman radiation initially propagates at an angle to the main beam. Thus, for example, this angle for the first anti-Stokes component can be found [54]: $\theta \approx \Omega_R \sqrt{\left(\partial^2 n_g / \partial \omega^2\right)}$, which amounts to $\sim 3 \cdot 10^{-4}$ rad for OIL radiation in atmospheric nitrogen.

Finally, one more reason for the change in the angular spectrum of Raman components is the stochastic nature of the Stokes signal. SRS develops from spontaneous noise of the medium and therefore has the random spatial structure. The spectrum of this partially coherent radiation is always wider than the diffraction one [47].

Figure 4.10 shows the dependence of the effective beam radius normalized to the initial radius $\bar{R}_e = R_e / R_0$ at SRS at a horizontal high-altitude path for the parameter G^*.

For comparison, Figure 4.10 shows also the values of R_e with diffraction neglected. One can see that, starting from threshold values of the gain coefficient, the size of both the main beam and the Stokes beam increases, and the more intense is the SRS process, the stronger is the effect of diffraction.

The increase of the dimensionless path length $\bar{L}$, corresponding to the increasing role of diffraction at the radiation propagation, is accompanied first by the increase of the cross size of the interacting beams and then by some decrease in comparison with the maximal achieved value. The calculations have shown that this maximum takes place at $\bar{L} \approx 0.17$. This correlates with the dependences of the power conversion coefficient shown in Figure 4.8. In other words, at the given level of the SRS gain increment, G^*-diffraction can both increase and decrease the efficiency of Raman interaction of waves.

As can be seen from Figure 4.11, the change of the tilt angle of the optical path to the horizon changes the efficiency of power conversion from the main radiation to the SRS components.

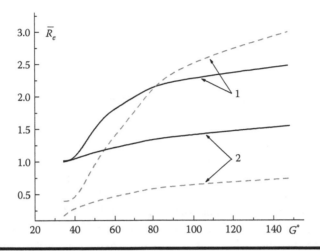

Figure 4.10 **Dependence of the effective beam radius of the main radiation (solid curve) and the first Stokes radiation (dashed curve) on the Raman gain parameter with diffraction taken into account (1) and neglected (2) at the propagation path ($\bar{L} = 0{,}17$; $\lambda_0 = 1.315$ μm, Gaussian beams).**

As follows from Equation 4.11, the effective atmospheric layer h_1, in which the active amplification of the Stokes signal occurs, decreases with an increase of the path tilt angle and amounts to only $h_1 \approx h^* \sim 7$ km for the zenith direction.

If the diffraction length for the pump radiation L_D is large (in comparison with the length h_1), then the diffraction has no time to interfere significantly the beam energy redistribution due to SRS, and the coefficient η_R decreases. In the

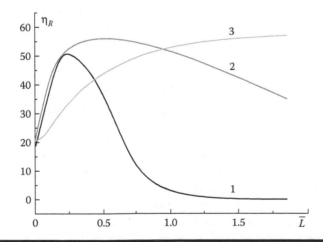

Figure 4.11 **Power conversion coefficient of the Gaussian beam as a function of the dimensionless optical path length at $G^* = 50$ and different zenith angles: $\theta = 90°$ (1); 75° (2); 0° (3).**

intermediate zone ($h_1 \leq L_D$), the influence of diffraction on η_R is positive and maximal, whereas at ($h_1 \geq L_D$) the beam diffraction begins to prevent the Raman amplification.

4.3.6 Cascade Excitation of SRS Components

In conclusion, a brief mention should be made of peculiarities of SRS manifestation with regard for the cascade excitation of the Stokes components in the medium. The results of numerical solution of the first three equation of system (4.54) for the power and the effective radius of the beam are shown in Figure 4.12. The parameters of calculation were chosen to correspond to the OIL radiation propagating along the horizontal high-altitude path at $G^* = 200$, $\overline{L} = 0.17$, and $\alpha_n = 0$.

It follows from Figure 4.12 that once some characteristic SRS length L_{R1}^*, which corresponds to the threshold G_1^*, is achieved, the marked growth of the power and radius of the first Stokes component begins with the simultaneous depletion of the main wave power. At the distance L_{R2}^*, the power of the Stokes signal at the frequency ω_{-1} attains the saturation with the level close to the limiting value of the conversion coefficient η_R^{max}. From this time, the amplification of the second Stokes component (ω_{-2}) occurs due to nonlinear polarization induced by the field of the first Stokes component, which, in its turn, begins to deplete, and so on.

As a result, the final value of the conversion coefficient η_R at the cascade amplification of the two Stokes components appears to be lower than that at the two-photon SRS due to energy losses in the medium to excitation of optical phonons first by the main radiation and then by the field of the first Stokes component. The limiting level of the SRS conversion, which can be achieved in this n-cascade process, is, obviously, $\eta_R^{max} = \omega_n / \omega_0$. In the considered case of only two Stokes components and the oxygen laser radiation, we get that $\eta_R^{max} \leq 41\%$.

4.4 Regularities of Nonstationary Self-Focusing of Profiled Laser Beams: Averaged Description

The averaged description of wave beams based on the formalism of effective (root-mean-square) parameters is widely used in linear optics for unified characterization of optical radiation with different transverse intensity profiles. Effective beam parameters were calculated from the information about the radiation intensity distribution at a chosen point of optical path. The most important effective characteristics of radiation include the effective radius, mean angular divergence or the beam quality coefficient, and the generalized wavefront curvature radius. In the linear medium, the effective radiation parameters are connected with each other by the simple evolution dependence, allowing the variation of mean beam characteristics to be predicted at any point of the optical path.

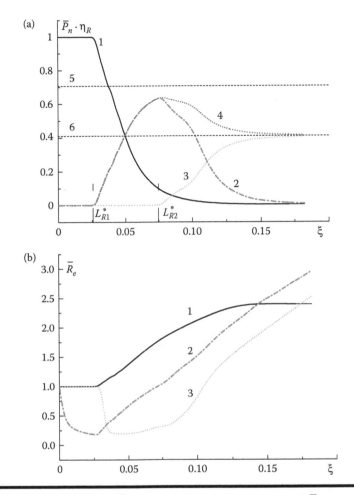

Figure 4.12 Relative power $\bar{P}_n = P_n/P_0$ (a) and effective radius $\bar{R}_e$ (b) of the main (1), first Stokes (2), and second Stokes (3) beams as functions of the normalized longitudinal coordinate ξ. Curve 4 is for the power conversion coefficient η_R. Lines 5, 6 correspond to the levels $\eta_R = 0.71$ and 0.41.

The similar averaged description of the process of laser radiation propagation is also used for the thermal defocusing of radiation in gases [2] and the stationary self-focusing in a medium with cubic Kerr nonlinearity [61]. In the last case, the law of variation of, for example, the effective beam radius takes into account the nonlinear response of the medium under the exposure of high-power electromagnetic field, which leads to progressive transverse compression of the optical beam and the following (in theory) beam collapse.

The propagation of high-power laser pulses of femtosecond (fs) duration through gas and condensed media is nonlinear and leads to significant changes not only in spatial and angular characteristics of the beam, but also in temporal and spectral ones [5]. In contrast to self-focusing of continuous-wave radiation, short laser pulses can carry power sufficient for partial ionization of the beam channel due to multiphoton absorption by medium molecules in the strong optical field. This process is accompanied by the avalanche-like growth of the free electron concentration in the medium, which leads to the appearance of nonlinear refraction and absorption in plasma, which, finally, limits the further growth of the radiation intensity and stabilizes the transverse size of the beam (see References 12, 62).

The spatial evolution of the effective parameters of femtosecond laser radiation under these conditions was studied theoretically in References 6, 7. For the beam with the Gaussian transverse intensity profile, analytical equations have been derived for the effective radius of femtosecond laser radiation in the single filamentation mode in three characteristic spatial zones, which can be separated conditionally based on the regularity of nonstationary radiation self-focusing.

4.4.1 Evolution Law of Effective Radius of the Optical Beam

The effective (root-mean-square) radius of the beam R_e (usually called instantaneous effective radius) is the second-order centered moment, and its square can be expresses as

$$R_e^2(z) = \langle R^2(z) \rangle - \langle R(z) \rangle^2 = 1/P(z) \iint\limits_{R_\perp} d^2 \mathbf{r}_\perp I(\mathbf{r}_\perp, z) \left(|\mathbf{r}_\perp|^2 - |\mathbf{r}_{gr}|^2 \right), \quad (4.73)$$

where $\mathbf{r}_\perp = \mathbf{e}_x x + \mathbf{e}_y y$ is the transverse coordinate; $\mathbf{e}_x$, $\mathbf{e}_y$ are the directing orths; $I = |U|^2 c n_0/8\pi$ is the radiation intensity; U is the strength of the electric field of wave; $P = \int_R \int_\perp d^2 \mathbf{r}_\perp I(\mathbf{r}_\perp, z)$ is the radiation power; n_0 is the (linear) refractive index of the medium; $\mathbf{R}_\perp$ is the region of determination of the transverse beam intensity profile

$$\mathbf{r}_{gr}(z) = \langle R(z) \rangle = 1/P(z) \iint\limits_{R_\perp} d^2 \mathbf{r}_\perp I(\mathbf{r}_\perp, z) \mathbf{r}_\perp$$

is the radius vector of the centroid. In contrast to the geometric radius of the beam measured at the given intensity level and reflecting the actual transverse scale of the beam only for the unimodal type of distribution, the effective radius determines the characteristic transverse size of the zone, where the most part of the radiation energy is concentrated for any beam intensity profile.

The solution of the equation of quasi-optics for the linear medium leads to the versatile evolution dependence of the effective radius of the focused optical beam at any point z of the optical path [61]:

$$R_e^2(z) = (\theta_D z)^2 + R_{e0}^2\left(1 - \frac{z}{F}\right)^2. \qquad (4.74)$$

Here

$$\theta_D = \frac{1}{k_0}\left[\frac{\displaystyle\iint_{R_\perp} d^2\mathbf{r}_\perp |\nabla_\perp U(\mathbf{r}_\perp, z=0)|^2}{\displaystyle\iint_{R_\perp} d^2\mathbf{r}_\perp |U(\mathbf{r}_\perp, z=0)|^2}\right]^{1/2} = \left[\frac{8\pi}{4cn_0 k_0^2 P_0}\iint_{R_\perp} d^2\mathbf{r}_\perp\left[\frac{\nabla_\perp I(\mathbf{r}_\perp, 0)}{I(\mathbf{r}_\perp, 0)}\right]^2\right]^{1/2}$$

$$(4.75)$$

is the effective diffraction divergence of the collimated beam of the transverse intensity profile; F is the initial wavefront curvature radius; $R_{e0} = R_e(z=0)$; $P_0 = P(z=0)$; $k_0 = 2\pi/\lambda_0$ is wave number; λ_0 is the radiation wavelength. It should be noted that the diffraction divergence θ_D is connected with the so-called beam quality parameter $M^2 = \theta_D k_0 R_{e0} = \theta_D/\theta_{Dg}$, demonstrating how many times the angular divergence of the studied beam is larger than the diffraction divergence of the Gaussian beam $\theta_{Dg} = 1/k_0 R_{e0}$ of the same initial effective radius R_{e0}.

At the stationary self-focusing of the laser beam in the medium with the cubic Kerr nonlinearity taking place at $P_0 > P_c$, where P_c is the critical power, there also exists the average description of the process similar to Equation 4.74, which is valid at least up to the point of transverse collapse of the beam z_N (nonlinear focus). In this case, for every particular type of the beam, it is sufficient to calculate few parameters (R_{e0}, θ_D, P_c) which depend only on the optical parameters of the medium and the transverse profile of the electric field strength of the wave, to determine the stationary effective beam radius R_e:

$$R_e^2(z) = (1 - \eta_0)(\theta_D z)^2 + R_{e0}^2(1 - z/F)^2, \quad z < z_N, \qquad (4.76)$$

where $\eta_0 = P_0/P_c$ is the self-focusing parameter, and critical self-focusing power is determined as

$$P_c = \frac{\lambda_0^2}{2\pi n_0 n_2}G(U(\mathbf{r}_\perp, 0));$$

$$G(U(\mathbf{r}_\perp,0)) = \frac{1}{2} \frac{\displaystyle\iint_{R_\perp} d^2\mathbf{r}_\perp \, |U(\mathbf{r}_\perp,0)|^2 \iint_{R_\perp} d^2\mathbf{r}_\perp \, |\nabla_\perp U(\mathbf{r}_\perp,0)|^2}{\displaystyle\iint_{R_\perp} d^2\mathbf{r}_\perp \, |U(\mathbf{r}_\perp,0)|^4} \qquad (4.77)$$

(n_2 is the nonlinear addition to the refractive index of the medium due to the Kerr effect). The parameters given by Equations 4.73, 4.75, and 4.77 describes the coefficients of radiation propagation in the nonlinear medium, and they depend on the laser beam profile.

We can transform Equation 4.76 to the following form:

$$\bar{R}_e^2(z) = (1 - \eta_0)\bar{z}^2 + (1 - \bar{z}/\bar{F})^2. \qquad (4.78)$$

Here, the dimensionless variables $\bar{z} = z/L_D$, $\bar{R}_e = R_e/R_{e0}$, $\bar{F} = F/L_D$, are introduced, and the parameter $L_D = R_{e0}/\theta_D$ has the meaning of the effective diffraction length of the beam with the given spatial intensity profile. It follows from Equation 4.78, in particular, which at $\eta_0 > 1$ the laser beam in general (in the sense of effective radius) can compress to a point at the distance:

$$z_N = z_K F/(z_K + F), \qquad (4.79)$$

where $z_K = L_D/\sqrt{\eta_0 - 1}$ is the coordinate of the point of transverse Kerr collapse for the collimated radiation. After the point of nonlinear focus z_N, Equation 4.78 makes no sense.

The type of transverse intensity distribution of an optical beam was not ever specified yet. This means that, in relative coordinates with allowance for the corresponding change of the relative radiation power η_0, Equation 4.78 does not lose its versatility for any type of beams. In the further discussions, we consider three most often practically used types of optical beams: Gaussian (GB), super-Gaussian (SGB), and ring beam (RB).

For the beam with the Gaussian initial profile, the electric field strength of optical wave is described by the function of the form

$$U_g(\mathbf{r}_\perp,0) = \exp\left(-|\mathbf{r}_\perp|^2/2R_0^2\right) \qquad (4.80)$$

with the geometric radius R_0 (at the $1/e$ level of intensity maximum). Then the initial values of the effective GB parameters are $R_{e0} = R_0$; $\theta_D \equiv \theta_{Dg} = 1/(k_0 R_0)$, $L_D \equiv L_{Dg} = k_0 R_0^2$ and $P_c \equiv P_{cg} = \lambda_0^2/(2\pi n_0 n_2)$

The beam of the super-Gaussian profile at the circular aperture

$$U_{sg}(\mathbf{r}_\perp,0) = \exp\left(-|\mathbf{r}_\perp|^{2q}/2R_0^{2q}\right), \quad q = 1,2,3,\dots \qquad (4.81)$$

is characterized by different values of the critical power, diffraction length, and divergence. And, finally, the third type of the beam intensity distribution—the ring beam—is described by the difference of two Gaussians

$$U_R(\mathbf{r}_\perp,0) = \exp\left(-|\mathbf{r}_\perp|^2/2R_0^2\right)\left(1 - \exp\left(-(p^{-2}-1)|\mathbf{r}_\perp|^2/R_0^2\right)\right)^{1/2}, \quad (4.82)$$

where $p = R_{sh}/R_0$ is the shadowing parameter; R_{sh} is the radius of the shadowed part of the beam at the e^{-1} intensity level.

Numerical estimates of the propagation coefficients for beams of different transverse profiles (4.80 through 4.82) have shown that the critical power and the quality parameter take minimal values of the Gaussian beam ($G = M^2 = 1$ at $q = 2, p = 0$), whereas the initial effective radius of SGB is always smaller than the beam radius of the Gaussian beam. Ring beams have the largest initial effective radius.

4.4.2 Effective Beam Radius at Nonstationary Self-Focusing

Self-focusing of a short laser pulse being a wave packet limited both in space and in time acquires the dynamic character. If we conditionally divide the temporal intensity profile into consecutive layers, then each layer is characterized by its own power level $P_i = P(t_i)$. Consequently, according to Equation 4.78, each layer has its own law of evolution of the effective radius $R_e(z; t_i)$ and position of the collapse point $z_N(t_i)$. The higher is power P_i of this temporal layer, the closer to the beginning of the path is this point. As a result, an observer being in the laboratory coordinate system sees the propagation of a high-power short pulse in the self-focusing mode as a series of local foci from each temporal layer moving with the speed of radiation propagation in the medium. This physical self-focusing model of Kerr is known in the literature as a model of moving foci and was proposed in Reference 63.

We introduce the integral effective radius of the beam R_{eg}, which characterizes the spatial size of the zone of concentration of the radiation energy density and is determined as an average of the array of squared local effective radii over the pulse power profile

$$R_{eg}^2(z) = \left\langle R_e^2(z,t) \right\rangle_P = \frac{1}{E(z)} \int\limits_{-\infty}^{\infty} P(z,t)R_e^2(z;t)dt$$

$$= \frac{1}{E(z)} \int\limits_{-\infty}^{\infty} dt \iint\limits_{R_\perp} d^2\mathbf{r}_\perp I(\mathbf{r}_\perp,z;t)\left(|\mathbf{r}_\perp|^2 - |\mathbf{r}_{gr}(t)|^2\right), \quad (4.83)$$

where $E(z) = \int_{-\infty}^{\infty} P(z;t)dt$ is the total radiation energy. Upon the substitution of Equation 4.78 into Equation 4.82, we obtain the law of variation of the squared integral effective radius at self-focusing

$$\bar{R}_{eg}^2(z) = (1 - \eta^*)\bar{z}^2 + (1 - \bar{z}/\bar{F})^2. \tag{4.84}$$

Here, $\eta^* = [E_0]^{-1} \int_{-\infty}^{\infty} \eta(t)P(0,t)dt$ is the pulse-averaged value of the self-focusing parameter. This equation predicts the transverse collapse of the collimate radiation pulse in general at the distance $\bar{z}_{Kg} = 1/\sqrt{\eta^* - 1}$ provided that $\eta^* > 1$.

Thus, the qualitative behavior of the integral effective beam radius at the nonstationary self-focusing can be described within the framework of the stationary theory with the use of Equation 4.78, but with the different modified self-focusing parameter η^*. For example, for the most widely used Gaussian temporal profile of the pulse

$$P(t) = P_0 \exp\{-\tau^2\},$$

where P_0 is the peak value; $\tau = t/t_p$ is dimensionless time; t_p is the pulse duration at the $1/e$ level, we obtain $\eta^* = \eta_0/\sqrt{2}$ $(\eta_0 = \eta(t = 0))$.

To check the applicability of Equation 4.83 under conditions of nonstationary self-focusing of laser radiation, we have carried out numerical calculations within the framework of the model of nonlinear Schrodinger equation (NSE) [64] for beams with different transverse intensity profiles. This equation describes the propagation of ultrashort laser radiation in a medium with allowance for the radiation diffraction, frequency dispersion of group velocity, and amplitude-phase self-modulation of the optical wave due to the Kerr effect and plasma nonlinearity. The change of the complex refractive index of the medium due to plasma generation in the radiation channel was taken into account by the Perelomov–Popov–Terentiev model of photoionization of air molecules [65]. The instantaneous density of free electrons in the medium was determined from the solution of the rate equation including the multiphoton and cascade mechanisms of ionization, as well as the decrease in the electron concentration due to electron recombination with ions.

The problem was solved numerically for beams with the Gaussian (4.80), super-Gaussian (4.81), and ring (4.82) transverse profiles of the envelope of optical field and the following initial parameters: pulse duration $t_p = 60$ fs, beam radius $R_0 = 1.0$ mm, and carrier wavelength $\lambda_0 = 800$ nm. The radiation was assumed to be initially collimated ($F = \infty$), the self-focusing parameter was set as $\eta_0 = 10$. For the Gaussian beam, this corresponded to the peak power $P_0 = 32$ GW ($P_{cg} = 3.2$ GW). For the super-Gaussian profile with the geometry parameter $q = 4$, according to Equation 4.77, the value of P_0 was approximately 2.4 times

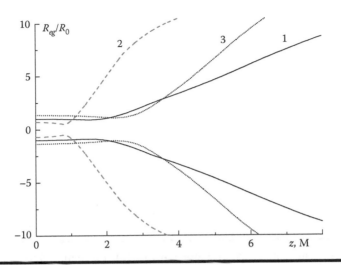

Figure 4.13 Evolution of the integral effective radius of the beams with the Gaussian (1), super-Gaussian (2), and ring (3) transverse profiles and the relative initial power $\eta = 10$ along the propagation path.

higher ($G \approx 2.4$) at $R_{e0} \approx 0.7R_0$. The ring beam at $p = 0.7$ had the initial power $P_0 = 80$ GW ($G = 2.5$) and $R_{e0} \approx 1.35R_0$. Since the initial radiation profiles were set in the form of ideal smooth functions, only one (axial) light filament was formed in the process of beam propagation.

Figure 4.13 shows the behavior of the integral effective beam radius normalized to the initial value for laser beams with different intensity profiles propagating in air.

The data are depicted as functions of the path distance in meters. The analysis of the plots allows to separate three characteristic spatial regions reflecting different stages of nonstationary self-focusing of radiation [66]: (1) zone of transverse compression of the beam to the global nonlinear focus due to self-focusing, (2) zone of sharp increase of the effective beam area after the nonlinear focus upon formation of a filament in its vicinity, and (3) zone of linear propagation of the radiation upon passage through the nonlinear medium.

As can be seen from Figure 4.13, the super-Gaussian beam, having the smaller initial effective radius, far earlier forms the nonlinear focus and has the far wider angular divergence than GB. At the same time, the beam with the ring profile, being the widest among the considered beams, is characterized by the slower rate of self-focusing, but strong divergence after the global focus.

Figure 4.14 shows the same data as in Figure 4.13, but for the normalized parameters: global effective radius $\bar{R}_{eg} = R_{eg}/R_{eg0}$ is normalized to its value at the path start $R_{eg0} = R_{eq}(z = 0)$, and the distance $\bar{z}$ is calculated for every beam in accord with its initial intensity profile.

In the considered case, this gives the relation of generalized diffraction beam lengths

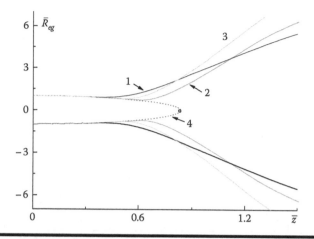

Figure 4.14 The same, as in Figure 4.13, but for generalized coordinates. Calculation by Equation 4.82 corresponds to curve 4; the circle shows the calculated position of beam collapse in general $(\bar{z}_{Kg})$.

$$L_D(\text{GB})/L_D(\text{SGB}) \approx 2.9 \quad \text{and} \quad L_D(\text{GB})/L_D(\text{RB}) \approx 0.9,$$

that is, the rate of diffraction of SGB is almost threefold higher than that of the Gaussian beam, whereas the beam with the ring profile, to the contrary, diffracts more slowly than GB.

It can be seen from Figure 4.14 that in the generalized coordinates the initial stage of spatial evolution of the effective radius (transverse compression) is similar for all the beams and corresponds to asymptotic dependence (4.84). Then each beam compresses and forms a global focal waist, whose path position $\bar{z}_g$ and cross size $\bar{R}_{eg}(\bar{z}_g)$ depend on the particular profile of radiation. Here, comparing Figures 4.13 and 4.14, SGB and RB in the new coordinates are characterized by the slower self-focusing at the path (longer $\bar{z}_g$) and have the narrower focal waist than the beam with Gaussian profile. This is a direct consequence of the quasi-uniform energy density distribution in the beams of the non-Gaussian profiles near the axis (Figure 4.15).

As a result, the super-Gaussian and ring beams self-focus under the Kerr effect and their energy concentrates closer to the beam axis, while in GB the intensity at the center increases faster in comparison with the periphery, which shows itself in the sharp gradient of the optical energy density profile at the central zone and, correspondingly, the larger effective radius of distribution at the point of nonlinear focus. The formation of the focal waist of effective radius of optical pulse is connected with the blocking effect of plasma nonlinearity of the medium, which restricts the peak value of intensity in the filamentation zone and exerts the effect on the refractive coefficient of gas opposing the Kerr effect.

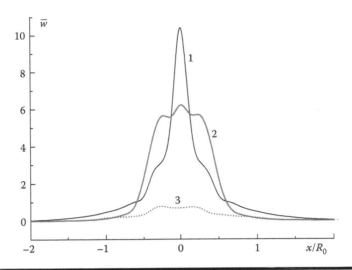

Figure 4.15 **Transverse profiles of energy density $\bar{w} = w(r,\ z)/w(r,\ z = 0)$ for GB (1), SGB (2), and RB (3) at the point of global nonlinear focus $\bar{z}_g$.**

The stronger transverse compression of the super-Gaussian and ring beams leads to the higher angular divergence after the global focus. However, the limiting radiation divergence formed at the linear stage of propagation, that is, after the passage through the layer of nonlinear medium, has close values for GB, SGB, and RB and far exceeds the initial, diffraction-driven level.

4.4.3 Estimating Equations for the Averaged Parameters of Optical Beam at the Nonlinear Focus

It seems important to estimate the coordinates of position of the global nonlinear focus of the beam z_g, because at this point the root-mean-square radiation energy density is maximal all over the path [67]

$$w_e(z) = E(z)/\left(\pi R_{eg}^2(z)\right).$$

As can be seen from the plots in Figure 4.16, which show the spatial evolution of the effective size for laser beams with different profiles, the behavior of the integral beam radius R_{eg} (from Equation 4.82) is closest to the behavior of the instantaneous effective radius R_e (from Equation 4.74) calculated at the time $\tau = 0$, that is, at the center of the temporal profile of the pulse. The evolution of the instantaneous effective radius in other temporal cross sections of the pulse demonstrates the qualitatively different behavior due to the lower power level in cross sections and influence of plasma (for details, see Reference 64).

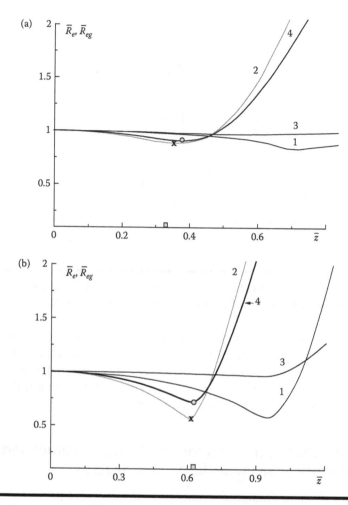

Figure 4.16 Variation of instantaneous (1–3) and integral (4) effective radii of GB (a) and SGB (b) along the propagation path at $\eta = 10$. The values of the instantaneous radius are calculated for time instants $\tau = -1$ (1), 0 (2), 1 (3). The dot shows the position of the global nonlinear focus of beam $\bar{z}_g$, the cross is for the nonlinear focus $\bar{z}_{l0}$, and the rectangle is for the filament start position $\bar{z}_{fl}$ of the central temporal cross section of the pulse.

The global focus of the entire pulse can be determined according to Equation 4.82 through the averaging of coordinates of local foci $\bar{z}_l$ over the radiation power distribution in temporal layers of the pulse:

$$\bar{z}_g = \langle \bar{z}_l(\tau) \rangle_P \approx \left[\int_{-T}^{T} P(0,\tau)d\tau \right]^{-1} \int_{-T}^{T} P(0,\tau)\bar{z}_l(\tau)d\tau, \qquad (4.85)$$

where T is the integration limit chosen in a special way. The sign of approximate equality in Equation 4.85 is caused by the choice (at averaging) of just the initial pulse power profile $P(0, \tau)$ without mutual influence of radiation in consequent temporal cross sections, which accumulate along the path.

The point of nonlinear focus $\bar{z}_l$ of the temporal layer (in the sense of instantaneous effective radius) is comparable with the start of filamentation in the layer at the distance $\bar{z}_{fl}$. Using the results of stationary theory, the coordinate $\bar{z}_{fl}$ can be determined approximately by the semi-empirical Marburger equation [68] modified by us to take into account a particular spatial profile of the beam

$$\bar{z}_{fl}(\tau) = \frac{0.367(L_{Dg}/L_{D^*})}{\sqrt{\left(\sqrt{G\eta(\tau)} - 0.852\right)^2 - 0.0219}}. \tag{4.86}$$

In this case, $\bar{z}_l(\tau) \approx \left(1/\bar{z}_{fl}(\tau) + L_{Dg}/(L_D\bar{F})\right)^{-1}$, and for unification the initial focal distance of the beam is normalized to the generalized diffraction length $\bar{F} = F/L_D$. As can be seen from Figure 4.16, the start of filamentation for the temporal layer of the pulse (shown by the rectangle in the plots) corresponds quite accurately to the position of focal waist of the effective radius. Now, the choice of the integration limit for T in Equation 4.85 becomes clear, that is, the following condition should be fulfilled:

$$\eta(\tau = \pm T) \geq G^{-1}\left(\sqrt{0.0219} + 0.852\right)^2 \approx G^{-1}.$$

At the Gaussian *temporal* profile of the pulse, we have $T \leq \sqrt{\ln[G\eta(\tau = 0)]}$.

The effective size of the global focal waist of the beam $R_{eg}(z_g)$ can be found by Equation 4.84 upon the obvious substitution of $\bar{z}_g$ for $\bar{z}$:

$$R_{eg}(z_g) = (1 - \eta^*)(\bar{z}_g)^2 + (1 - \bar{z}_g/\bar{F})^2,$$

and the normalized effective focal pulse energy density can be determined as

$$\bar{w}_{eg} = w_e(\bar{z}_g)/w_e(0) = 1/\pi R_{eg}^2.$$

References

1. Zuev, V. E., *Propagation of Visible and Infrared Waves in the Atmosphere*. Moscow: Sov. Radio, 1970, 496 p.
2. Zuev, V. E., Zemlyanov, A. A., Kopytin, Yu. D., and Kuzikovskii, A. V., *High-Power Laser Radiation in Atmospheric Aerosols*. Dordrecht: D. Reidel, 1984, 291 p.

3. Zuev, V. E., Zemlyanov, A. A., and Kopytin, Yu. D., *Nonlinear Optics of the Atmosphere, Current Problems of Atmospheric Optics*, vol. 6. Leningrad: Gidrometeoizdat, 1989, 256 p.

4. Geints, Yu. E., Zemlyanov, A. A., Zuev, V. E., Kabanov, A. M., and Pogodaev, V. A., *Nonlinear Optics of Atmospheric Aerosol*. Novosiborsk: SB RAS Publishing House, 1999, 260 p.

5. Akhmanov, S. A., Vysloukh, V. A., and Chirkin, A. S., *Optics of Femtosecond Pulses*. Moscow: Nauka, 1988, 312 p.

6. Belyaev, E. B., Godlevskii, A. P., and Kopytin, Yu. D., Laser spectrochemical analysis of aerosols, *Kvant. Elektron.*, vol. 5, No. 12, pp. 2594–2601, 1978.

7. Zuev, V. E. Ed. *Sensing of Physical-Chemical Parameters of the Atmosphere with the Use of High-Power Lasers: Collection of Papers*. Tomsk: IOA SB AS, 1979, 220 p.

8. Belyaev, E. B., Kopytin, Yu. D., Godlevskii, A. P., Krasnenko, N. P., Muravskii, V. P., and Shamanaeva, L. G., On character of generation of acoustic radiation at laser break-down of gas-disperse media, *Pis'ma v ZhTF*, vol. 8, No. 6, pp. 333–337, 1982.

9. Godlevskii, A. P., Kopytin, Yu. D., Korol'kov, V. A., and Ivanov, Yu. V., Spectrochemical lidar for analysis of the elemental composition of atmospheric aerosol, *Zh. Prikl. Spektroskopii*, vol. 39, No. 5, pp. 734–740, 1983.

10. Bochkarev, N. N., Zemlyanov, A. A., Zemlyanov, Al. A., Kabanov, A. M., Kartashov, D. V., Kirsanov, A. V., Matvienko, G. G., and Stepanov, A. N., Experimental investigation into interaction between femtosecond laser pulses and aerosol, *Atmos. Ocean. Optics*, vol. 17, No. 12, pp. 971–975, 2004.

11. Bochkarev, N. N., Zemlyanov, A. A., Zemlyanov, Al. A., Kabanov, A. M., Kartashov, D. V., Kibitkin, P. P., Matvienko, G. G., and Stepanov, A. N., Fluorescence of a drop with dye excited by femtosecond laser pulses, *Russian Physics Journal*, vol. 48, No. 4, pp. 344–348, 2005.

12. Couairon, A., and Myzyrowicz, A., Femtosecond filamentation in transparent media, *Phys. Reports*, vol. 441, No. 2–4, pp. 47–189, 2007.

13. Schweiger, G., Raman scattering on single aerosol particles and on flowing aerosols: A review, *J. Aerosol Sci.*, vol. 21, No. 4, pp. 483–509, 1990.

14. Ledneva, G. P., Nonstationary generation in a spherical microparticle, *Optics Spectrosc.*, vol. 76, No. 3, pp. 506–509, 1994.

15. Lin, H.-B., Eversole, J. D., and Campillo, A. J., Continuous-wave stimulated Raman scattering in microdroplets, *Opt. Lett.*, vol. 17, pp. 828–830, 1992.

16. Vehring, R., and Schweiger, G., Threshold of stimulated Raman scattering in microdroplets, *J. Aerosol Sci.*, vol. 26, No. 1, pp. 235–236, 1995.

17. Datsyuk, V. V., and Izmailov, I. A., Optics of microdroplets, *Physics-Uspekhi*, vol. 44, No. 10, pp. 1061–1073, 2001.

18. Belokopytov, G. V., and Pushechkin, N. P., Threshold of resonance parametric striction excitation in droplets by optical pumping, *Radiophys. Quantum Electron.*, vol. 35, No. 6–7, pp. 498–510, 1992.

19. Serpengüzel, A., Chen, G., Chang, R. K., and Hsieh, W.-F., Heuristic model for the growth and coupling of nonlinear processes in droplets, *J. Opt. Soc. Amer. B.*, vol. 9, No. 6, pp. 871–883, 1992.

20. Hill, S. C., Leach, D. H., and Chang, R. K., Third-order sum-frequency generation in droplets: Model with numerical results for third-harmonic generation, *J. Opt. Soc. Amer. B.*, vol. 10, No. 1, pp. 16–33, 1993.

21. Bohren, C. F., and Huffman, D. R., *Absorption and Scattering of Light by Small Particles*. New York: Wiley, 1998, 544 p.
22. Geints, Yu. E., Effect of surface deformations of spherical microparticles on Q-factor of their resonance modes: Geometric optics approach, *Atmos. Ocean. Optics*, vol. 15, No. 7, pp. 579–584, 2002.
23. Sukhorukov, A. P., *Nonlinear Wave Interactions in Optics and Radiophysics*. Moscow: Nauka, 1988, 232 p.
24. Khanin, Ya. I., *Dynamics of Quantum Generators, Quantum Radiophysics*, vol. 2. Moscow: Sov. Radio, 1975, 496 p.
25. Biswas, A., Latifi, H., Armstrong, R. L., and Pinnik, R. G., Double-resonance stimulated Raman scattering from optically levitated glycerol droplets, *Phys. Rev.*, vol. 40, No. 12, pp. 7413–7416, 1989.
26. Geints, Yu. E., Zemlyanov, A. A., and Chistyakova, E. K., Threshold of stimulated Raman scattering (SRS) in transparent drops, *Atmos. Ocean. Optics*, vol. 8, No. 10, pp. 1480–1487, 1995.
27. Zemlyanov, A. A., and Geints, Yu. E., Stimulated Raman scattering in spherical particles, *Atmos. Ocean. Optics*, vol. 10, No. 4–5, pp. 313–321, 1997.
28. Zemlyanov, A. A., and Geints, Yu. E., Resonance excitation of light field in weakly absorbing spherical particles by a femtosecond laser pulse. Peculiarities of nonlinear optical interactions, *Atmos. Ocean. Optics*, vol. 14, No. 5, pp. 316–325, 2001.
29. Qian, S.-X., Snow, J. B., and Chang, R. K., Coherent Raman mixing and coherent anti-Stokes Raman scattering from individual micrometer-size droplets, *Optical Lett.*, Vol. 10, No. 10, pp. 499–501, 1985.
30. Pinnick, R. G., Biswas, A., Pendleton, J., and Armstrong, R. L., Aerosol-induced laser breakdown thresholds: Effect of resonant particles, *Appl. Opt.*, vol. 31, No. 3, pp. 311–317, 1992.
31. Basov, N. G., Danilychev, V. A., Rudoi, I. G., and Soroka, A. M., Kinetic self-focusing of CO_2 laser radiation in air, *Dokl. Akad. Nauk SSSR*, vol. 284, No. 6, pp. 1346–1349, 1985.
32. Strohbehn, J. W. Ed., *Laser Beam Propagation in the Atmosphere*. Berlin: Springer, 1978.
33. Vorob'ev, V. V., *Thermal Self-Action of Laser Radiation in the Atmosphere. Theory and Model Experiment*. Moscow: Nauka, 1987, 200 p.
34. Belyaev, E. B. et al., *Nonlinear Optical Effects in the Atmosphere*. Tomsk: IOA SB RAS, 1987, 224 p.
35. Banakh, V. A., Karasev, V. V., Konyaev, Yu. N., Sazanovich, V. M., and Tsvyk, R. Sh., Defocusing of a laser beam under conditions of thermal blooming, *Atmos. Ocean. Optics*, vol. 6, No. 12, pp. 881–883, 1993.
36. Konyaev, P. A., and Lukin, V. P., Thermal distortions of focused laser beams in the atmosphere, *Izv. Vysh. Uchebn. Zaved. SSSR, Fizika*, vol. 26, No. 2, pp. 79–89, 1983.
37. Zemlyanov, A. A., and Sinev, S. N., Multimode partially coherent laser beam thermal blooming in regular medium, *Atmos. Ocean. Optics*, vol. 1, No. 8, pp. 44–50, 1988.
38. Kravtsov, Yu. A., and Orlov, Yu. I., *Geometric Optics of Inhomogeneous Media*. Moscow: Nauka, 1980, 304 p.
39. Zemlyanov, A. A., and Sinev, S. N., *Self-Action of Partially Coherent Beam at Large Nonlinear Parameters*. Tomsk: IOA SB RAS, 1984, 26 p.
40. Netesov, V. V., Influence of the kinetics of molecular absorption of radiation on the propagation of a pulse with $\lambda = 10.6$ μm in the atmosphere, *Zh. Prikl. Mehk. and Tekhn. Fiz.*, No. 4, pp. 3–8, 1986.

41. Zemlyanov, A. A., and Sinev, S. N., Propagation of partially coherent radiation pulses along vertical atmospheric path, in *Nonlinear Optics and Opto-Acoustics of the Atmosphere*, Zuev, V. E. Ed. Tomsk: IOA SB RAS, 1988. pp. 13–21.

42. Kolosov, V. V., and Kuznetsov, M. F., Stationary thermal defocusing of partially coherent beams, *Radiophys. Quantum Electron.*, vol. 30, No. 9, pp. 1099–1105, 1987.

43. Zemlyanov, A. A., and Martynko, A. V., Divergence of a laser beam in a regular non-linearly refractive medium, *Atmos. Ocean. Optics*, vol. 4, No. 11, pp. 834–838, 1991.

44. Banakh, V. A., and Smalikho, I. N., Laser beam propagation along extended vertical and slant paths in the turbulent atmosphere, *Atmos. Ocean. Optics*, vol. 6, No. 4, pp. 377–385, 1993.

45. Shalyaev, M. F., and Sadovnikov, V. P., Stimulated Raman scattering of a focused pulsed laser beam in the atmosphere, *Atmos. Ocean. Optics*, vol. 4, No. 11, pp. 781–783, 1991.

46. Lukin, V. P., *Atmospheric Adaptive Optics*. Novosibirsk: Nauka, 1986, 248 p.

47. Henesian, M. A., Swift, C. D., and Murray, J. R., Stimulated rotational Raman scattering in nitrogen in long air paths, *Optical Lett.*, vol. 10, No. 11, pp. 565–577, 1985.

48. Ignat'ev, A. B., and Morozov, V. V., Influence of stimulated Raman scattering on the high-power laser beam propagation through the atmosphere, *Atmos. Ocean. Optics*, vol. 14, No. 5, pp. 413–417, 2001.

49. Martin, W. E., and Winfield, R. J., Nonlinear effects on pulsed laser propagation in the atmosphere, *Appl. Optics*, vol. 27, No. 3, pp. 567–573, 1988.

50. Kolomiets, Yu. N., Lebedev, S. S., and Semenov, L. P., The effect of stimulated Raman scattering on the propagation of laser beams having different profiles in the atmosphere, *Atmos. Ocean. Optics*, vol. 5, No. 1, pp. 38–41, 1992.

51. Konstantinov, K. K., Starodumov, A. N., and Shlenov, S. A., Influence of rotational SRS on the angular spectrum of laser radiation in the atmosphere, *Atmos. Ocean. Optics*, vol. 2, No. 12, pp. 1291–1297, 1989.

52. Sevruk, B. B., Numerical simulation of SRS of cylindrical wave beams, *J. Appl. Spectrosc.*, vol. 70, No. 4, pp. 513–521, 2003.

53. Betin, A. A., Pasmanik, G. A., and Piskunova, L. V., Stimulated Raman scattering of light beams under saturation conditions, *Quantum Electron.*, vol. 2, No. 11, pp. 2403–2411, 1975.

54. Arbatskaya, A. N., Investigation of the angular distribution of stimulated Raman scattering of light, *Trudy FIAN*, vol. 99, pp. 3–48, 1977.

55. Nathanson, B., and Rokni, M. J., The effect of Stokes-antiStokes coupling on the gain of resonant stimulated Raman scattering, *Phys. D.*, vol. 24, pp. 233–236, 1991.

56. Losev, L. L., and Lutsenko, A. P., Parametric Raman laser with a discrete output spectrum equal in width to the pump frequency, *Quantum Electron.*, vol. 20, No. 11, pp. 1054–1062, 1993.

57. Sogomonian, S., Grigorian, G., and Grigorian, K., Parametric suppression of Raman gain in coherent Raman probe scattering, *Optical Commun.*, vol. 152, pp. 351–354, 1998.

58. Matveev, V. S., Approximate representations of absorption coefficient and equivalent widths of lines with Voigt profile, *J. Appl. Spectrosc.*, vol. 16, No. 2, pp. 228–233, 1972.

59. Vorontsov, M. A., and Shmalgauzen, V. I., *Principles of Adaptive Optics*. Moscow: Nauka, 1985, 336 p.

60. Shen, Y. R., *The Principles of Nonlinear Optics*. New York: Wiley, New York, 1984, 560 p.

61. Vlasov, V. N., Petishchev, V. A., and Talanov, V. I., The averaged description of wave beams in linear and nonlinear media (the method of moments), *Radiophys. Quantum Electron.*, vol. 14, No. 9, pp. 1353–1363, 1971.
62. Kandidov, V. P., Kosareva, O. G., Mozhaev, E. I., and Tamarov, M. P., Femtosecond nonlinear optics of the atmosphere, *Atmos. Ocean. Optics*, vol. 13, No. 5, pp. 394–401, 2000.
63. Lugovoi, V. N., and Prokhorov, A. M., A possible explanation of the small scale self-focusing filaments, *JETP Lett.*, vol. 7, No. 5, pp. 117–119, 1968.
64. Geints, Yu. E., and Zemlyanov, A. A., Conditions of nonstationary self-action of tightly focused high-power femtosecond laser pulse in air, *Atmos. Ocean. Optics*, vol. 21, No. 9, pp. 688–696, 2008.
65. Perelomov, A. M., Popov, V. S., and Terent'ev, M. V., Ionization of atoms in an alternating electric field, *Zh. Eksp. Teor. Fiz.*, vol. 50, pp. 1393–1397, 1966.
66. Zemlyanov, A. A., and Geints, Yu. E., Evolution of effective characteristics of laser beam of femtosecond duration upon self-action in a gas medium, *Optics Spectrosc.*, vol. 104, No. 5, pp. 852–864, 2008.
67. Zemlyanov, A. A., and Geints, Yu. E., Integral parameters of high-power femtosecond laser radiation at filamentation in air, *Atmos. Ocean. Optics*, vol. 18, No. 7, pp. 514–519, 2005.
68. Marburger, J. H., Self-focusing: Theory, *Prog. Quant. Electr.*, vol. 4, pp. 35–110, 1975.

Chapter 5

Peculiarities of Propagation of Ultrashort Laser Pulses and Their Use in Atmospheric Sensing

Gennady G. Matvienko and Alexander Y. Sukhanov

Contents

5.1 Lidar Control in Problems of Atmospheric Sensing by Ultrashort Pulses

Permanent monitoring of the state of the atmosphere, as well as technogenic pollutions, dangerous types of bioaerosol, which are indicators of chemical and biological contamination, as was mentioned in Chapter 1, is one of the first-priority problems in atmospheric optics because information of this kind is necessary for the prediction and improvement of the environmental situation both in particular regions and on the planetary scale. The existing global tendencies imply continuous development of remote sensing methods as the most efficient technology for monitoring of the atmospheric state with the use of ground-based, airborne, and spaceborne stations.

Methods of lidar sensing occupy a particular place in remote sensing. The understanding of the fact that the radiation with wavelengths different from the wavelength of laser radiation carries information about the composition of matter in the target region was very important for application of lasers in remote sensing. Such specific features of lasers, as high power, monochromaticity, short pulse duration, and high directivity of an optical beam, have been tested, in the first turn, in atmospheric sensing. Results of laser sensing of the atmosphere have shown that lidar systems can be used to detect and measure parameters of atmospheric constituents of both natural and anthropogenic origin [1].

At the end of the twentieth century, large-scale investigations of high-power femtosecond laser pulses with aerosol-gas constituents of the atmosphere were started to provide for the long-distance propagation of optical radiation and to develop new methods for sensing of gas and aerosol characteristics [2–5]. Most interesting phenomena arise at the propagation of radiation with femtosecond pulse duration and power exceeding some critical level. In this case, the spectral, temporal, and spatial characteristics of laser radiation change significantly, accompanying the formation of light and plasma channels—filaments, in which the intensity of laser radiation is concentrated up to the level of optical breakdown in air ($\sim 10^{14}$ W/cm^2). The phenomenon of filamentation is accompanied by generation of supercontinuum radiation, that is, broadband radiation extending over the range from 0.3 to 4 μm in the wavelength scale. A pulse of this "white" radiation is considered as a promising source for laser sensing and environmental monitoring.

Physical methods for detection, identification, and quantitative estimation of various gas and aerosol constituents of the atmosphere and atmospheric parameters with the use of existing optical effects, in particular, laser-induced fluorescence, spontaneous and stimulated Raman scattering, Doppler shift, spectral absorption, anisotropic scattering at atmospheric inhomogeneities, and effect of appearance of conical emission of broadband radiation at the propagation of femtosecond pulses are continuously improved.

The effect of an extremely high-power femtosecond optical pulse on aerosol particles was studied under laboratory conditions. The scattering matrix and the scattering phase function of an aerosol particle in an extremely high-power optical field were studied experimentally. Field studies of physical characteristics and the structure of high-power femtosecond laser radiation at open-air paths till several kilometers long were carried out [6].

The Teramobile international research group [5] was the first to conduct the atmospheric sensing by femtosecond pulses up to heights of 10–15 km. The new field—femtosecond atmospheric optics—has been actively developed. Within the framework of this research field, physical principles of remote monitoring of microphysical and optical characteristics of atmospheric aerosol by femtosecond pulses with the use of intra-atmospheric broadband source of radiation (supercontinuum) are actively developed.

During the recent decades, this field was also intensively developed in Russia, which has a wide experience in atmospheric sensing and developed methods for solution of inverse atmospheric-optical problems, in particular, in multifrequency sensing [7–9]. Methods for control of the filament structure of radiation with the aid of spatial and temporal focusing of a laser beam, as well as with the aid of systems generating phase-modulated pulses, were tested [10]. In References 11–13, one can find a review of the state of the art in investigation of the phenomena of filamentation of high-power femtosecond laser radiation in transparent media. In addition, these works study the mechanisms of supercontinuum generation and spatial distribution of the sources of supercontinuum radiation during the propagation of a high-power femtosecond laser pulse in liquid and gas.

In Reference 14, a model of coherent scattering of high-power laser radiation at an ensemble of water aerosol particles oriented at problems of femtosecond nonlinear optics was considered. It is shown that the appearance of maxima in the intensity distribution of a laser beam at coherent scattering of radiation at aerosol particles can lead to generation of a random set of filaments.

In References 15–18, the additional effect of aerosol particles on an increase of the local field in the incident radiation was considered, thresholds of optical breakdown of a transparent particle were determined, and effects of the interaction of laser pulses with liquid-droplet aerosol were demonstrated.

In the series of works (see References 19–23), the problems of propagation of a high-power laser pulse in the atmosphere were considered with allowance for the multiple scattering, and the efficiency of using the white-light lidars for sensing of the molecular atmosphere and microphysical parameters of cloudiness was assessed.

The use of the effect of filamentation with formation of an intra-atmospheric broadband source of pulsed sensing radiation (supercontinuum) turns out to be the most attractive for the solution of problems of remote sensing. Approaches for the use of this kind of radiation in the laser sensing of the atmosphere are developed to estimate the characteristics of molecular and aerosol components. Problems of atmospheric sensing by femtosecond pulses require revision of the approach for the analysis of echo signals: it is necessary to invoke practically all known sensing methods both active and passive and, correspondingly, formulate new sensing equations.

The form of the lidar equation depends on the type of laser radiation interaction with the atmospheric medium. The used form of the lidar equation is that applicable to measurements with high spatial resolution [1,6].

As is known, if the sensing is carried out by pulses in an ordinary mode with the intensity level lower than the critical one and the pulse duration exceeding the level at which it is necessary to take into account the dependence of optical characteristics of the atmosphere on the pulse duration, then it is possible to solve the lidar equation in the known forms, namely, scattering, differential absorption, and fluorescence.

As an ultrashort radiation pulse having the high energy density propagates in the atmosphere, effects of nonlinear optics manifest themselves. The nonlinearity is most pronounced at the laser wavelength and at wavelengths of neighboring spectral ranges. Spatially, the nonlinearity is pronounced at the initial part of the path. High intensities in the filamentation zone turn on the processes of multiphoton absorption and phase self-modulation, which leads to spectral broadening, that is, supercontinuum generation. A part of the pulse energy is distributed over a wide, consisting of several spectral octaves, range, but to the edges of the distribution the energy decreases sharply, by several orders of magnitude.

The long-wavelength (i.e., Stokes) part of the spectrum is localized near the filament axis and takes part in the process of self-focusing, while the short-wavelength
(i.e., anti-Stokes) component (generation of conical emission) is partially localized
in the axial zone and partially experiences divergence with the linear wavelength
dependence of the angle. The spectral components are coherent with each other. In
the short-wavelength part, a ring structure is observed and the interference interaction of spectral components of different filaments is pronounced. The pulse continues to propagate in the direction of sensing. With allowance for the broadening
and energy loss with time, the supercontinuum generation continues until the pulse
intensity becomes lower than some critical value.

The spectral composition of the radiation scattered toward an observer is determined by the spectrum of the initial pulse and the composition of the propagation
medium. The possibility of analyzing the absorption of broadband supercontinuum
radiation develops the idea of a multicomponent laser. The broadband feature of
the radiation of an intra-atmospheric source assumes the possibility of applying the
known and specifically developed laser sensing methods to the determination of
qualitative and quantitative composition of the atmosphere.

The spectrum recorded by a femtosecond lidar can be conditionally divided
into the short-wavelength part (laser spectrum) and the long-wavelength part. The
spectrum of the backscattered lidar signal in the laser wavelength range contains
the scattered radiation of the femtosecond pulse and an addition due to supercontinuum. The short-wavelength and long-wavelength parts contain the supercontinuum spectrum. The lidar equation in its traditional form should be complemented
with an additional term responsible for radiation of the intra-atmospheric source.

From the viewpoint of experimental geometry, the path of sensing by extremely
short pulses can be conditionally divided into parts by the "linear-nonlinear" attribute. These parts are localized in length: the field of view of the recording system
can include the section from a radiator to the filamentation zone, filamentation
and generation of conical emission, the section illuminated by supercontinuum
radiation, and the linear sensing pulse with possible partial overlapping. At these
sections, the probability of illumination by the supercontinuum radiation varies
from 0 to 1. At the initial section, the probability that the lidar system receives
the supercontinuum radiation depends on the geometric form-factor and the ratio
of the pulse intensity and the critical intensity in the conditional filamentation
channel. This probability should be close to unity at all points of the path located
farther than the point of supercontinuum generation. In the presence of several
filamentation channels, the energy fractions distributed over the sensing pulse
spectrum are summed. It is important to point out that the component at the
laser wavelength is presented in the backscattered signal from all parts of the path:
before the filamentation zone, and in the filamentation zone—in the nonlinear
sensing mode, from farther parts—the rather powerful sensing pulse in the linear
mode. It is necessary to take into account that the transition to the linear mode
is caused not only by the energy loss for filamentation, but also by the temporal

broadening of the initial pulse. After the boundary of the zone of conical emission generation, the supercontinuum radiation is presented in the backscattered signal with the unit probability as an additive component to the attenuated radiation at the laser wavelength.

The experience of the Teramobile international research group shows that the most significant results are obtained at the combination of the powerful femtosecond system and the powerful astronomical telescope. For example, in Reference 24, the telescope with 2-m mirror was used as a receiving system in the bistatic lidar scheme with a separation of 30 m between the transmitter and receiver.

For the linear case, the laser radiation power scattered toward an observer (subscript *L*) recorded as the pulse with duration τ_L and initial energy E_0 passes the distance *R* can be presented in the traditional form:

$$P(\lambda, R) = P_L \frac{A_0}{R^2} \xi(\lambda) \beta(\lambda_L, \lambda, R) \xi(R) \frac{c\tau_L}{2} e^{-\int_0^R k(R)\,\mathrm{d}R} \qquad (5.1)$$

where

$k(R)$ is the vertical profile of the sum of extinction coefficients in the atmosphere for the laser and detectable wavelengths, the two-side (two-pass) extinction coefficient, $k(R) = k(\lambda_L, R) + k(\lambda, R)$;

τ_L is the laser pulse duration;

$\xi(R)$ is the geometrically justified probability that the radiation from an object located at the distance *R* reaches the detector (geometric form-factor);

$\xi(\lambda)$ is the spectral transmission coefficient of the optical system;

A_0 is the effective area of the optical receiving system (account for the form-factor);

$\beta(\lambda_L, \lambda, R)$ is the volume backscattering coefficient;

$P_L = E_L/\tau_L$ is the power of laser radiation.

For the case of elastic (Mie or Rayleigh) scattering, the observation wavelength coincides with the laser wavelength. Just the linear version of the lidar equation is used in Reference 25 with the only difference that the equation is written for the photon number, and the supercontinuum formation point is introduced to the integration limits in determination of the extinction coefficient. This form is used in analysis of the results of sensing of path points lying above the generation point by the supercontinuum radiation with ignoring of nonlinear effects. The detection time achieves the values from 1.5 to 15 μs. The overlapping of the field of view of the receiving telescope (i.e., a mirror of 40 cm, frequency –*f*/3, bistatic lidar scheme, focusing to the end of a waveguide 1 mm in diameter) and the divergence angle of the supercontinuum radiation was provided by the geometry of the experiment, and thus the probability of overlapping of these angles was believed to be equal to unity.

The way to take into account nonlinear effects in the lidar equation that was proposed in Reference 26, in which the emphasis was at the transformation of the backscattering component for consideration of nonlinear effects in determination of aerosol particle size spectra. Peculiarities of propagation, such as self-focusing, were not considered. Consequently, the extinction component remained in its original form, which is unreal from the physical point of view because of the high power of radiation.

The lidar equation for sensing of supercontinuum radiation (SCR) was considered in Reference 27. Particular attention was paid to consideration of the influence of the function of overlapping of the solid angles of field of view of the recording and the receiving instrumentation on the decrease of signal power under conditions of SCR self-focusing and self-channeling in the atmosphere. High intensities achieved in filaments also lead to nonlinear extinction of pulses. With allowance made for these conditions, the modified equation was presented, which takes into account the influence of the form of the sensing beam and the extinction due to multiphoton ionization on nonlinear extinction and backscattering components. The assumption of the low influence of pulse energy loss on supercontinuum generation is used, and the strong narrowing of the spectrum on the both sides of the main wavelength is taken into account.

With allowance made for the spatial characteristics of the spectrally distributed signal, we come to the conclusion that the geometric form-factor of lidar (probability of overlapping of the field of view angle of the receiving system and the divergence angle of the sensing pulse) is significantly different for the short-wavelength and long-wavelength spectral ranges: the initial divergence of the collimated laser pulse can be equal to fractions of milliradian, while divergence in the short-wavelength supercontinuum spectral range is 0.1–0.2°. Consequently, it is necessary to take into account the dependence of the geometric form-factor on the distance and on the wavelength. In particular, this dependence can be minimized through introduction of at least two channels for receiving of the backscattered sensing signal to the lidar scheme. Each channel should have the possibility of recording of the spectrally distributed signal and the corresponding form-factors. The requirement of temporal resolution appears to be specific. As a rule, photomultiplier tubes with the temporal resolution (for detection) from fractions of microseconds to few microseconds are used as measurement converters. With allowance for the observation experience (Teramobile), the supercontinuum channel operates in the accumulation mode with times up to several tens and even hundreds of seconds. It is natural that the signal in this case is received from a chosen part of the path, which is implemented either through time gating in the receiver or in the bistatic scheme through fixation of the angle between optical axes of the receiver and the radiator. In any case, the problem of influence of multiple scattering remains nonsolved. With allowance for temporal restrictions at the reception of optical systems by measurement converters employing the photoeffect (fractions of nanosecond are the best samples), we come to the conclusion that modern lidars can provide for the effective reception of backscattered signals in the accumulation (time averaging) mode from path sections located farther than the filamentation zone.

Thus, it is possible to adjust the lidar so that the signals from the path part with predominance of nonlinear effects (intensities much higher or equal to the critical value) fall within the dead zone. The proper choice restricts the spectral range of reception and the form-factor value. Then, backscattered signals from the point of supercontinuum generation R_{SC} to the control point $R > R_{SC}$ and in the backward direction—from the entire path to the receiver of radiation at the point $R = 0$ will be assigned to the results of sensing. In general, the lidar equation can be reduced to the linear one, but the scattering signals will be influenced by nonlinear effects at the initial parts of the path. The paradoxicality of the situation is that the generation of supercontinuum radiation has the nonlinear character, while backscattered signals can be considered as linear signals with allowance for the spatial-temporal scales of a filament. In turn, at organization of the channel of parallel sensing by long pulses, it is possible to compare results and to draw conclusions about, for example, the relation of the extinction coefficients in the linear and nonlinear modes [28].

To estimate comparative characteristics of the femtosecond lidar, analogous to that described in experiments of the Teramobile group, and the traditional nanosecond lidar [29], the amplitude of the backscattered signal was calculated for the set of wavelengths from the spectral range of supercontinuum and the wavelength of sensing by the nanosecond lidar (Figure 5.1). The calculation was made with the use of reconstructed profiles of the extinction coefficient, the scattering and backscattering coefficient, and the lidar ratio. The calculation logics and algorithms correspond to the traditional linear approach, from which follows [30–32]:

> …With an accuracy sufficient for practical needs, the process of laser beam propagation in the aerosol atmosphere can be described with only three optical characteristics available: the extinction coefficient, scattering coefficient, and scattering phase function.

For the nanosecond lidar, the following parameters are specified: pulse energy of 10 mJ, pulse duration of 10 ns, wavelength of 0.53 μm, and receiver's aperture of 0.305 m. The range of 1000 m is determined for the practically cloudless atmosphere (visibility range >30 km) for the signal-to-noise ratio equal to 10. For the conditional femtosecond lidar, the following characteristics were chosen: wavelength of 0.8 μm, pulse energy of 10 mJ, pulse duration of 100 fs, pulse repetition frequency of 10 Hz, and receiver's aperture of 1 m. Photodetector (conditional) is of mark FEU-83 PMT. The point of supercontinuum generation was taken at a distance of 0.1 km from the source, and the total coefficient of energy transfer into the energy of supercontinuum is 0.23 of the energy of femtosecond pulse at this point of the path. For the chosen supercontinuum wavelengths, the energy transfer

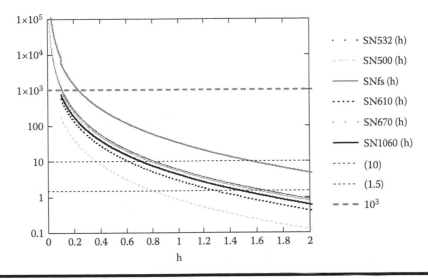

Figure 5.1 Vertical profiles of the ratio of the backscattered signal amplitude to the detector (FEU-83) noise—signal-to-noise ratio, current mode of operation. Designations: SN532—for the nanosecond lidar, wavelength of 532 nm (control version); SN500, SN610, SN670, SN1060—for supercontinuum wavelengths of 500, 610, 670, 1060 nm, respectively; *h* is distance in km. In the top part: solid line—femtosecond lidar, dots—nanosecond lidar. Horizontal dashed lines show the range of satisfactory values of the signal-to-noise ratio from 1.5 to 10 and conditional level of 1000, limiting the reception of signals from the near zone of the path.

coefficients were taken to be 0.01, 0.04, 0.08, and 0.1 for wavelengths of 0.5, 0.61, 0.67, and 1.06 μm, respectively [22].

Figure 5.1 shows vertical profiles of the signal-to-noise ratio, the results of model calculation of the amplitudes of backscattered signal from different points of 2-km path for the following conditions: meteorological visibility range of 5 km at the ground level (weather type: summer period, stable haze, visibility range of more than 4 km).

One can see from the figure that, in general, for the given initial conditions, the recording of spectral signals from supercontinuum is possible for path points from 0.8 km (for wavelength of 0.5 μm) to 1.6 km (for wavelengths of 0.67 and 1.06 μm). The range of the femtosecond lidar in the visible and near-IR range is comparable with the range of the traditional nanosecond lidar and determined by the energy transferred into the supercontinuum spectrum in the process of filamentation of the initial pulse and by the parameters of the receiving system.

In the generalized form, the lidar signal from the path points, from R_{SC} to R, recorded by the lidar receiving system in the case of illumination by the supercontinuum radiation, can be represented as a sum of two components:

$$P(R,\lambda,\Delta\lambda) = P_1(R,\lambda_L) + P_{SC}(R,\lambda,\Delta\lambda) \tag{5.2}$$

where $P_1(R, \lambda) = P_1(R)$ is the ordinary lidar equation at the wavelength of laser radiation $\lambda = \lambda_L$:

$$P_1(R) = \frac{E_1}{\tau_L} \frac{c\tau_L}{2} \frac{A}{R^2} \beta(\lambda_L, R) \exp\left[-2\int_{R_{SC}}^{R} k(\lambda_L, x)\,dx - \int_0^{R_{SC}} k(\lambda_L, x)\,dx\right] \tag{5.3}$$

where E_1 is the residual energy of the laser pulse after generation of supercontinuum, R_{SC} is the distance to the point of appearance of supercontinuum, $R > R_{SC}$, A is the effective receiver area, $A = \xi(R)A_0$, A_0 is the aperture of the receiving system, ξ is the geometric form-factor of the lidar, c is the speed of light, β is the vertical profile of the volume backscattering coefficient, k is the vertical profile of the extinction coefficient, and λ_L is laser wavelength. This equation allows ordinary procedures of inversion for the backscattering coefficient or extinction coefficient.

Keeping in mind the peculiarity of radiation generation in a wide spectral range—the generation starts at the remote (from the receiver) path point R_{SC}—the echo signal from supercontinuum radiation can be described as follows:

$$P_{SC}(R,\lambda_{SC},\Delta\lambda) = \frac{E_{SC}}{\tau_{SC}} \frac{c\tau_{SC}}{2} \frac{A}{R^2} \beta(R,\lambda_{SC}) \exp\left[-2\int_{R_{SC}}^{R} k(x,\lambda_{SC})\,dx - \int_{R_0}^{R_{SC}} k(x,\lambda_{SC})\,dx\right] \tag{5.4}$$

where

$$E_{SC} = \int_{\lambda}^{\lambda+\Delta\lambda} \varepsilon_{SC}(\lambda)\,d\lambda$$

is the energy of supercontinuum radiation pulses in the wavelength range $\Delta\lambda$, ε_{SC} is the spectral density of supercontinuum energy; $\Delta\lambda$ is the band of wavelengths detected by the receiving system of the femtosecond lidar; $k(x, \lambda) = k(x, \lambda_L) + k(x, \lambda)$ is the total extinction coefficient at the sensing wavelength and the detected wavelength.

Equation 5.4 describes the result of scattering of supercontinuum radiation in the atmosphere.

Let us analyze the possibility of inversion of Equation 5.4 for determination of the concentration of gas components. We use the approach of the well-known differential absorption and scattering (DAS) method, which is based on detection of four signals: $P_{SC}(R, \lambda_{on})$; $P_{SC}(R + \Delta R, \lambda_{on})$; $P_{SC}(R, \lambda_{off})$; $P_{SC}(R + \Delta R, \lambda_{off})$. Here, λ_{on} lies within the profile of the absorption line of the sought gas, λ_{off} lies in a wing or outside the absorption line; $\Delta\lambda$ is chosen based on the widths of absorption bands of detected gases. For simplicity, we assume that the system is well aligned, that is, $\xi(R) = 1$.

We can write the atmospheric extinction coefficient with the absorption by the analyzed gas separated in the explicit form $K(\lambda) = \sigma(\lambda) + \alpha(\lambda)$, where $\alpha(\lambda) = N \cdot \Sigma(\lambda)$, N is the concentration, and $\Sigma(\lambda)$ is the absorption cross section per one molecule. Then we can formulate the following equations:

$$\ln[P_{SC}(R,\lambda_{on})] - \ln[P_{SC}(R+\Delta R,\lambda_{on})] = \ln\left(\frac{R+\Delta R}{R}\right)^2 + \ln\left[\frac{\beta(R,\lambda_{on})}{\beta(R+\Delta R,\lambda_{on})}\right]$$
$$+2\int_R^{R+\Delta R}\sigma(x,\lambda_{on})\,dx + 2\int_R^{R+\Delta R} N\cdot\Sigma(x,\lambda_{on})\,dx$$

$$(5.5)$$

$$\ln[P_{SC}(R,\lambda_{off})] - \ln[P_{SC}(R+\Delta R,\lambda_{off})] = \ln\left(\frac{R+\Delta R}{R}\right)^2 + \ln\frac{\beta(R,\lambda_{off})}{\beta(R+\Delta R,\lambda_{off})}$$
$$+2\int_R^{R+\Delta R}\sigma(x,\lambda_{off})\,dx + 2\int_R^{R+\Delta R} N\cdot\Sigma(x,\lambda_{off})\,dx$$

$$(5.6)$$

We introduce now the average values $\bar{\sigma}$ and $\bar{N}$ in the range ΔR and solve this system of equations for $\bar{N}$:

$$\bar{N} = \frac{1}{2\Delta R[\Sigma(\lambda_{on}) - \Sigma(\lambda_{off})]}\left\{\ln\left[\frac{P_{sc}(R,\lambda_{on})\cdot P_{sc}(R+\Delta R,\lambda_{off})}{P_{sc}(R,\lambda_{off})\cdot P_{sc}(R+\Delta R,\lambda_{on})}\right]\right.$$
$$\left. + \ln\left[\frac{\beta(R,\lambda_{on})\cdot\beta(R+\Delta R,\lambda_{off})}{\beta(R,\lambda_{off})\cdot\beta(R+\Delta R,\lambda_{on})}\right] - 2\Delta R[\bar{\sigma}(\lambda_{on}) - \bar{\sigma}(\lambda_{off})]\right\} \qquad (5.7)$$

As a rule, λ_{on} and λ_{off} are close and, from the viewpoint spectral dependences of β and σ, we can believe that $\beta(\lambda_{on}) = \beta(\lambda_{off})$ and $\sigma(\lambda_{on}) = \sigma(\lambda_{off})$. Then

$$\bar{N}(R) = \frac{1}{2\Delta R \Delta \Sigma} \left\{ \ln \left[\frac{P_{sc}(R,\lambda_{on}) \cdot P_{sc}(R + \Delta R, \lambda_{off})}{P_{sc}(R,\lambda_{off}) \cdot P_{sc}(R + \Delta R, \lambda_{on})} \right] \right\} \tag{5.8}$$

We have derived the equation analogous to that for ordinary lidar employing the DAS method. An advantage of the femtosecond lidar is that the spectrally wide supercontinuum radiation overlaps absorption bands of many gases, thus allowing the multigas analysis of the atmosphere. In Equation 5.8, there is no dependence on the spectral density of supercontinuum energy. This is valid only in the case when the signals $P_{SC}(R, \lambda_{on})$ and $P_{SC}(R + \Delta R, \lambda_{on})$, as well as $P_{SC}(R, \lambda_{off})$ and $P_{SC}(R + \Delta R, \lambda_{off})$, are detected from the same generation pulse. The coefficient of energy conversion into supercontinuum radiation determines the potential of the white-light lidar.

In many problems, it is needed to determine the profile of the extinction coefficient of the atmosphere related to the variability of the atmospheric aerosol composition [33]. In this case, it is possible to use the algorithms of mutifrequency analysis [34], which provide information about the microphysical properties of particles. The supercontinuum radiation is ideal for the application of these algorithms for some classes of atmospheric aerosol. Consider the solution of lidar Equation 5.4 for the extinction coefficient with the emphasis, for example, at the Klett method [35]. In Equation 5.4, one can pass to logarithmic and square-amplified return signals $S(R, \lambda) = \ln [P_{sc}(R, \lambda) \cdot R^2]$ and, in addition, $S(R)$ is taken to be known at the limited range R_n (e.g., the value is close to the noise level). Then the difference

$$S(R,\lambda) - S(R_n,\lambda) = \ln \left[\frac{\beta(R,\lambda)}{\beta(R_n,\lambda)} \right] + 2 \int_R^{R_n} k(x,\lambda)\,dx \tag{5.9}$$

depends only on the atmosphere, and the solution can be written in the form

$$K(R,\lambda) = \frac{\exp\{-[S(R_n,\lambda) - S(R,\lambda)]/g\}}{K^{-1}(R_n,\lambda) + \dfrac{2}{g} \displaystyle\int_{R_{SC}}^{R_n} \exp\{-[S(R_n,\lambda) - S(R,\lambda)]/g\}\,dR} \tag{5.10}$$

where g is the parameter of relation between the backscattering and extinction coefficients for the elastic scattering (g depends on the lidar wavelength and specific properties of scattering centers; it varies in the range 0.67–1.0). The value of g can also be estimated from the condition of stability of solutions [1].

Thus, we can note that the inversion of the lidar equation with supercontinuum does not impose any specific formal differences on the signal processing algorithms. However, to be noted are possible limitations of the presented equations attributed to the contribution from multiple scattering, which is proportional to the field-of-view angle of the lidar receiving system. The point is that the necessity of complete interception of the supercontinuum radiation cone (with an angle degree exceeding 0.1°) requires the corresponding values of the fields of view of the receiving system, which explains the schemes of most successful atmospheric experiments with the use of astronomical telescopes having the high light-gathering power and small field-of-view angles. The consideration of the influence of multiple scattering and backscattered signals from the zone of development of nonlinear effects on signals of the femtosecond lidar is a future task.

5.2 Effect of Splitting of the Femtosecond Pulse in the Linear Transfer Mode

One of the required problems of physical optics consists in obtaining optical images of objects hidden in the optically dense disperse medium [37–49]. This problem arises, first of all, in the development of efficient methods of transmission optical tomography and optical imaging [36,49], laser sensing of the atmosphere and ocean [37], confocal microscopy [38], etc. The recent progress in the technology of femtosecond lasers [39] has given a new impulse to these investigations, allowing measurements with the ultrahigh spatial resolution close to the diffraction limit. However, in this case, one faces some methodological difficulties connected with the information content of the transmitted and backscattered laser radiation, which, in contrast to X-ray and microwave radiation, interacts actively with the disperse medium, losing the coherence and forming the multiple-scattering noise. In addition to the diffusion broadening of the lidar pulse, multiple scattering can be a source of a false signal leading to uncontrolled error of tomographic or lidar measurements. For the first time, this was noticed in early experiments by R. R. Alfano, one of the founders of femtosecond optics, with colleagues [40–42]. The experiments dealt with the use of ultrashort radiation pulses for imaging of small objects in dense scattering media. In some situations, as the duration of the initial pulse decreased, the time splitting of the transmitted signal into two barely distinguishable components was observed. These components, in opinion of the authors of Reference 40, owe their origin to the coherent interaction of ballistic and diffusely scattered photons. The observation of this effect requires the precision recording of the temporal envelope of the signal. Therefore, the number of works, in which this effect was noted, is limited. Only in the recent time, some experiments were reported [43,44], in which the splitting effect was confirmed with high accuracy. At the same time, these experiments have shown that the short duration of a pulse is necessary, but not sufficient condition for appearance of the bimodal configuration of the envelope of transmitted signal. In References 45–47, it is shown that the shape and position of the secondary peak caused by the diffuse component of the signal depends on the optical parameters of the disperse medium (first of all, the scattering coefficient) and its optical thickness. In Section 5.3, based on the numerical solution of the radiation transfer equation, we make an attempt to specify the parametric range, in which the effect of splitting of an ultrashort optical pulse takes place.

5.2.1 Theoretical Prerequisites of the Issue

Recent theoretical and experimental investigations dealing with the analysis of peculiarities of propagation of ultrashort laser pulses in disperse media [40–47] have led to a certain qualitative classification of photons forming the spatiotemporal distribution of the transmitted radiation. Some photons, virgin of collisions with medium particles and, consequently, keeping their original direction and state of polarization, fall in the category of so-called ballistic photons [40,46]. Other photons enter the statistical series of random scattering events, deflect from the initial direction, and lag behind the ballistic photons. The main part of scattered photons is represented by chaotically polarized diffuse photons. In media with the strong scattering anisotropy, the intermediate group of photons, which have experienced a few number of scattering events and kept the nearly axial direction of propagation [46], plays an important role. In the international literature, this nearly axial group of photons is referred to as *snake* photons [42,43]. The recognition of validity of this classification leads to new interpretation of the problem of nonstationary radiation transfer with allowance made for the existence of different types of photons.

As is well known, the radiation transfer in an inhomogeneous disperse medium is described by the phenomenological integro-differential equation, which takes the following form for the case of a pulsed radiator:

$$\left[v^{-1}\frac{\partial}{\partial t} + \Omega\nabla + \sigma(\mathbf{r}) \right] I(\mathbf{r},\Omega,t) = 1/4\pi \int_{4\pi} G(\mathbf{r},\Omega',\Omega)I(\mathbf{r},\Omega',t)d\Omega' + S(\mathbf{r},\Omega,t)$$

$$(5.11)$$

where $S(\mathbf{r}, \Omega, t)$ is the source function; $I(\mathbf{r}, \Omega, t)$ is the density of photon flux at the point $\mathbf{r}$ in the propagation direction Ω at the time t; $x = (\mathbf{r}, \Omega, t)$ is a point of the phase space, $x \in X$, $X = \{(\mathbf{r}, \Omega, t): \mathbf{r} \in R \subset R^3, \Omega \in W = \{(a, b, c) \in R^3: a^2 + b^2 + c^2 = 1\}, t \in T\}$ or $X = R \times W \times T$ is the seven-dimensional phase space; $G(\mathbf{r}, \Omega, \Omega') = \sigma_s(r) g(\mu, \mathbf{r}')$ is the volume coefficient of directed elastic scattering, mostly, the Mie scattering in the direction (Ω, Ω'); $g(\mu, \mathbf{r}')$ is the scattering phase function; $\mu = \cos \theta$; θ is the scattering angle; $\sigma(\mathbf{r})$ is the extinction coefficient, that is, $\sigma(\mathbf{r}) = \sigma_a(\mathbf{r}) + \sigma_s(\mathbf{r})$, where $\sigma_a(\mathbf{r})$ and $\sigma_s(\mathbf{r})$ are respectively the absorption and scattering coefficients of the disperse medium; and v is the speed of light in the medium.

The first successful attempts of analytical description of the process of formation of the bimodal structure of a short laser pulse in a strongly scattering medium (SSM) based on the approximate solution of the transfer equation were undertaken in References 45–47. The authors used the well-known Kubelka–Munk two-flux model [48] generalized for the case of nonstationary radiation transfer processes. As a result, the system of two interrelated differential equations was obtained in the 1D approximation

$$v^{-1}\frac{\partial F_+}{\partial t}+\frac{\partial F_+}{\partial x}=-m_e F_+ + m_s F_- \tag{5.12}$$

$$v^{-1}\frac{\partial F_-}{\partial t}+\frac{\partial F_-}{\partial x}=m_s F_+ - m_s F_- \tag{5.13}$$

where $F_+(t, x)$ and $F_-(t, x)$ are the radiation fluxes propagating in the direction of the x-axis and in the opposing direction; m_s and m_e are, respectively, the specific dimensionless scattering and extinction coefficients [48,49].

Using the methods of theory of generalized functions for solution of systems (5.12) and (5.13), it is possible to obtain, following References 46, 47, a rather compact equation for radiation fluxes $F(t, x)$ at the time t at the depth x in the form

$$F_+(t,x)=U_0 m_e v \delta(m_e vt - m_e x)\exp(-m_e x)+U_0 \eta(m_e vt - m_e x)$$
$$\times vm_s x J_1\left(m_s\sqrt{(vt)^2 - x^2}\right)\exp\left(-m_e vt/\sqrt{(vt)^2 - x^2}\right), \tag{5.14}$$

where U_0 is the energy of the initial pulse, $\delta(\cdot)$ is the Dirac delta, $J_1(\cdot)$ is the modified first-kind Bessel function, and $\eta(\cdot)$ is the Heaviside function.

The first term in Equation 5.14 describes the ballistic component of the radiation, while the second one is the diffuse component. Equation 5.14 gives the qualitatively adequate pattern of the temporal pulse intensity distribution of radiation passed through a scattering layer of finite thickness. Unfortunately, it does not allow, by the definition of the two-flux approximation, the rigorous consideration of microphysical and optical properties of the medium, first of all, the scattering phase function. At the same time, the factor of anisotropy of the scattering phase function plays, as the further estimates show, the decisive role in establishment of the pulse splitting mode.

In problems of transmission optical tomography, the diffusion approximation is traditionally used for solution of the radiation transfer equation. The diffuse approximation is based on the assumptions that the photon flow in SSM undergoes scattering at a large ensemble of particles and the scattering phase function is close to isotropic [48]. Under these assumptions, it is sufficient to multiply (5.11) by the unit direction vector e_r and to integrate over the solid angle Ω. After the usage of known transformations [48,49], we can obtain the nonstationary diffusion equation for the photon flow $F(t, r)$

$$v^{-1}\frac{\partial F(t,r)}{\partial t} - D\nabla^2 F(t,r) + \sigma_a(r)F(t,r) = S_d(t,r),\qquad(5.15)$$

where $D = \{3[\sigma_a(r) + (1 - g_{HG})\sigma_s(r)]\}^{-1}$ is the diffusion coefficient; g_{HG} is the average cosine of the scattering angle; $S_d(t, r)$ is the photon source. In contrast to $S(r, \Omega, t)$ in Equation 5.11, here $S_d(t, r)$ is the isotropic function.

In some canonic situations, Equation 5.15 permits the analytical solution. In particular, in Reference 50, the equation was derived for the temporal intensity distribution at the beam axis at the depth z in the case of a point unidirectional source illuminating a semi-infinite homogeneous medium

$$\begin{aligned}
T(z,t) = (4\pi Dv)^{-1/2}t^{-3/2}\exp(-\sigma_a vt)\Bigg\{&(z-z_0)\exp\left[-\frac{(z-z_0)^2}{4\pi Dv}\right]\\
-(z+z_0)\exp\left[-\frac{(z+z_0)^2}{4\pi Dv}\right] &+ (3z-z_0)\exp\left[-\frac{(3z-z_0)^2}{4\pi Dv}\right]\\
-(3z+z_0)\exp\left[-\frac{(3z-z_0)^2}{4\pi Dv}\right]&\Bigg\}
\end{aligned}\qquad(5.16)$$

where $z_0 = [(1 - g_{HG})\sigma_s]$.

If we know the anisotropy factor g_{HG}, then Equation 5.16 determines, with the high accuracy, the shape of the pulse passed through a layer of homogeneous SSM having significant optical thickness, that is, at establishment of the asymptotic deep mode. This is illustrated by the results of the comparative analysis performed in Reference 50 for the optical thickness of the scattering layer $\tau = \sigma_e vt \approx 60.0$. On the other hand, the diffusion approximation fails to describe the behavior of ballistic photons having the highest information content (by definition) and does not allow consideration of the inhomogeneous optical structure of the medium, and cannot provide solutions under complex boundary conditions characteristic of the actual geometry of optical sensing. Thus, for the precision quantitative prediction of the shape of a short optical pulse in the SSM volume, it is necessary to solve the complete radiation transfer equation of type (5.11) without simplifying physical assumptions. Under complex boundary conditions taking place in actual experiments of transmission spectroscopy and laser sensing, the only promising approach to solution of this class of problems is the Monte Carlo technique.

5.2.2 Mathematical Model of Transfer: Monte Carlo Technique

The transformation of the shape of an optical signal upon the propagation through an optically dense scattering medium is one of traditional problems demonstrating the advantages of the Monte Carlo technique. The first attempts of computer simulation of an actual experiment connected with the study of regularities of temporal blooming due to multiple scattering were undertaken in Reference 51. However, neither in this study nor in the following studies [52–55], the effect of splitting was not revealed. This is likely caused by the imperfect algorithmic technique of statistical simulation and the well choice of optical parameters, not falling, as is shown below, within the parametric range of existence of the effect. Speaking about the Monte Carlo technique, we would like to note that in recent time a tendency of using simplified, so-called analog, simulation algorithms is observed. Increasing capabilities of computer system justify this approach to a certain extent. Nevertheless, for the solution of multidimensional radiation transfer problems in complex boundary conditions corresponding to an actual experiment, it is worth using *weighted* Monte Carlo methods [43,44], allowing a significant decrease of the variance of sought estimates. The construction of weighted algorithms requires prior analysis of the integral radiation transfer equation or, more exactly, peculiarities of its kernel. It was shown repeatedly [43–46] that, in the most cases, integral-differential Equation 5.11 can be transformed into the canonical form of the second-kind Fredholm equation. In the operator form, it is usually written for the density of collisions $f(x) = \sigma(x)I(x)$ as follows:

$$f = Kf + \varphi, \tag{5.17}$$

where the integral operator K is defined as

$$[Kf](r,\Omega,t) = \int_R \int_W \int_T k[(r',\Omega',t') \rightarrow (r,\Omega,t)]f(r',\Omega')\,dr'\,d\Omega'\,dt' \tag{5.18}$$

The kernel of nonstationary integral Equation 5.17, following Reference 56, has the form

$$k[(r',\vec{\Omega}',t') \rightarrow (r,\vec{\Omega},t)] = \frac{\Lambda(\mathbf{r}')\sigma(r)g(\mu,\mathbf{r}')\exp[-\tau(\mathbf{r}',\mathbf{r})]}{2\pi\,|\mathbf{r}-\mathbf{r}'|}$$

$$\times \delta\left(\Omega - \frac{\mathbf{r}-\mathbf{r}'}{|\mathbf{r}-\mathbf{r}'|}\right)\delta\left[t' - \left(t + \frac{|\mathbf{r}'-\mathbf{r}|}{c}\right)\right], \tag{5.19}$$

which is the probability density of photon transition from the state $(\vec{r}', \vec{\Omega}', t')$ to the state $(\vec{r}, \vec{\Omega}, t)$;

$$\varphi(r, \Omega, t) = \int_R k[(r', \Omega_0, t_0) \to (r, \Omega, t)]\sigma(r')\exp[-\tau(r', r_0)]p(t_0)\,dr' \quad (5.20)$$

is the density of initial collisions (source function); $\tau(\mathbf{r}', \mathbf{r}) = \int_0^l \sigma(\mathbf{r}, l')\,dl'$ is the optical length of the section $l = |\mathbf{r}' - \mathbf{r}|$; $\Lambda(\mathbf{r}') = \sigma_s(r')/\sigma(\mathbf{r}')$ is the survival probability of a quantum. We assume that the initial point, $r_0(x_0, y_0, z_0) \in S$, $S \subset R$, is illuminated by the incident ray in the direction $\Omega_0(a_0, b_0, c_0)$ at the time t_0.

In many cases corresponding to actual optical experiments, in particular, in this problem, it is needed to estimate the spatiotemporal characteristics of a short optical signal under conditions of collimated illumination and detection. This assumes the presence of localized detectors and sources of radiation. The domain of localization is $D^* \ll R$. The probability of informative events with $h \neq 0$ is minimal. To solve the problem with these boundary conditions, it is worth constructing another Markovian chain such that the transition density $k[(r', \Omega', t') \to (r^*, \Omega^*, t^*)]$ will include the function $\delta(\Omega^* - r^* - r'/|r^* - r'|^2)$.

In the theory of the Monte Carlo technique [57], this artificial approach is called "the method of local estimation." Shortly speaking, it consists essentially in the following. Let us assume that $x^*(r^*, \Omega^*, t) \in D$, where D is the phase volume of some optical detector, $\{D \subset R, D \ll R\}$. We rewrite Equation 5.12 in the form

$$f(x) = \int_X k(x' \to x)f(x')\,dx' + \varphi(x), \quad (5.21)$$

assuming that $x = x^*$, $\varphi(x^*) = 0$, $\sigma_s(r) = \sigma(r)$ in Equation 5.21. Then Equation 5.21 can obviously be written as follows:

$$I(x^*) = \int_X \frac{k(x' \to x^*)}{\sigma(r^*, \lambda)} f(x')\,dx' \quad (5.22)$$

Thus, the spectral flux density $I_\lambda(x^*)$ is formally represented as a linear functional of the collision density. However, the kernel $k(x' \to x^*)$ contains generalized δ-functions. To remove them, we integrate consecutively over some limited ranges of the directions $\Omega_i \in W$ and the time of photon registration $T_j \in T$ ($\Omega_i \ll W$ and $T_j \ll T$). As a result, we get the statistical estimate of the radiation flux in the region of the localized detector $D_{ij} = \Omega_i T_j$ in the form

$$\tilde{I}(r^*) = \int_{D_{ij}} I(r^*, \Omega^*, t^*) d\Omega^* dt^* = \iint_{T_j \Omega_i} I(r^*, \Omega^*, t^*) d\Omega^* dt^*$$

$$= \int_X \xi_{ij}(x', x^*) f(x') dx' = M \sum_{n=0}^{N} q_n \xi_{ij}(x_n, x^*), \qquad (5.23)$$

where

$$\xi_{ij}(x_n, x^*) = \frac{\exp[-\tau(r_n, r^*)] g(\mu^*, r)}{2\pi |r_n - r^*|^2} \Delta_i(s^*) \Delta_j(t^*). \qquad (5.24)$$

Here, $\Delta_i(s^*)$ and $\Delta_j(t^*)$ are, respectively, indicators of the areas Ω_i and T_j;

$$\Delta_i(s^*) = \begin{cases} 1, s^* \in \Omega_i, \\ 0, s^* \bar{\in} \Omega_i, \end{cases}; \quad \Delta_j(s^*) = \begin{cases} 1, s^* \in T_j, \\ 0, s^* \bar{\in} T_j, \end{cases}; \quad s^* = \frac{r_n - r^*}{|r_n - r^*|}; \quad t^* = \left(t + \frac{|r^* - r|}{c}\right);$$

$$\mu^* = (\Omega_n, s^*);$$

M in Equation 5.23 is the symbol of mathematical expectation; $n = 1, 2, \ldots N$ is the number of random photon collision; q_n is the statistical weight of photon compensating the fictive character of the transitions $k(x_n \to x^*)$.

The weighted methods have demonstrated high efficiency in problems of remote sensing of the environment. Partial technical algorithms for simulation of geometric trajectory of photon packet in a homogeneous and layered-inhomogeneous medium are commonly known (see, e.g., Reference 56), and we do not dwell on them. If it is needed to study the dependence of sought functional on the set of parameters, the simple modification of the method of local estimation of the flow, given in Section 5.2.5, appears to be useful.

5.2.3 Results of Model Estimation

As a physical prototype of our numerical experiment, we have taken the scheme of experimental measurements used in recent works [43,58]. This choice is caused by the fact that these articles report the results illustrating the dynamic pattern of formation of the bimodal structure of ultrashort pulse having passed through a layer of model SSM. The results are obtained with the good temporal resolution (about 2 ps) and the record signal-to-noise ratio of 1000:1. Based on these data, the following optical-geometric conditions of numerical experiment were taken.

The pulsed laser radiation with the wavelength $\lambda = 790$ nm (wavelength of femtosecond Ti:Sa laser) is directed at the model chamber having the transverse dimension of about 5 cm and the longitudinal thickness of 3 cm. The input radius of the optical beam is 0.05 cm; the angular divergence is $\varphi_s = 1$ μrad. The initial pulse shape satisfies the Gauss distribution $I(t-t_0)=\exp[-(4\ln 2)(t-t_0)^2/\tau_p^2]$ with the modal value $t_0 \cong 35\,\text{fs}$ and the half-width $\tau_p \cong 50\,\text{fs}$ (herein, time is in femtoseconds). The minimal angle of the receiving fiber-optic system is $\varphi_d = 0.6$ m rad; the radius of the receiver entrance hole is 0.6 cm.

Optical properties of the model medium correspond to the water suspension of intralipid of variable concentration [43]. It should be noted that intralipid is widely used as a phantom of biological tissues, because their optical properties are extremely close [59]. In particular, to estimate the influence of anisotropy of the phase function $g(\mu)$ on the shape of the diffusely transmitted pulse, the Henyey–Greenstein approximation, $g(\mu)=1-g_{HG}^2/2(1+g_{HG}^2-2g_{HG}\mu)^{3/2}$, was used [36,48], which includes the anisotropy factor g_{HG} in the explicit form; for the chosen radiation wavelength $\lambda = 790$ nm in 20% intralipid suspension $g_{HG}(\lambda = 790HM) = 0.6$. Model values of the volume scattering and absorption coefficients are, respectively, $\sigma_s = 3.6$ cm^{-1} and $\sigma_a = 0.00001$ cm^{-1}. For convenience, all basic parameters are summarized in Table 5.1.

It should be noted that the values of the basic parameters correspond to the optimal conditions of observation of the bimodal configuration of the shape of ultrashort pulse for the chosen optical model of SSM. Numerical simulation consisted in the consecutive variation of one of the basic parameters at the fixed values of other parameters.

Now, let us pass to the results of model estimates. The character of amplitude envelope of the optical pulse propagated through the SSM layer depends on the contribution of photons, which underwent different number of scattering events. At the same time, the scatter in delay time takes place among scattered photons. The contribution of different numbers of scattering events to the detected signal, as well as the photon path distribution due to single scattering, depends on the shape of the scattering phase

Table 5.1 Basic Parameters Determining Optical Characteristics and Conditions of Illumination of the Model Medium

Parameter	Value
Scattering coefficient, σ_s(cm^{-1})	3.6
Absorption coefficient, σ_a(cm^{-1})	0.00001
Geometric thickness of the medium, cm	3.0
Anisotropy factor, $g(\mu)$	0.6
Pulse duration, τ_p(fs)	50.0
Beam radius, cm	0.1
Receiver radius, cm	0.6

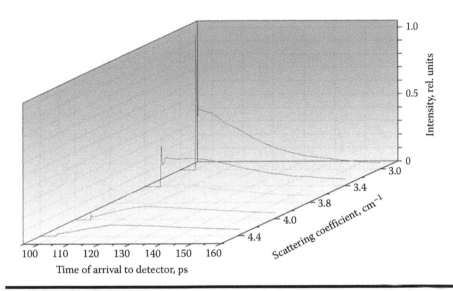

Figure 5.2   Influence of the volume scattering coefficient on the shape of femtosecond pulse transmitted through a layer of strongly scattering medium (SSM). All other basic parameters are as in Table 5.1.

function, macroscopic cross sections of scattering, and absorption of an elementary volume of the medium. Figures 5.2 and 5.3 show the influence of, respectively, the volume scattering and absorption coefficients on the shape of the femtosecond laser pulse transmitted through the SSM layer of finite geometrical thickness.

It is obvious that with increase of σ_s the ballistic component forming the first mode of the signal becomes weaker, and its energy is transferred to scattered photons. It can be seen from presented illustrations that recommendations on the increase of the contribution of ballistic photons through an increase of the scattering coefficients of SSM proposed in References 40, 41 and based, to a great extent, on incorrect assumptions on the character of decrease of a fraction of scattered photons in SSM, are not confirmed by our calculations (see Figure 5.2). An increase of the absorption coefficient (Figure 5.3) leads to a systematic decrease of the signal level and practically does not influence the two-mode configuration of the signal. The effect from an increase of the geometric thickness of the layer at constant σ_s (Figure 5.4) is equivalent, as expected, to the effect from increase of σ_s at the constant layer thickness shown in Figure 5.2. The influence of the shape of the scattering phase function on the temporal structure of the signal can be judged from the results shown in Figure 5.5. For a medium with the high anisotropy factor $g_{HG} \geq 0.7$, the role of quasi-ballistic or snake photons increases. These photons smooth out the dip between the ballistic and diffuse components, thus making the output signal unimodal. This effect explains why in many computational and experimental studies attributed to transmission spectroscopy the bimodal structure of the transmitted signal does not show itself. The pulse duration is the no less

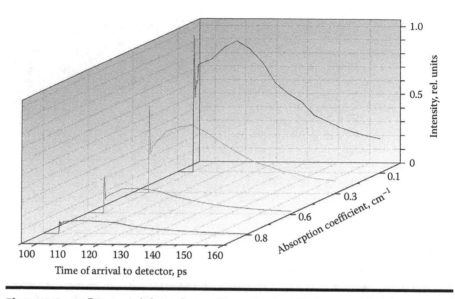

Figure 5.3 **Influence of the volume absorption coefficient on the shape of femtosecond pulse transmitted through a layer of strongly scattering medium (SSM). All other basic parameters are taken from Table 5.1.**

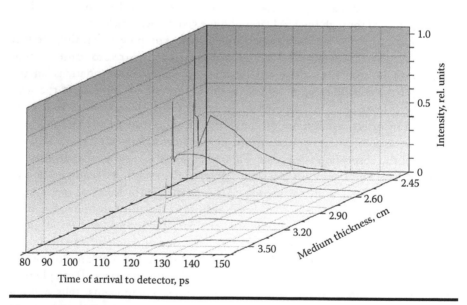

Figure 5.4 **Influence of the geometric thickness on the shape of femtosecond pulse transmitted through a layer of strongly scattering medium (SSM). All other basic parameters are as in Table 5.1.**

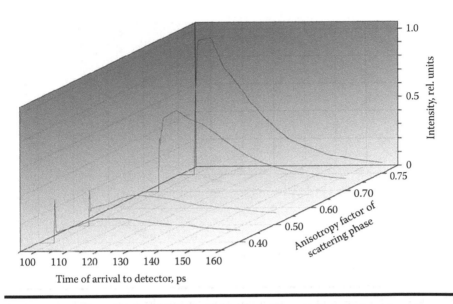

Figure 5.5 **Influence of the anisotropy factor of the scattering phase function on the shape of femtosecond pulse transmitted through a layer of strongly scattering medium (SSM). All other basic parameters are taken from Table 5.1.**

important condition for the effect of pulse splitting. Not incidentally, this effect was discovered in the studies [40–42], in which femtosecond laser pulses were practically used for the first time. It is obvious that, in the case of a wide (in duration) pulse, ballistic photons delaying from the source fill the time gap between the ballistic and diffuse components, thus removing the bimodal structure. This phenomenon is confirmed by the time sweeps of the signal shown in Figure 5.6. At pulse duration of about 3 ps, bimodality practically disappears. It should be noted that these, rather strict, estimates of the boundary values of parameters determining the range of existence of the pulse splitting effect for an ultrashort pulse are not ultimate. They may change in different media; this issue calls for further investigations. The situation is even more complicated, because not only the optical model of the medium, but also geometric conditions of medium illumination and signal detection determine the conditions for appearance of the signal splitting mode. If insignificant variations of the diameter of incident laser beam exert practically no effect (see Figure 5.7) on the time configuration of the received signal, then the boundary conditions of detection affect significantly. At small dimensions of the entrance hole of the receiver, the considerable part of the diffuse component is not detected, and the second mode of the signal disappears (see Figure 5.8). On the other hand, at the significant increase of the detection area, it becomes possible to lose the informative signal of the ballistic component against the background of unlimited increase of the multiple-scattering noise. The same effect also follows from variation of the detector field-of-view angle (Figure 5.9). The larger the

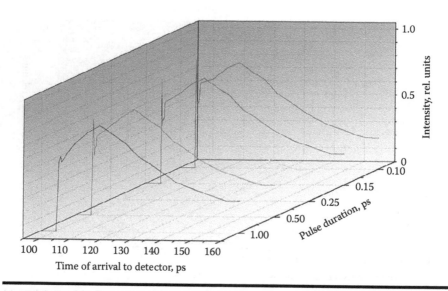

Figure 5.6 **Influence of the pulse duration on the shape of femtosecond pulse transmitted through a layer of strongly scattering medium (SSM). All other basic parameters are taken from Table 5.1.**

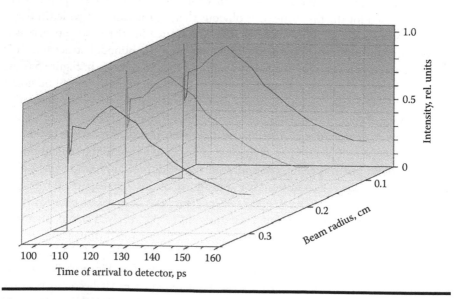

Figure 5.7 **Influence of the laser beam radius on the shape of femtosecond pulse transmitted through a layer of strongly scattering medium (SSM). All other basic parameters are as in Table 5.1.**

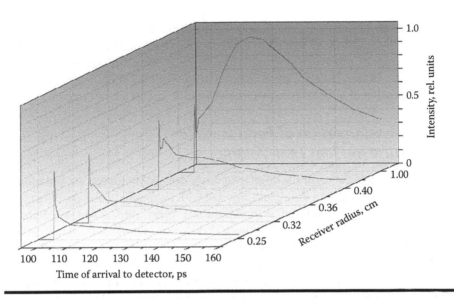

Figure 5.8 **Influence of the receiver radius on the shape of femtosecond pulse transmitted through a layer of strongly scattering medium (SSM). All other basic parameters are as in Table 5.1.**

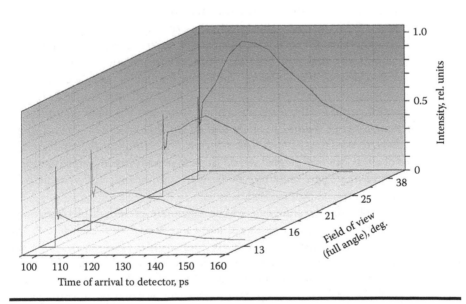

Figure 5.9 **Influence of the detector acceptance angle on the shape of femtosecond pulse transmitted through a layer of strongly scattering medium (SSM). All other basic parameters are as in Table 5.1.**

acceptance angle, the larger the fraction of the diffusion component of the signal, which finally leads to the complete distortion of signal informativeness from the viewpoint of transmission tomography. The results presented suggest that in practical problems of transmission spectroscopy and tomography the attention should be paid to selection of optimal parameters and angular resolution of detectors.

5.2.4 Transformation of Femtosecond Pulse in Plant Leaf Volume

As a practical application of the developed methodology of numerical simulation to real SSMs, we consider an example connected with the prediction of possible transformation of a femtosecond pulse in volume of a plant leaf. As was noted above, the modern methods of transmission optical tomography and confocal microscopy are widely used for analysis of the inner structure, metabolic processes, and physiological state of biological tissues, in particular, plant bodies [60–62]. The informativeness of these methods depends, to a large extent, on the a priori knowledge of possible multiple-scattering noise inevitable in SSMs, which include, in particular, a plant leaf.

Average size of a leaf of higher plants is much larger than the wavelength of optical radiation. Thus, we can consider a leaf as an extended object of disperse structure containing optically active microelements of various physical nature. In Reference 63, it was found that the optically active fraction of leaf mesophyll particles has, at least, the three-mode structure. The average radius of particles of the first mode is $r_s \approx 0.15$ μm. This value falls in the size range of chlorophyll particles (thylakoid grana in biology) observed in laboratory experiments. The second mode has a maximum in the range $r_l \approx 2.0$ μm, which is characteristic of chloroplast size in representation by equivalent spheres. The coarse fraction of organic compounds of leaf mesophylls, in the first turn, cells with the characteristic size $r_c \geq 8$–12 μm, does not contribute markedly to the absorption spectrum, but plays an important role in formation of the field of scattered radiation. Therefore, it is worth including a polydisperse structure of air bubbles with $r_b \approx 9.5$ μm into the model of the lower layer of a leaf (spongy parenchyma). Thus, we pass to the multiphase system containing three aggregate ensembles of particles significantly different in their microphysical and optical characteristics.

Regularities of radiation scattering and absorption in such multiphase medium are determined by specific features of radiation interaction with each phase. The optical characteristics of mesophyll of dicotyledonous plant leaf necessary for the solution of the radiation transfer equation were calculated in Reference 63 for the spectral range 400–800 nm. The boundary and initial conditions for the numerical solution of the problem of propagation of a femtosecond pulse through the leaf volume correspond to actual conditions of the laboratory physical experiment [62]. Reference 62 presents the results of a unique, in its way, experiment, in which the envelope of the ultrashort pulse ($\tau_p \simeq 100$ fs) propagated through leafs samples of different plants were recorded with high temporal resolution ($\Delta t \simeq 3$ ps). The uniqueness of the experiment consists in the fact that the supercontinuum radiation was used in it as a source for the first time, which allowed recording of signal characteristics for the set of wavelengths, in particular: $\lambda = 550$, 670, and 740 nm. This approach extends significantly the information capabilities of sensing, allowing the solution of simplest inverse problems. In the numerical experiment, to estimate the possibility of manifestation of the pulse splitting effect, we have chosen the wavelength $\lambda = 740$ nm, because the first two wavelengths fall within the rather strong chlorophyll absorption bands [63]. According to the model estimates presented above (see Figure 5.3), the manifestation of the effect is insignificant. Figure 5.10 shows the results of comparison of experimental and calculated data for the *Phaseolus vulgaris* plant leaf with the geometric thickness of about 0.02 cm taken as an example. Calculation by the Monte Carlo technique was performed for the following parameter values: $\tau_p = 200$ fs; $\sigma_s = 100$ cm^{-1}; $\sigma_a = 0.5$ cm^{-1}; $g_{HG} = 0.7$. The temporal resolution of the recorded signal was taken to be at the level of the initial pulse duration.

It can be easily seen that the results of simulation, in general, reflects well the qualitative pattern of temporal transformation of the ultrashort pulse at the expense of multiple scattering. Moreover, based on them, we can predict that the effect of splitting of the pulse into ballistic and diffuse components will manifest itself with the further increase of the receiving system resolution.

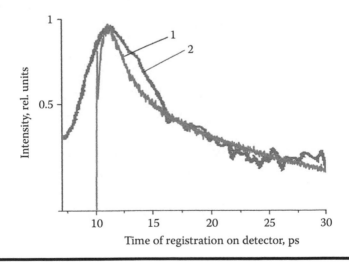

Figure 5.10 Transformation of the shape of femtosecond pulse passed through the layer of Phaseolus vulgaris plant leaf: curve 1 – results of numerical simulation at the values of the basic parameters $\tau_p = 200$ fs; $\sigma_s = 100$ cm⁻¹; $\sigma_a = 0.5$ cm⁻¹; $g_{HG} = 0.7$; curve 2 – data of field experiment. (From Johanson, J. et al., *Photochem. Photobiol.*, Vol. 69, pp. 242–247, 1999.)

5.2.5 The Main Formulation of Local Estimations

In Monte Carlo calculations, it is often, as in this case, necessary to study the solution as a function of some or other parameter of the optical model of the scattering medium (extinction coefficient, quantum survival probability, and others). In this case, to reduce stochastic variance of sought estimates and to increase the efficiency of calculations, it is worth using the same sample of random numbers. Below we show simple technical approaches allowing the variant parameters to be factor out from brackets of the simulation algorithm in the method of local estimation.

1. Consider the case of homogeneous scattering medium, whose extinction coefficient σ takes a set of given values. For a detector located at the medium boundary

$$\tilde{I}(r^*) = \frac{1}{N_i} \sum_{i=1}^{N_i} \sum_{n=1}^{N} \frac{\Lambda_{i,n} \exp(-\tau_{i,n}) g(\mu_{i,n})}{2\pi(-\tau_{i,n}/\sigma)}, \qquad (5.25)$$

where $i = 1, 2, \ldots N_i$ is the number of random photon trajectory.

It is reasonable to assume that for every particular scheme of numerical experiment at a finite value the optical thickness of the scattering layer, the range of $\tau_{i,n}$ values in Equation 5.25 is limited from above, that is, $\tau_{i,n} \in (0, \tau_m)$, where τ_m is the given parameter. Divide this range into m parts $0, \tau_1, \tau_2, \ldots, \tau_j, \tau_{j+1}, \ldots, \tau_m$ and introduce the parameter η_j so that $\eta_j = 1$ if $\tau_j \leq \tau_{i,n} \leq \tau_{j+1}$, and $\eta_j = 0$ otherwise. Let n be so that at $\eta_j = 1$

$$|\tau_{i,n} - \tau_j| \leq \varepsilon \quad \text{and} \quad |\tau_{i,n} - \tau_{j+1}| \leq \varepsilon, \varepsilon \geq 0, \varepsilon < 1 \qquad (5.26)$$

If Equation 5.26 is true, we can believe that $\tau_{i,n} \cong \overline{\tau}_j$, where $\overline{\tau}_j$ is the average value of τ in the jth interval. Then Equation 5.25 takes the form

$$\tilde{I}(r^*) = \frac{1}{N_i} \sum_{i=1}^{N_i} \sum_{n=1}^{N} \sum_{j=1}^{m} \eta_j \frac{\Lambda_{i,n} \exp(-\tau_{i,n}) g(\mu_{i,n})}{2\pi(\tau_{i,n}/\sigma)^2}$$

$$\cong \sum_{j=1}^{m} c_j \frac{\exp(-\tau_j)}{2\pi(\tau_j/\sigma)^2}, \qquad (5.27)$$

where

$$c_j = \frac{\eta_j}{2\pi N_i} \sum_{i=1}^{N_i} \sum_{n=1}^{N} \Lambda_{i,n} g(\mu_{i,n}) \qquad (5.28)$$

Thus, to estimate the functional dependence $\tilde{I} = \tilde{I}(\sigma)$, all other parameters fixed, it is sufficient to calculate a set of c_j values for $\sigma = 1$ and to process the results according to Equations 5.27 and 5.28.

2. The no less important optical parameter determining the spatiotemporal structure of the radiation field in the scattering medium is the quantum survival probability. If the values of $\Lambda(r)$ are independent of the spatial coordinates $r(x, y, z)$, and the number of informative photon collisions is $n \leq n_m$, where the parameter $n_m = const$, that is, it is limited from above, then the statistical estimate

$$\tilde{I}(r^*) = \frac{1}{N_i} \sum_{i=1}^{N_i} \sum_{n=1}^{N} \frac{\Lambda_{i,n} \exp(-\tau_{i,n}) g(\mu_{i,n})}{2\pi \, |r_{i,n} - r^*|^2} \cong \sum_{n=1}^{n_m} \Lambda^n c_n, \qquad (5.29)$$

where

$$c_n = \frac{1}{2\pi N_i} \sum_{i=1}^{N_i} \frac{\exp(-\tau_{i,n}) g(\mu_{i,n})}{|r_{i,n} - r^*|^2} \cong \sum_{n=1}^{n_m} \Lambda^n c_n \qquad (5.30)$$

Estimate (5.29) is unbiased, if n_m actually is the maximal possible number of scattering events. This is admissible, if the scattering layer has a finite optical thickness.

5.2.6 Conclusions

The results of numerical Monte Carlo calculations have confirmed the reality of the sometimes experimentally observed effect of splitting of femtosecond laser pulse at propagation through a layer of an optically dense medium. At the same time, it was found that the effect is observed in a limited range of parameters characterizing optical properties of the medium, conditions of medium illumination, and signal detection. At the optimal selection of the parameters, the temporal configuration of the calculated signals is in a good qualitative agreement with the known measured data, for example, References 43, 59. The calculations were based on the linear radiation transfer, which suggests that the only physical reason for appearance of the bimodal structure of the pulse is the different effective speed of propagation of the ballistic and diffuse components of the signal. This important circumstance was noticed by A. Ishimaru in a classic radiation transfer theory [48]. The assumption that the bimodal structure owes its origin to coherent effects, that is, interference of the primary and scattered waves [41], does not lack support, since these effects were not included in the calculation algorithms. Refinement of the parametric range of existence of the ultrashort pulse splitting effect under real disperse formations calls for further investigations. The consideration of the fine structure of a transmitted signal and the multiparameter dependence of the expected effect has required certain modification of the method of local estimation, which is described above in Section 5.2.5.

5.3 Estimation of the Efficiency of the Use of Promising White-Light Lidar for Sensing of Optical and Microphysical Parameters of Cloud Aerosol

The advent of broadband laser radiation sources allows us to find new application of them in problems of remote sensing of the atmosphere, in particular, problems of determination of microphysical parameters of the atmosphere, including cloudiness. The effect of filamentation and the following generation of directed radiation with the narrow divergence angle of the beam having the wide spectrum of wavelengths, which extends from 0.4 to 2 μm, covers a lot of absorption lines of atmospheric gases and bands falling within atmospheric windows. This, in turn, opens possibilities of selecting the wavelengths for sensing of atmospheric aerosols. In References 22, 84, absorption spectra of atmospheric gases were analyzed and four wavelengths free of influence of the absorption of atmospheric gases were selected: 1.28, 1.56, 1.61, and 2.13 μm. Then the numerical experiment on lidar sensing of a layered-inhomogeneous model of the transfer model by the Monte Carlo technique was carried out. For reconstruction of optical characteristics of clouds, more exactly, vertical profiles of the extinction and backscattering coefficients, the parameterized iterative algorithm was used [21]. Then, the backscattering coefficients were used for reconstruction, with the use of the genetic algorithm, of the particle size distribution function also in dependence on the depth of penetration into clouds. This approach requires two stages of operation, and the iterative method of reconstruction of the backscattering coefficients diverges at the end of the path and requires the knowledge of the basic backscattering coefficients at the initial point of iterations and some simplifications based on a priori information. In addition, the applied genetic algorithm in its initial version is a rather slow procedure, reconstructing the data based on the results of the previous stage and depending on the accuracy of the iterative method.

As is well known, optical characteristics of radiation scattering are sources of information about microstructure and microphysical properties of cloud aerosol at the laser sensing of the atmosphere (see References 64–79 and bibliography therein). At the same time, the optical characteristics themselves at extended paths cannot be measured directly. In an actual experiment, some functionals of the backscattering signal are usually measured, and the profiles of the coefficients of optical coefficients are determined through the solution of the corresponding inverse problem. In this case, the character and complexity of inverse optical problems are naturally determined by the structure of functional dependences between measured and sought parameters in a given optical experiment. It is obvious that the formulation of inverse problems in the theory of broadband radiation scattering should start from the analysis of initial functional equations determining the informativeness and relationships between measured and estimated parameters in a particular experiment. In the real situation of laser sensing of atmospheric aerosol, revealing of information about scattering characteristics is complicated by the circumstance that received optical signals depend at least on two optical parameters. For example, at the monostatic laser sensing of an aerosol layer under conditions of single scattering, the functional relation between the spatially resolved profile of the backscattered signal $P(z, \lambda)$ at a given wavelength λ and optical characteristics β_{sct}, β_π (respectively, scattering coefficient and backscattering coefficient) can be written in the implicit form as $F(P, \beta_{ext}, \beta_\pi, z, \lambda) = 0$. Obviously, this equation is completely undefined, if the task is to find characteristics of radiation scattering. To have the possibility of simplest quantitative interpretation of lidar measurements, it is necessary to define additionally this equation by specifying, for example, the parameter of relation between the total scattering and backscattering coefficients or the so-called lidar ratio $b(z, \lambda) = \beta_\pi(z, \lambda)/\beta_{sct}(z, \lambda)$. This leads to the problem of a priori estimation of $b(z, \lambda)$ from additional measurements or model estimation. There is the following way to overcome this uncertainty: to increase the amount of measured information, on the one hand, and to use finer mathematical methods for the inversion, on the other hand. This has led to the known methodology of laser sensing [34,64–67], whose further development is the subject of this study.

5.3.1 Optical Equations of the Theory of Multiwavelength Laser Sensing

The theory of multiwavelength (or multi-frequency) laser sensing as an optical method for remote determination of optical and microphysical parameters of the Earth's atmosphere and hydrosphere is based on the nonstationary radiation transfer equation and operators of polydisperse radiation scattering at particle systems. These equations and operators were derived in fundamental investigations by I. E. Naats [34,64,65]. At the multifrequency sensing, the radiation transfer equation written in the single scattering approximation for a set of spectral ranges reduces to the system of lidar equations of the form [64,67]:

$$P(\lambda_i, z) = P_0(\lambda_i)\beta_\pi(\lambda_i, z)G(\lambda_i)z^{-2} \exp\left[-2\int_0^z \beta_{ext}(\lambda_i, z')dz'\right], \quad (5.31)$$

where $P_0(\lambda_i)$ is the power of the emitted signal at the wavelength λ_i; $G(\lambda_i)$ is the instrumental constant determined by the area of the receiving system of the lidar and by the transmittance of optical elements at the wavelength λ_i; $\beta_{ext}(\lambda_i, z)$ is the extinction coefficient.

For simplicity, we assume below that under conditions of clouds the molecular scattering in the near-IR range can be neglected. For solution of the problem on reconstruction of microphysical parameters of clouds, first of all, the droplet size distribution function $f(r)$, (r is the droplet radius), system of Equation 5.31 should be complemented with the known integral relations:

$$\beta_{ext}(\lambda_i, z) = \int_{r_1}^{r_2} K_{ext}(\lambda_i, z, r) s(z, r) \, dr \qquad (5.32)$$

$$\beta_\pi(\lambda_i, z) = \int_{r_1}^{r_2} K_\pi(\lambda_i, z, r) s(z, r) \, dr, \qquad (5.33)$$

where $s(r) = \pi r^2 f(r)$; r_1, r_2 are the boundaries of the size spectrum $f(r)$; K_{ext}, K_π are factors of efficiency of, respectively, extinction and backscattering.

Since optical measurements with the chosen spectral range Λ are conducted for the discrete set of wavelengths $\lambda_i \in \Lambda$, the operator form of integral relations (5.32) and (5.33) is often used [34]:

$$\beta_\pi(\lambda_i) = (K_\pi s)(\lambda_i), \; \beta_{ext}(\lambda_i) = (K_{ext} s)(\lambda_i) \qquad (5.34)$$

With allowance for Equation 5.34, initial system of Equations 5.31 through 5.33 can be rewritten in the following form:

$$F(P_i, \beta_{\pi,i}, \beta_{ext,i}, z, \lambda_i) = 0,$$
$$K_\pi s = \beta_\pi, K_{ext} s = \beta_{ext}, \qquad (5.35)$$

where the component values of the vectors $\boldsymbol{\beta}_\pi$ and $\boldsymbol{\beta}_{ext}$ are determined by the relations $\beta_{\pi,i} = \beta_\pi(\lambda_i)$, $\beta_{ext,i} = \beta_{ext}(\lambda_i)$, $i = 1, 2, \ldots n$.

This system is obviously defined with respect the vector $\vec{s}$, which reflects the sought distribution $s(z, r)$ for all z along the sensing path. Actually, there are equations for determination of the vector **s**, whose right-hand sides $\boldsymbol{\beta}_\pi$ and $\boldsymbol{\beta}_{ext}$, although are unknown directly, but related component-to-component with each other by lidar equations. This proves finally the validity of the multifrequency lidar sensing as an optical method for determination of aerosol microstructure.

5.3.2 Iterative Method for Solution of the System of Lidar Equations

Let us dwell briefly on issues associated with construction of an algorithm for numerical solution of system of Equation 5.35. For this purpose, we use the formal operator of mutual transition introduced in References 34, 64 as $W = K_{ext} K_\pi^{-1}$ or $W = K_\pi K_{ext}^{-1}$. In this case, system (5.34) can be rewritten as

$$F(P_i, \beta_{\pi,i}, \beta_{ext,i}, z, \lambda_i) = 0,$$
$$\beta_{ext} = W\beta_\pi, i = 1, 2, \dots n \tag{5.36}$$

Then we transform (5.36) into the discrete form, assuming that digitization of the lidar signal at every wavelength $\lambda_i \in \Lambda$ yields a set of equidistant readouts $P_{k,i} = P(z_k, \lambda_i)$ with a step $\Delta z = z_{k+1} - z_k$, ($k = 1, 2, \dots Nz$). Then, for every k we have a system of lidar equations

$$S_{k,i} = \beta_{\pi,k,i} \exp\left\{ -2\Delta z \sum_{j=1}^{k} \omega_j \beta_{ext,j,i} \right\}, \tag{5.37}$$

where $S_{k,i} = P_{k,i} z_k^2 / P_{0i} G_i$ is the square-amplified backscattering signal; ω_j are quadrature coefficients.

If system (5.37) is solved consecutively starting from $k = 1$, then for the point z_k all the previous values of optical characteristics up to $\beta_{\pi,k-1,i}$ and $\beta_{ext,k-1,i}$ are known. In this case, it is more convenient to represent Equation 5.37 in the form [64,68]:

$$\beta_{\pi,k,i} \exp(-2\tau_{k,i}) = S_{k,i} / T_{k-1,i}, \tag{5.38}$$

where

$$T_{k-1,i} = \exp\left[-2\Delta z \sum_{j=1}^{k-1} \omega_j \beta_{ext,j,i} \right], \quad k \geq 2, \quad T_{1,i} = 1$$

$\tau_{k,i} = \Delta z (\beta_{ext,k-1,i} + \beta_{ext,k,i})/2$ is the optical thickness of the k-th layer of the sensed cloud.

To solve system (5.37) along with vector Equation 5.36 for vectors β_π and β_{ext}, an iterative algorithm was proposed and mathematically justified in Reference 68. The scheme of this algorithm can be represented in the following analytical form:

$$\beta_{\pi,k,i}^{(m)} = \frac{S_{k,i}}{T_{k-1,i}} e^{2\tau_{k,i}^{(m-1)}}$$
$$\beta_{ext,k}^{(m-1)} = W\beta_{\pi,k}^{(m-1)}, \quad i = 1, 2, \dots n; \quad k = 1, 2, \dots \tag{5.39}$$

where m is the iteration number; the operator W is the matrix analog of the operator $K_{ext} K_\pi^{-1}$.

For sensing of optically dense scattering media, in particular, stratified low-level clouds, the increased resistance of the algorithms to the multiple-scattering noise is needed. Based on these requirements, the parametric modification of iterative scheme (5.39) was proposed in Reference 20. It was used in further calculations of the vectors $\beta_\pi(\lambda_i)$ and $\beta_{ext}(\lambda_i)$ for the chosen wavelengths of laser sensing λ_i.

5.3.3 Selection of Informative Wavelengths of Laser Sensing

It is obvious that the number of sensing wavelengths n in the given spectral range Λ depends on how close is the separation between the chosen neighboring wavelengths, for example, λ_i and λ_{i+1}, at the given accuracy of the measurement instrumentation. Denote the smallest relative error of measurements in the spectral range Λ as ε and consider the problem on the acceptable closeness of two wavelengths at the sensing of atmospheric aerosol. Assume that the values of the optical characteristic $\beta(\lambda)$ are measured experimentally in the range Λ. If $\Delta\beta = \beta(\lambda_{i+1}) - \beta(\lambda_i)$ is the increment of the optical characteristic β in the vicinity of the point λ_i, then the smallest separation between λ_i and λ_{i+1} should satisfy the obvious inequality $|\Delta_i\beta| \geq \varepsilon\beta(\lambda_i)$. In References 34, 64, based on the analysis of polydisperse integrals, the validity of the following inequality was demonstrated:

$$\min\left|1 - \frac{\lambda_i}{\lambda_{i+1}}\right| \geq \varepsilon A, \quad i = 1, 2, \ldots, n, \tag{5.40}$$

where the minimum is taken for the pair of neighboring λ_i and λ_{i+1} at all possible divisions of the sensing range Λ. Here, A is a constant depending only on properties of the aerosol distribution ($0 < A < 1$). Thus, if λ is some sensing wavelength, then the next wavelength should be chosen at a distance exceeding the value $|\Delta\lambda| \geq \varepsilon\lambda A$. Then, it is easy to estimate the acceptable number of wavelengths totally for the range Λ. Denote the left and right boundaries of Λ as λ_{min} and λ_{max}, respectively. Then the following relations can be used:

$$\lambda_i = (1 + \varepsilon A)^{i-1}\lambda_{min}, \quad i = 1, 2, \ldots, n$$
$$n = 1 + 1g\left(\frac{\lambda_{max}}{\lambda_{min}}\right) / 1g(1 + A\varepsilon), \tag{5.41}$$

of which the former determines the sequence of the values λ_i, while the latter is the maximum allowable number of readouts. These equations show that, regardless of the properties of microstructure, with an increase of λ the readouts should be more and more rare, and the value of the corresponding interval $\Delta\lambda$ is directly proportional to λ. To use these equations, it is necessary to know A. Formally, A is the functional of the distribution $s(r)$ and is determined by the equation

$$A = S \left\{ \int\limits_{R} | (rs') | \, dr \right\}^{-1} \qquad (5.42)$$

which can be used for numerical estimates. In the case of the particle size distribution in the form of the gamma function [69] (see also Chapter 1):

$$s(r) = ar^{\alpha+2} \exp(-br) \qquad (5.43)$$

we can show the validity of the approximate equality $A \approx 1/(\alpha + 3)$, where α is the distribution parameter. It should be noted that the equations resented are approximate, but they are sufficient in the case when the assumption of the unimodal form of the sought distribution $s(r)$ is allowable, for example, for the droplet spectrum in low-level stratified clouds. In the absence of a priori information on the character of $s(r)$, the more rigorous estimation of the number of independent measurements can be performed following References 70, 79 within the framework of the Fourier analysis of continuous measurements.

5.3.4 Analytical Methods for Inversion of the Integral Equation

Formally, the particle distribution by size $f(r)$ or by cross sections $s(r)$ can be determined from the spectral behavior of any optical characteristics, that is, coefficients of optical interaction (extinction, scattering, absorption, backscattering) or components of the scattering matrix. In this chapter, keeping in mind the lidar measurement scheme, the backscattering coefficients $\beta_{\pi}(\lambda_i, z_k)$ are preferred, because they, among other things, have the higher information content [66]. Reconstructed values of the backscattering coefficients $\beta_{\pi}(\lambda_i, z_k)$ formally satisfy first-kind Fredholm integral Equation 5.33, whose kernel includes the function $s(r)$ to be estimated.

Assume that the function $s(r)$ describes the size distribution of the geometric cross section of aerosol particles in the unit volume of the scattering medium. In problems of laser sensing, the form of the function $s(r)$ can change depending on the spatial coordinates. The backscattering efficiency factor $K_{\pi}(\lambda, r)$ for a particular particle of the radius r at the wavelength λ depends also on the complex refractive index of the aerosol particulate matter $m - i\kappa$. The complex refractive index and the particle size boundary in Equation 5.33 can also be unknown in the general case. Under these conditions, the inversion of Equation 5.33 is an ill-posed problem [34,64].

The construction of approximate solutions of ill-posed inverse problems resistant to small variations in initial data requires the use of specialized mathematical methods [64,65,71]. The theory and efficient methods for solution of this class of problems are developed now. The comparative analysis of the analytical methods for inversion of the first-kind Fredholm integral equation in the application to problems of atmospheric optics can be found in References 66, 72, 73.

Methods and algorithms for solution of the inverse problems of laser sensing of atmospheric aerosol developed in IAO SB RAS have passed the successful practical testing for a long time [34,64,65]. These methods are mostly based on the rigorous mathematical approach of the regularization theory [71]. For unimodal distribution spectra characteristic of cloud aerosol, the method of optimal parameterization has shown good results as a method of optimal parameterization, simplest in the algorithmic aspect [64,68,74]. The simplification is connected with the a priori specification of the analytical form of the distribution function, for example, in the form of generalized γ-distribution [69,75].

The inverse problem is solved by the variation method in the form of optimal estimates of the parameters $\{a, b, \alpha, \gamma\}$ characterizing the microstructure, on average, in the given range of solutions $(r_1, r_2) \in R$. The numerical algorithm for determination of the sought parameters is based, as a rule, on minimization of the square form

$$F(P) = \sum_{i=1}^{Nl} \{\beta(\lambda_i) - \beta_m(\lambda_i, a_1, a_2, ... a_m)\}^2 \qquad (5.44)$$

where β_m is a set of model optical characteristics formed by the operator K and the model function $s_m(r, a_1, a_2, ... a_m)$ in some limited domain Ω of the space of solutions.

It should be noted that model particle distribution functions, as a rule, depend nonlinearly on the sought parameters and, as a result, function (5.44) to be minimized can have more than one extreme in the domain Ω, which leads to the solution uncertainty in the estimation of a large number of parameters. With allowance for this circumstance, it is important to conduct the previous numerical analysis of the inverse problem of aerosol radiation scattering for selection of the smallest number of independent parameters, data measured in a particular optical experiment are most informative for.

5.3.5 Statistical and Intellectual Methods for Solution of Inverse Problems

It should be noted that the regularization method requires fitting of the regularization coefficient, which determines, to a great extent, the accuracy of reconstruction of the sought parameters of aerosol particle microstructure. There exist various techniques for automatic fitting of this coefficient, but they significantly decrease the speed of the algorithm because of the need to multiply large matrices many times (see, e.g., References 49, 79 and bibliography therein). In addition, the regularization method quickly loses its stability with an increase of the multiple-scattering background noise in the received signal.

This problem becomes governing in the sensing of optically dense atmospheric hydrometeors, including, in the first turn, low-level stratified clouds, whose monitoring is rather urgent from the viewpoint of flight safety. In connection with the arising difficulties of quantitative interpretation of optical sensing data, the interest in statistical methods of solution of ill-posed problems [76–78] and methods using the artificial intelligence technology [49,79–86] have increased in the recent time.

5.3.5.1 Monte Carlo Technique

One of the methods of problem solution is based on the use of the stochastic algorithm or the random search algorithm [76,80,81]. The method consists in the following: the expected size range $[r_1, r_2]$ is divided into equal parts, and then the iterative search for the distribution density in each part is performed with minimization of functional (5.44), that is, the value of the distribution density in the parts at every iteration step is corrected to some random value distributed by the uniform law with zero mean. If this leads to a decrease of the functional, then the corrected value is saved, otherwise the previous value at this part is taken:

$$\sum_{i=1}^{Nl} \left(\beta_\pi(\lambda_i) - \int_{r_1}^{r_2} f(r) K_\pi(\lambda_i, r) \mathrm{d}r \right)^2 \rightarrow \min \qquad (5.45)$$

Other method of reconstruction is based on application of the Monte Carlo technique. In this case, a sample of radius values (of the given dimension) distributed in the range $[r_1, r_2]$ is generated. Then the iterative procedure starts, in which one of the values selected in a random way is replaced with the new random value in the case if this leads to a decrease of the functional

$$\sum_{i=1}^{Nl} \left(K_\pi(\lambda_i) - \frac{S \sum_{j=1}^{N_v} K_\pi(\lambda_i, r_j)}{N_v} \right)^2 \rightarrow \min \qquad (5.46)$$

where $K_\pi(\lambda_j)$ is the backscattering coefficient at the j-th wavelength; N_v is the sample dimension; S is the scaling coefficient; Nl is the number of wavelengths, the sensing is carried out at; r_i is the radius of the i-th particle in the sample. This procedure is repeated until the preset accuracy or the given number of iterations is achieved. The scaling coefficient S also changes randomly, the decision on the change of this coefficient is taken at every stage of the iterative process according to the preset probability.

After the procedure of minimization, the normalized histogram in the array of radii is constructed. A feature of this method is that it does not use a priori information on the form of the distribution function. The conducted numerical experiments [84] have shown that the use of the Monte Carlo technique for correct reconstruction of the distribution function, the large number of wavelengths (more than 10) is needed, and the second version of the method has a disadvantage that the number of possible solutions of Equation 5.44 increases drastically in connection with the fine tuning of every particle separately. That is why it was decided to use the genetic algorithm and the method of neural networks.

5.3.5.2 Genetic Algorithm

The genetic algorithm [63,87–90] is used for minimization of functional of form (5.45). In this case, the distribution function $f(r)$ can be taken in any form, in particular, in form (5.43). The search is performed over the space of parameters of the distribution function. The genetic algorithm of search for the minimum is based on simulation of the process of natural evolution and falls in the category of the so-called evolution search methods. That is why it is described by terms borrowed from biology. Operations used in the practical implementation of this method are based on analogs from the living world, such as mutation and crossing. Operations are carried out with a set of solutions (population), and their results are descendants (new solutions), which are also included in the population. Along with creation of descendants, in adaptive forms (worst solutions) are removed from the population. As a result of multiple repetition of these operations, the population consists of the best adapted forms, and thus the optimal solution can be achieved.

More rigorously, the genetic algorithm can be determined as the following object $GA(P^0, r, Fit, sl, cr, m, sc)$, where P^0 is the initial population; r is the number of elements in the population; Fit is the fitness function (utility function); sl is selection of solutions for creation of a new solution; cr is the "crossover" operator determining the possibility of obtaining a new solution; m is the mutation operator; sc is the selection operator. The block diagram of our algorithm is shown in Figure 5.11.

The initial population P^0 is a set of initial solutions belonging to the space of solutions $X \in D$. The population P^0 is generated either in a random way or based on a priori data about the sought solution. The fitness function is determined by the minimizing functional. The selection of solutions is necessary for creation of a new solution. It is worth selecting the solutions corresponding to the optimal and most widely different solutions.

In other words, the algorithm strategy can consist in the equiprobable selection of the solutions, for which $Fit(X_k) < Fit_{cp}$, where Fit_{cp} is the average value of the fitness function. Upon selection of the generating solutions (ancestors), they are subjected to the crossing-over operation for creation of a new solution (descendant). The "crossover" operation can be performed with the aid of recombination of vector elements of ancestor solutions [63] or by the following equation:

$$x_i^n = x_{k1,i} \cdot q + x_{k2,i} \cdot (1 - q),$$

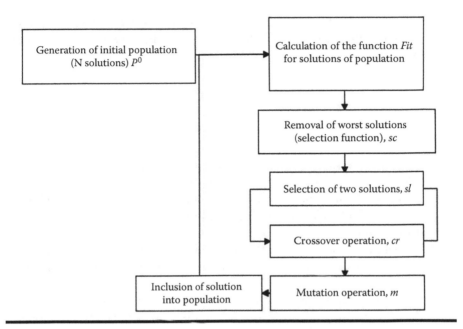

Figure 5.11   Block diagram of the genetic algorithm.

where x_i^n is the i-th element of the new solution vector, q is the uniformly distributed value from 0 to 1, $x_{k1,i}$, $x_{k2,i}$ are the generating solutions. Once the new solution is created, it is subjected to the mutation operation, which can be carried out in the following way:

$$x_i = x_i + \delta, \quad i \in [1, Np]$$

where Np is the dimension of the solution vector, δ is a small random value, i are selected from the given range, and the probability of change of a large number of elements is small. After this operation, the solution is added to the population. Then the good solutions are selected and worst solutions are rejected from the population. This process can proceed for one solution or several solutions simultaneously. For example, by the condition $Fit(X_k) < Fit_{cp}$ all solutions are rejected, and then the new set of solutions is created with the aid of overcrossing and mutations.

In the problem of lidar sensing of the microstructure of a cloud layer on the assumption that the droplet spectrum satisfies the generalized γ-distribution of form (5.43), the fitness function can be represented in the form:

$$Fit(\vec{p}) = \sum_i \left| \beta_\pi(\lambda_i) - \int_0^\infty \left(\sum_{j=0}^{Ng} A_j \exp\left(\frac{-(r - r_j)^2}{2\mu_j^2} \right) \right) K_\pi(\lambda_i, r, m(\lambda_i)) \, dr \right|^2, \quad (5.47)$$

where $\vec{p} = \left(Ng, A_0, r_0, \mu_0, A_1, r_1, \mu_1 \ldots \ldots A_{x-1}, r_{x-1}, \mu_{x-1} \right)$.

The number of functions is also an unknown parameter of the search, whose value at the mutation operation is selected every time by the exponential distribution law

$$F(x) = \begin{cases} q \exp(-qx), & x > 0 \\ 0, & x \leq 0 \end{cases}. \tag{5.48}$$

The random number generator is described by the equation $Ng = [-\ln (1 - F/q) + 1]$, where the square brackets mean the rounding off to an integer number. The amplitude and center of every γ-function are also introduced as unknown parameters. The half-widths μ are variable parameters at $\mu \ll \max(r)$, where r is the particle radius.

In general, the proposed algorithm can be described in the following way. For every sought parameter, the search range is specified based on our knowledge of the function $f(r)$.

At the first stage, the initial array of solutions is determined. Random values from the corresponding range are assigned to the array parameters for every possible solution. For every solution (vector of sought parameters $[a_1, a_2, \dots, a_n]$ for an arbitrary function or $[a, \alpha, b, \gamma]$ for the function of γ-distribution), fitness function (5.47) is calculated. The smaller the value of functional (5.47) (the smaller accuracy of parameter determination), the better the solution. Two or more solutions are selected in the array of solutions, and then the crossing-over operation is carried out. It is followed by the mutation operation, that is, change of one or more descendant parameter by a small value or complete change of several parameters. Once the descendant is created, it is determined whether this descendant is better than the most inadaptive member of the population. If it is so, the inadaptive member is removed and replaced with the descendant. The operation is repeated until the acceptable solution is obtained.

5.3.5.3 Method of Artificial Neural Networks

The other proposed method is based on the application of artificial neural networks [86,91–94]. Neural networks can be considered as some device allowing construction of the necessary functional dependence by the given learning examples. In our case, in implementation of the method of neural networks, it is important to choose the learning method, number of layers, and type of neurons, as well as to create the learning sample. Creation of the learning sample is a separate complicated problem. It is necessary for this sample to be complete, most informative, and not very large. At its creation, it is necessary to model the process of lidar sensing for different atmospheric conditions.

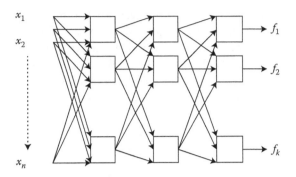

Figure 5.12   Structural scheme of three-layer neural network.

For solution of the problem of lidar sensing, several constructions of neural networks different in the form of activation functions, number of layers, combination of input parameters, and number of separate neural networks for reconstruction of the concentration profile can be used.

As was shown in References 93, 94, for problems of optical sensing, the most efficient network is the three-layer neural network of the multilayer perceptron type (see Figure 5.12).

Mathematically, the functioning of the three-layer neural network can be described by the following equation:

$$f_i = \varphi_{3,i}\left\{\sum_{j1=0}^{Nr-1}\omega_{i,j1}^3\cdot\varphi_{2,j1}\left[\sum_{j2=0}^{Nin-1}\omega_{j1,j2}^2\cdot\varphi_{1,j2}\left(\sum_{j3=0}^{Nl-1}\omega_{j2,j3}^1\cdot x_{j3}\right)\right]\right\} \quad (5.49)$$

where

$\varphi_{i,j}$ is an activation function of the j-th neuron of the i-th layer;
$\omega_{i,j}^n$ is the j-th weighting coefficient of the i-th neuron of the n-th layer;
f_i are output values of the neural network, corresponding to the output of the i-th neuron of the last layer;
x_i are input values of the neural network (NN), their number is equal to the number of inputs of every input neuron of the neural network;
Nr is the number of discrete values of the function $f(r)$.

To give the needed functionality to a neural network, it is previously trained with the use of examples. In our case, input data for NN can be represented both by signals received at different wavelengths and by values of optical characteristics, in particular, backscattering at several wavelengths $\beta_\pi(\lambda_i)$; ($i = 1, 2, \ldots, n$). The values of the distribution function $f(r, z_j)$, $j = 1, 2, \ldots$, served input data for NN.

For creation of examples, the imitation direct problem of sensing of a cloud layer was solved by the Monte Carlo technique. The fitting coefficients of backscattering factors were calculated by the Mie theory with the refractive indices for water at their modification up to 5%. The distribution function in the below example was calculated, as earlier, based on the γ-distribution (5.43) with different parameters. The parameters were chosen in the following stochastic way:

$$\alpha = [5 - 1.5(\xi - 0.5)]$$
$$b = 2 + (\xi - 0.8)$$
$$\gamma = 1 + 0.5(\xi - 0.1) \tag{5.50}$$
$$a = \left(1 + \frac{0.1\xi}{(\alpha/(\gamma \cdot b))^{2\gamma}}\right) b^3$$

where $\xi \in (0, 1)$ is the uniformly distributed random value, the square brackets mean the integer part of a value.

The scheme of training used in this problem consists, briefly, in the following. Let, at the initial time, the weighting coefficients $w_{i,j}^n$ of all synaptic links of three-layer perceptron, being in one-to-one relation with the sought value of the parameters of Equation 5.50, take some random values. For NN training, some number of learning examples, consisting of input signals $x_1, x_2, \ldots, x_n$, and the corresponding desirable responses at the network output $d_1, d_2, \ldots, d_n$ is used. Consecutive learning assumes presenting the input vectors one by one to NN and the correction of weights in accordance with the chosen rule. The error signal of the output j-th neuron at the m-th iteration (corresponding to the m-th learning pair x_m and d_m) can, obviously, be determined as $e_j(m) = d_j(m) - y_j(m)$. Sometimes [86], the thermodynamic analogy such as the error energy of the j-th neuron defined as $e_j^2(m)/2$ is introduced. Then the total error energy of the network is determined as

$$E(m) = \sum_{j=1}^n e_j^2(m)/2. \tag{5.51}$$

This equation can be treated as a goal function for minimization (or the cost function) depending on all NN weighting coefficients. We can introduce the energy of the root-mean-square error, that is, averaged energy (5.51) over all samples of the learning set

$$\bar{E} = \frac{1}{N} \sum_{m=1}^N E(m) = \frac{1}{N} \sum_{m=1}^N \sum_{j=1}^n e_j^2(m)/2, \tag{5.52}$$

which has the meaning of the measure of NN learning efficiency.

Consider the *j*-th neuron in the second layer of the network at the *m*-th iteration (i.e., as the *m*-th sample of the learning set is applied to the network input). The argument of the activation function of this neuron is

$$g(m) = \sum_{i=0}^{N_i} \omega_{ij}(m) y_i(m),$$

where $y_i(m)$ are signals from outputs of neurons of the previous layer. Then the functional signal at the output of the considered *j*-th neuron is determined as $y_j(m) = \varphi_j[g(m)]$.

The backpropagation algorithm (BPA) [86], which has shown good results in problems of fluorescent spectroscopy, was chosen as an efficient learning scheme. This algorithm consists, essentially, in application of the correction $\Delta \omega_{i,j}(m) = -\eta \partial E(m)/\partial \omega_{i,j}$, proportional to the error gradient, to the weight $\omega_{i,j}(m)$. The learning is carried out by the gradient descent method, that is, at every iteration the weight is changed as

$$\omega_{i,j}(m+1) = \omega_{i,j}(m) - \eta \frac{\partial E}{\partial \omega_{i,j}}, \tag{5.53}$$

where η is the parameter determining the learning rate. The error function in the explicit form does not depend on weights $\omega_{i,j}$. Therefore, we can use equations of implicit differentiation of the complex function

$$\frac{\partial E}{\partial \omega_{i,j}} = \frac{\partial E}{\partial y_j} \frac{\partial y_j}{\partial S_j} \frac{\partial S_j}{\partial \omega_{i,j}}, \tag{5.54}$$

where S_j is the weighted sum of input signals. In this case, the factor $\partial S_j/\partial \omega_{i,j} = x_i$, where x_i is the value of the *i*-th input. Determine the first factor of Equation 5.54:

$$\frac{\partial E}{\partial y_j} = \sum_k \frac{\partial E}{\partial y_k} \frac{\partial y_k}{\partial S_k} \frac{\partial S_k}{\partial y_j} = \sum_k \frac{\partial E}{\partial y_k} \frac{\partial y_k}{\partial S_k} \omega_{k,j}^{(n+1)} \tag{5.55}$$

where *k* is the number of neurons in the $(n+1)$-th layer. Introduce the auxiliary variable $\delta_j^{(n)} = (\partial E/\partial y_j)(\partial y_j/\partial S_j)$, then we can find the recursive equation for determination of $\delta_j^{(n)}$ of the *n*-th layer, if the value of this variable is known for the next $(n+1)$-th layer

$$\delta_j^{(n)} = \frac{\partial y_j}{\partial S_j} \sum_k \delta_k^{(n+1)} \omega_{k,j}^{(n+1)}. \tag{5.56}$$

Determination of $\delta_j^{(n)}$ for the last layer of the neural network presents no difficulties, because the vector of the values, the network should output at the given input vector, is known a priori:

$$\delta_j^{(n)} = \left(y_j^n - d_j\right)\frac{\partial y_j}{\partial S_j} \tag{5.57}$$

As a result of all transformations, we obtain the following equation for calculation of the increment of link weights in the neural network:

$$\Delta w_{i,j} = -\eta\delta_j^{(n)}x_i^n, \tag{5.58}$$

$$w_{i,j}(m+1) = w_{i,j}(m) + \Delta w_{i,j}. \tag{5.59}$$

Thus, we can formulate the complete algorithm of learning by the BPA method (structure scheme of the algorithm is shown in Figure 5.13):

1. Apply the next input vector from the learning sample to the input of the neural network and determine the outputs of neurons in the output layer.
2. Calculate $\Delta w_{i,j}$ with the use of (5.57) and (5.58) for the output layer of NN.

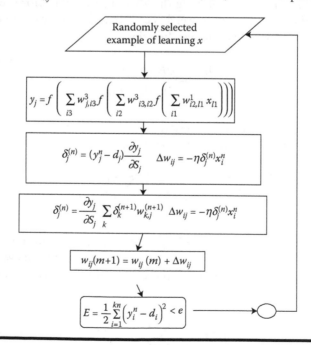

Figure 5.13 **Structural scheme of the backpropagation algorithm.**

3. Calculate $\Delta\omega_{i,j}$ with the use of Equations 5.56 and 5.58 for the other layers.
4. Correct all weights of NN by Equation 5.59.
5. If the error is significant, return to step 1. Otherwise, complete the learning procedure.

The described BPA algorithm in its canonical form [86] is not efficient enough in the case, when derivatives with respect to different weights differ widely, which often takes place in the considered class of problems, because of the oscillating character of amplitude Mie coefficients. Therefore, to increase the convergence of the method, a hybrid scheme based on inclusion of the error of elements of the genetic search considered above into the backpropagation algorithm is implemented in this study. The genetic algorithm itself has a disadvantage that it is necessary to calculate the function of network error for all examples in the sample, which significantly decreases the algorithm operation speed. It is reasonable to use this algorithm for a small part of the sample including example with the widest spread, then the algorithm speed increases considerably. Finally, the resultant algorithm consists in the alternative use of the genetic algorithm and the BPA method. If the BPA method reduces the error by very small value, then the control is given to the genetic algorithm, which, in its turn, returns the control to the BPA method after a certain number of cycles.

5.3.6 Comparative Analysis of the Efficiency of the Monte Carlo Technique and Intellectual Methods for the Reconstruction of Microstructural Parameters of Cloud Aerosol in the Scheme of Multi-Wavelength Lidar Sensing

Earlier in Reference 22, we have carried out the comparative analysis of the efficiency of the inverse Monte Carlo method and intellectual methods for reconstruction of microstructural parameters of cloud aerosol in the scheme of multiwavelength lidar sensing. The analysis was performed within the framework of the closed computational experiment. The direct problem of estimation of backscattering signals within the scope of the nonstationary linear radiation transfer theory was calculated by the semianalytical Monte Carlo method allowing the boundary conditions of the real physical experiment to be taken into account quite rigorously. In this case, the physical prototype of the lidar was the mentioned mobile French–German Teramobile system, whose operational and technical characteristics are considered in detail in the literature [6]. The results of solution of the direct problem, that is, time scans of signals, obtained for the given set of wavelengths and satisfying the requirements of the informativeness, and the a priori known model of the cloudy atmosphere, were used for the reconstruction of optical and microphysical parameters of the atmosphere included in the model. The calculations were performed for four sensing wavelengths 1.28, 1.56, 1.61, and 2.13 μm, which fall within the near-IR micro-windows, according to our estimates. The middle-cyclic continental aerosol model was chosen as an optical model of the atmosphere [75]. A 100-m cloud layer was taken to be at a height of 200 m, and microphysics of the stratified-inhomogeneous cloud corresponded to commonly used test Cloud C1 model by Deirmendjan [69].

Examples of reconstruction of microphysical parameters of a cloud layer by different inversion methods are shown in Figures 5.14 through 5.16.

Figure 5.14 shows the vertical cross section of the cloud droplet size distribution function of the particle number density $N(r)$ reconstructed with the use of the inverse Monte Carlo method. This method does not invoke a priori information about the form of the distribution function. Consequently, numerous solutions can be obtained, because they are not limited by one class of functions. To limit the number of possible solutions, the initial sample of radii is generated in a random way according to some probability density having the preset modal radius. Results obtained with the use of the Monte Carlo method show that this method begins to produce large deviations of large errors in backscattering coefficients. In addition, in real experiments, the modal radius is known only approximately, and, what is most important, to increase the accuracy of the method, it is needed to use the much greater number of sensing wavelength, which is instrumentally impossible now.

Then the genetic algorithm was used for solution of the problem, and the obtained results of reconstruction of the distribution function $N(r)$ are shown in Figure 5.15. This algorithm has allowed an increase in accuracy of reconstruction of the distribution function. However, it should be noted that at the strong distortion

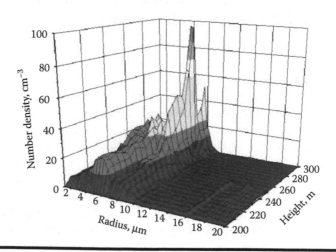

Figure 5.14 **Vertical profile of the particle size distribution function of number density $N(r)$ reconstructed by the inverse Monte Carlo method.**

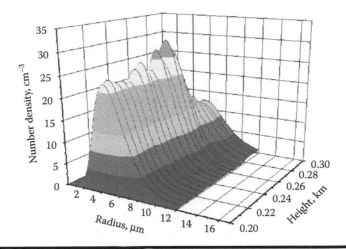

Figure 5.15   Vertical profile of the particle size distribution function of number density *N(r)* reconstructed by the genetic algorithm.

of reconstructed profiles of the backscattering coefficient it is rather difficult to select the proper criterion for termination of the iteration process of the genetic algorithm, because the obtained solutions are limited to the class of gamma functions. In addition, a certain disadvantage of the genetic algorithm is its rather slow convergence, which complicates the use of this algorithm in real time. Nevertheless, the accuracy of reconstruction is rather high, because the model dependence of the backscattering coefficient on the wavelength is limited to some class of functions. This method can be successfully used for processing of actual data, because

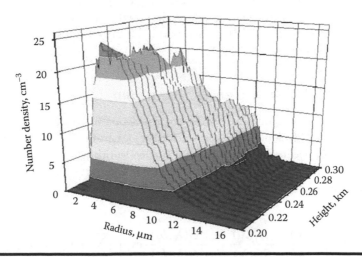

Figure 5.16   Vertical profile of the particle size distribution function of number density *N(r)* reconstructed by the method neural network.

unimodal distribution functions $N(r)$ in the most cases are well approximated by the generalized gamma distribution. In the case of complex composite distributions $N(r)$, the larger number of sensing wavelengths should be invoked. However, numerical experiments have shown that, even with the use of many wavelengths in the limited class of functions $N(r)$, several solutions of problem (5.45) can be obtained. In this case, it is unclear which solution corresponds to the actual one. It is often proposed [34,64] to obtain the averaged solution or to refine the solution with the aid of available a priori ideas about the distribution function.

Figure 5.16 shows vertical profiles of the distribution function $N(r)$ for the analogous experimental model as reconstructed with the aid of the method of artificial neural networks. It can be said that the method of neural networks appeared to be somewhat more sensitive to errors of input values of backscattering coefficient profiles in comparison with the genetic algorithm, which is caused, to a large extent, by the retraining of NN with particular examples, as well as by the strong distortion of the backscattering coefficient. In addition, capabilities of the neural network are limited to the domain of learning examples. Therefore, if real or test values are outside the learning sample, then the network can yield inadequate or unreal solutions. However, this problem can be solved to a large extent through creation of the most complete learning sample. The neural network can solve the problem of reconstruction in real time, as well as it automatically solves the problem of multiple solutions of problem (5.45) by averaging them. Alternatively, this problem is solved at the stage of construction of the neural network and creation of the learning sample through introduction of the additional input of the neural network for a priori information (e.g., specification of the approximate value of the modal radius or the distribution half-width). A common disadvantage of the intellectual methods employing vertical profiles of backscattering coefficients $\beta_\pi(\lambda_i, h)$, $i = 1, \dots ,$ 4 reconstructed with the increasing error as input information, as follows from the analysis of Figures 5.15 and 5.16, remains the domain of stable solutions insufficient in the cloud depth.

5.3.7 Modification of Genetic Algorithm

It is worth joining the procedure for the reconstruction of optical and microphysical characteristics of clouds and aerosol formations from signals obtained from several optimally selected wavelengths in a single algorithm.

It turns out that the equation of functional for solution of the problem of optimization with unknown parameters can be written in the following form:

$$\sum_{i=1}^{M}\sum_{j=2}^{N}\sum_{n=1}^{j-1}\left(\frac{S_j(\lambda_i)}{S_n(\lambda_i)} - \frac{\beta_{\pi j}(\lambda_i)\cdot(h_n)^2 G(\lambda_i,h_j)}{\beta_{\pi n}(\lambda_i)\cdot(h_j)^2 G(\lambda_i,h_n)} \cdot \exp\left(-\sum_{k=n}^{j-1}(\beta_{ext\,k} + \beta_{ext\,k+1})\Delta h\right)\right)^2$$

$$(5.60)$$

where $\beta_{\pi j}(\lambda_i) = \int_{r_1}^{r_2} f_j(r) K_\pi(r, n(\lambda_i), \lambda_i) dr$ is the backscattering coefficient at the wavelength λ_i at the height h_j on the assumption of stratified-inhomogeneous structure.

$\beta_{ext\,j}(\lambda_i) = \int_{r_1}^{r_2} f_j(r) K_{ext}(r, n(\lambda_i), \lambda_i) dr$ is the extinction coefficient $S_j(\lambda_i)$ is the signal at the wavelength λ_i obtained at the height h_j, $f_j(r)$ is the particle size distribution function at the height h_j, $K_\pi(r, n(\lambda_i), \lambda_i)$ and $K_{ext}(r, n(\lambda_i), \lambda_i)$ are the backscattering and extinction coefficients for the spherical particle of the radius r at the refractive index of the particulate matter $n(\lambda_i)$.

As was noted above, the genetic algorithm is a heuristic algorithm modeling the process of natural evolution for solution of optimization problems; it allows us to find the global optimum. The general principle of the genetic algorithm consists in formation of the set of solutions referred to as a population. From this population, two solutions are selected and then subjected to the operation of recombination—crossing-over, which usually corresponds to exchange of bit sequences in elements of the solution vector or the following operation: $x_i = x1_i a + x2_i(1 - a)$, where a is a random value from 0 to 1, x_i, $x1_i$, $x2_i$ are elements of the solution vector and parent pairs. Then the new solution is subjected to the operation of mutation, usually consisting in the change of bit representation or the change of an element of the solution vector x_i. The obtained solution is included in the population of solutions. The operation of selection of parent pairs and creation of a new member can take a certain number of steps. After a certain number of iterations, one member or several members, being the worst solutions, are removed from the population or the worst solutions are removed with the maximal probability. The worst solution is determined with the aid of the so-called fitness function. The role of this function is usually played by the optimized functional. In the canonical form, this algorithm often leads to the situation that a local minimum is taken for the global one, which corresponds to the degeneration of the population of solutions. The convergence in this case is rather slow. We decided to modify the fitness function, as well as the genetic algorithm itself for more efficient convergence to the sought solution. The particle size distribution function represented in the form of the sum of three Gauss functions specified for every height level

$$f_i(r) = \sum_{k=1}^{3} A_{i,k} \exp\left(-\frac{(r - r_{i,k})^2}{2\sigma_{i,k}^2}\right), \tag{5.61}$$

was taken as an unknown solution. The refractive index of water at four sensing wavelengths was taken as the refractive index of the medium. In this case, the coefficients of optical interaction $\beta_{opt\,j}(\lambda_i)$ can be written as

$$\beta_{opt\,j}(\lambda_i)_j(\lambda_i) = \sum_{l=1}^{L} \sum_{k=1}^{3} A_{i,k} \exp\left(-\frac{(r - r_{i,k})^2}{2\sigma_{i,k}^2}\right) K_{opt}(\eta, \lambda_i) \Delta r, \tag{5.62}$$

where r_l is the uniform grid from 0.001 to 15 μm with a step Δr and the number of elements L. The fitness function in our case is given in the following form:

$$fit(A_{i,j}, r_k, \sigma_k) = \sum_{i=1}^{M} \sum_{j=2}^{N} \sum_{l=1}^{j-1} \left(\frac{(Sr_{i,j,l} - Sa_{i,j,l}) \cdot 2}{Sr_{i,j,l} + Sa_{i,j,l}} + 1 \right)^4,$$

$$i \in 1...M = 4, j \in 1...N, k \in 1...3, \tag{5.63}$$

where

$$Sr_{i,j,l} = \frac{S_j(\lambda_i)}{S_l(\lambda_i)}, \quad l \in 1...j-1 \tag{5.64}$$

$$Sa_{i,j,l} = \frac{\beta_{\pi j}(\lambda_i) \cdot (h_l)^2}{\beta_{\pi l}(\lambda_i) \cdot (h_j)^2} \cdot \exp\left(-\sum_{p=l}^{j-1} (\beta_{ext\,p}(\lambda_i) + \beta_{ext\,p+1}(\lambda_i)) \Delta h \right), \tag{5.65}$$

It is assumed that the lidar's geometric factor $G(\lambda_i, h)$ varies insignificantly with height. Correspondingly, the ratio of geometric factors in Equation 5.60 is close to 1. For improvement of the solution and normalization of the coefficients of optical interaction at a priori known coefficients at some height, we can complement the fitness function with an additional term:

$$\sum_{i=1}^{M} \left(\frac{2(\beta_{opt0}(\lambda_i) - \beta_{opt}a(\lambda_i))}{\beta_{opt0}(\lambda_i) + \beta_{opt}a(\lambda_i)} + 1 \right)^4, \tag{5.66}$$

where $\beta_{opt}a$ is the known value of the coefficient of optical interaction at the height h_0.

The high degree causes the sharper descent to the sought solution than the use of low degrees, which leads to the possible degeneration of population in the process of iterations. The parameters $A_{i,j}$, r_k, σ_k of the fitness function, being, essentially, parameters of the particle size distribution are considered as a sought solution. Modification of the genetic algorithm consists in modification of the "crossover" operator. In this case, the following operation is used:

Three solutions $x1$, $x2$, and $x3$ are selected randomly from the set of solutions. The solution $x1$ is selected among them as an optimal one. The new solution results from the transformation $x = x1 - (x3 - x2) \cdot sign(fit(x3) - fit(x2)) \cdot 0.001$, which is, with the high probability, a step toward an increase of the function.

The mutation operator is the following stochastic procedure for the arbitrary number of arbitrary elements of the vector formed through overcrossing, namely, $x_v = x_v \cdot a \cdot 2$, where a is a random value distributed uniformly from 0 to 1.

Besides, there are two addition populations. The first one includes the solution, whose fitness functions differ from all available in the population by 1%. Such members replace the most inadaptive members of the population. A solution having the better adaptation than the least adapted member of the population and the larger separation from all the existing members is included in other population. With the small probability, members of these two populations are also selected for the crossingover operation. It is necessary for the population to have the high level of variety of members and does not degenerate into local minima or worst solutions.

5.3.8 Numerical Experiment on Sensing of Optical and Microphysical Parameters of Clouds Based on the Modified Genetic Algorithm

To solve the direct problem of multiwavelength lidar sensing of cloud aerosol, the technique of statistical simulation identical to References 21, 22 was used. The genetic algorithm, to whom we followed, is fully described in Reference 95. As before, initial and boundary conditions for solution of integral radiation transfer equation correspond to the known scheme of the monostatic white-light lidar [68]. It was assumed that the source emits an ultrashort optical pulse in the directional cone $2\pi(1 - \cos\varphi_s)$, where $\varphi_s = 0.1$ m rad is the total divergence angle of the source. The return signal is recorded by a detector in angular cones $2\pi(1 - \cos\phi_d^k)$, where ϕ_d^k is a set of total detection angles, $k = 1, 2, \ldots$ A 0.5-km thick cloud layer was taken at a height of 1.0 km. For solution of the direct problem, the optical parameters of the chosen cloud model, including the scattering, extinction, and backscattering coefficients, quantum survival probability, and scattering phase function, for optical sensing wavelengths of 1.285, 1.557, 1.629, and 2.130 μm were calculated previously by the Mie equations. The set of wavelengths is caused, as was noted above, by the requirement of lidar operation in atmospheric windows [22,75] and by the informativeness of this set of wavelengths for the sought parameters of cloud aerosol. Traditional optical characteristics of the atmosphere, including the cloud layer, were set in this example as piecewise linear functions of height h.

The backscattering signals calculated by the Monte Carlo technique for the chosen scheme of the model computer experiment are shown in Figure 5.17 for one wavelength λ as functions of the field of view angle of the assumed detector ϕ_d^i.

The values of actual signals including the multiple-scattering noise are then used as input parameters for solution of the inverse problem, that is, reconstruction of discrete vertical profile of a priori preset optical and microphysical characteristics of cloud aerosol based on the proposed genetic algorithm. This is the principal difference of the new approach. An intermediate stage of reconstruction of $\beta_\pi(\lambda_j, h)$ profiles, which is most susceptible to the influence from the multiple-scattering noise, is removed, and, as follows from Equation 5.60, just the values of signals

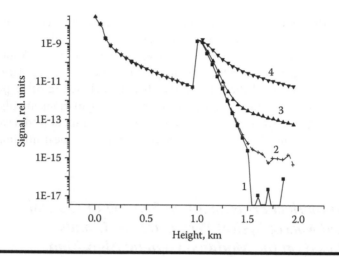

Figure 5.17 **Lidar signals at wavelength** $\lambda = 1.285$ **μm calculated by the Monte Carlo technique as functions of the detector field-of-view angle** $\varphi_d = 0.0001$, **0.001, 0.01, and 0.1 rad (curves 1–4, respectively).**

$S_j(\lambda_i)$ become input functional in the procedure a genetic search, Moreover, the use of the signal ratio in Equation 5.60 allows, as is well known, a significant decrease in the multiple-scattering noise. The reconstructed values of the extinction and backscattering coefficients for the most realistic value $\varphi_d = 0.01$ rad are shown in Figure 5.18 in comparison with the model values. The integral optical thickness of

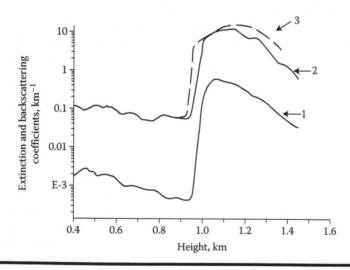

Figure 5.18 **Example of reconstruction of backscattering** $\beta_\pi(\lambda, h)$ **and extinction** $\beta_{ext}(\lambda, h)$ **coefficients at wavelength** $\lambda = 1,285$ **μm at the detector field-of-view angle** $\varphi_d = 0.01$ **rad by the modified genetic algorithm.**

a cloud is large and equal to $\tau = 5 - 6$. Consequently, the results of reconstruction of $\beta_{ext}(\lambda_i, h)$ and $\beta_\pi(\lambda_i, h)$ are quite justifiable.

The model vertical (within the cloud) profile of the cloud droplet size distribution of number density $N(r)$ in this experiment was specified in the form of the generalized γ-distribution [70] with constant modal radius $r_m = 4$ μm and the vertically variable function $N(r)$, whose values are reflected indirectly in the model values of the extinction coefficient $\beta_{ext}(\lambda_i, h)$ shown in Figure 5.18. The results of $N(r)$ reconstruction by the genetic search method (5.60) through (5.66) are depicted in Figure 5.19.

Thus, we can state that the proposed algorithm reconstructs, to a sufficient extent, the unimodal character of the preset $N(r)$ profile in the class of Gauss functions (5.61). At the same time, with an increase of the sensing depth, the estimated value of the modal radius experiences marked shifts, whose value increases gradually. However, in general, the proposed modification of the genetic algorithm has demonstrated the increased, in comparison with the known methods, resistance to the multiple-scattering noise, allowing us to reach the record ($\tau \leq 5$) optical sensing depth. The further increase of the algorithm efficiency can be attained through reasonable, within the scope of technical feasibilities, increase of the number of sensing wavelengths.

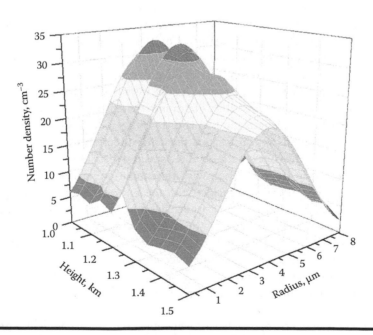

Figure 5.19 **Vertical profile of the particle size distribution of the number density $N(r)$ obtained with the modified genetic algorithm.**

5.3.9 Summary

The analytical methods for solution of some ill-posed inverse problems in application to the problem of multifrequency laser sensing of aerosol atmosphere has been reviewed briefly. It has been shown that under conditions of significant multiple-scattering noise, which takes place in the case of sensing of cloud aerosol, it is preferable to use flexible, so-called intellectual, algorithms for reconstruction of the microphysical parameters of aerosol. The hybrid scheme, in which the learning procedure in the method of artificial neural networks is optimized based on the genetic search, appears to be the most efficient. The ways for further modification of the genetic algorithm with the use of specific features of the multi-wavelength lidar sensing technology have been described. The results of closed numerical simulation shown in Figures 5.15, 5.16, and 5.19 demonstrate the rather high quality of reconstruction of the a priori preset spectra of cloud droplet size distribution up to optical depths 1.5–2 times exceeding the estimates known until now.

5.4 Lidar Experiment on the Study of Microphysical Characteristics of Artificial Aerosol at Short Paths

Within the framework of development of femtosecond atmospheric optics, a field experiment of lidar sensing of an artificial aerosol cloud at a short (85 m) atmospheric path was conducted in 2014. Backscattering signals from artificial aerosol (ethylene glycol) with a particle diameter of 1 μm under the exposure to supercontinuum radiation from the zone of laser filamentation were recorded. In the described experiment, the nonlinearity of the propagation medium manifests itself: supercontinuum radiation is generated at the initial part of the path in air, the energy of the initial pulse redistributes both into the short-wavelength (anti-Stokes) and into the long-wavelength (Stokes) spectral ranges. Amplitude values of spectral components of the optical backscattered (in direction of 174.5°) signal from supercontinuum at the edges of the obtained spectral distribution decrease by three orders of magnitude in comparison with the amplitude of the pulse at the laser wavelength (800 nm), which coincides with the known results of investigation of scattering of femtosecond pulses at atmospheric air. The spectral distribution of backscattered signals was detected by the Hr4000 spectrophotometer (usually used in Ocean Optics) in the range 0.2–1.1 μm. Once energy losses for filamentation and accompanying effects are taken into account, the result of sensing from path parts located after the filamentation zone can be described by linear sensing equations with invoking of the methodology developed for the multifrequency lidar sensing [17].

The scheme of the computer experiment copied the conducted field experiment, thus being the biaxial lidar scheme. The optical aerosol model included the coefficients of optical interaction for ethylene glycol with concentration of 100 cm^{-3}

calculated by the Mie theory with regard for the linear propagation of supercontinuum radiation. In this case, the aerosol scattering coefficient exceeded background values by two to ten times. As the initial supercontinuum spectrum, we took the experimental spectra recorded at the end of the sensing path at a distance of 85 m from the source. Then the spectrum was divided by the model transmittance along the path. The spectral angular divergence was obtained from Reference 96 and the analysis of experimental color patterns obtained at propagation of filaments. The Monte Carlo simulation consisted in solution of the radiation transfer equation with the use of the local estimation method [97].

5.4.1 Solution of Inverse Problem with the Genetic Algorithm

As was already said, the genetic algorithm falls in the category of methods dealing with search for the solution of the optimization problem. It is borrowed from natural evolution mechanisms and consists in formation of a set of solutions selected with the use of the fitness function, as well modification of solutions through the operations of crossing-over and mutation [22,95,98]. Heuristically, this leads to the finding of an optimal solution. Commonly recognized problems of genetic algorithms are the slow convergence to the solution and the long operation because of the need in multiple calculation of the selection function. In our case, this function was represented as

$$
f(a,b,r,g)
$$

$$
= \sum_{i=1}^{n} \frac{\left[I_e(\lambda_i) - a \cdot I_s(\lambda_i) \left(b \int_{r1}^{r2} M \exp\left(-\frac{(r'-r)}{2g} \right) \beta_{174.5}(r,\lambda_i) dr' + \beta_{174.5}^{a+m}(\lambda_i) \right) \right]^2}{I_e(\lambda_i)^2 \exp\left(\int_0^{85} \beta_{ext}^{a+m}(\lambda_i,h) dh \right)}
$$

(5.67)

where $f(a, r, g, b)$ is the fitness function depending on the signal scale parameters a and the coefficient of contribution b from the artificial aerosol to scattering at an angle of 174.5° (in fact, it corresponds to the relative contribution in comparison with the molecular and aerosol atmospheric components and the concentration of ethylene glycol particles), r and g are the mode and the half-width of the Gauss distribution density, M is the coefficient of normalization to the unit distribution density, $\beta_{174.5}(r, \lambda_i)$ is the coefficient of scattering at the corresponding angle for the ethylene glycol particle with the radius r, $\beta_{174.5}^{a+m}(\lambda_i)$ is the total coefficient of scattering by aerosol and molecular atmosphere, $\beta_{ext}^{a+m}(\lambda_i,h)$ is the extinction coefficient, $I_e(\lambda_i)$ and $I_s(\lambda_i)$ are the backscattered signal spectrum and the spectrum of supercontinuum spectrum coming to the detector (results from multiplication by transmittance). The Gauss distribution was taken as the initial particle size spectrum, because now we consider rather narrow spectra. In the process of calculation, this function requires long computational time. For faster calculations, the dependence of scattering on the wavelength and the particle radius was calculated previously. In addition, the integral for averaging over the distribution function is calculated only in the vicinity of the mode, which results in computation time of about 1 h at modern processors without parallelization.

5.4.2 Discussion

The particle distribution functions were reconstructed for particles with radii of 0.5, 1, 1.5, and 2 μm. Figure 5.20 shows the examples of fitting for particles of these radii with deviations of about 0.001–0.01 μm. The mode reconstruction is accurate enough, even with fitting discrepancy and the level of signal-to-noise (S/N) ratio of 30 on average over the spectrum (minimal S/N ratio in some cases achieved 0.5).

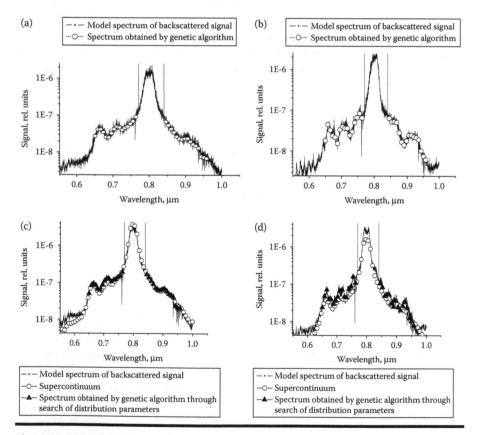

Figure 5.20 Fitting to model spectra obtained by the Monte Carlo technique: (a) 1 μm, (b) 1.5 μm, (c) 0.5 μm, and (d) 2 μm.

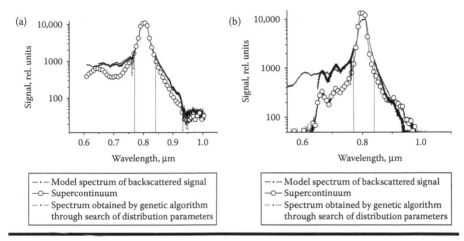

Figure 5.21 Experimental spectra and fitting by genetic algorithm.

For the experimental spectrum with particle diameter of 1 μm and the profile as in Figure 5.21a, the reconstruction yielded the mode with the radius 0.498 μm. For Figure 5.21b, if we take the spectrum averaged over the set of realizations, then two modes with the radii of 0.27 and 0.6 μm were reconstructed.

Thus, particular attention should be paid to the issue of obtaining the initial spectrum. Recommendations include the obtaining of the spatially resolved spectrum before the cloud or the use of two receivers in the bistatic scheme, when one receiver is directed to the aerosol cloud, while another is directed to the space before the cloud. Analysis of possibilities of reconstruction for aerosol distributions with wider half-width and nearly lognormal forms requires further investigations in this field.

Thus, the distribution functions have been reconstructed from experimental data in accord with the narrow particle size distribution of the artificial aerosol formation based on ethylene glycol, and the possibilities of reconstruction of narrow distributions with a mode in the range 0.5–2 μm with the genetic algorithm have been studied.

5.4.3 Assessment of Feasibility of Reconstructing Microphysical Characteristics of Mist with the Use of White-Light Lidars at Short Paths

The further investigations are aimed at assessment of reconstruction feasibilities for lognormal distribution functions with a half-width larger than 0.01 μm. A disadvantage of the genetic algorithm consists in the rather long search for unknown parameters. In addition, the conditions for its termination cannot always be selected unambiguously. We have considered standard approaches to solution of the inverse problem of reconstruction of the particle size distribution functions, as well as approaches based on solution of the problem of optimization from the parameters of the distribution function and scaling parameters:

$$\sum_{i=1}^{n} \frac{\left(\left| I_e(\lambda_i) - a \cdot I_s(\lambda_i) \left[b \int_{r1}^{r2} \phi(r', M, \alpha, \bar{r}, \bar{g}) \beta_{174.5}(r, \lambda_i) dr' + \beta_{174.5}^{a+m}(\lambda_i) \right] \right|^2 \times \left(\frac{1}{2\beta(\lambda_i)} (1 - \exp(-2\beta(\lambda_i)\Delta)) \right) \cdot \exp\left(-2 \int_0^{85} \beta_{ext}^{a+m}(\lambda_i, h) dh \right) + \int_0^5 \beta_{ext}^{a+m}(\lambda_i, h) dh \right)}{I_e(\lambda_i)^2} \to \min$$

(5.68)

where a is the signal scale parameter, b is the coefficient of contribution to scattering from artificial aerosol at an angle of 174.5° (in fact, it accounts for the ethylene glycol particle concentration and the relative contribution in comparison with the molecular and aerosol atmospheric components), r and g are the vectors of modes and half-widths of the Gauss distribution densities, M is the coefficient of normalization to the unit distribution density, $\beta_{174.5}(r, \lambda_i)$ is the coefficient of scattering at the corresponding angle for the ethylene glycol particle with the radius r, $\beta_{174.5}^{a+m}(\lambda_i)$ is the total coefficient of scattering by aerosol and molecular atmosphere, $\beta_{ext}^{a+m}(\lambda_i, h)$ is the extinction coefficient, Δ is the depth of aerosol formation, $\beta(\lambda_i) = \int_{r1}^{r2} \phi(r', M, \alpha, \bar{r}, \bar{g}) \beta(r, \lambda_i) dr' + \beta_{atm}(\lambda_i)$ is the extinction in a cloud, $I_e(\lambda_i)$ and $I_s(\lambda_i)$ are the backscattered signal spectrum and the supercontinuum spectrum coming to the detector (results from multiplication by transmittance), $\beta_{atm}(\lambda_i)$ is the background atmospheric extinction. In general, the considered gradient methods and regularization methods failed to provide a positive result. Then we have proposed other approach, which is considered below.

If the model transmittance or transmittance calculated by model data is used in Equation 5.68, then, with the standard aerosol and molecular model, this equation can be simplified and transformed into a system of equations for wavelengths λ_i. The coefficients α_1 and α_2, accounting for the contribution of the background (model) scattering component in the direction of 174.5° and the contribution of the scattering by mist play the role of scaling coefficients

$$\frac{I_e(\lambda_i)}{I_s(\lambda_i)T(\lambda_i)} \approx \alpha_1 \int_{r1}^{r2} \phi(r', \bar{r}, \bar{g}) \beta_{174.5}(r, \lambda_i) dr' + \alpha_2 \beta_{174.5}^{a+m}(\lambda_i). \quad (5.69)$$

The particle size distribution function is lognormal and depends on two parameters (Figure 5.22):

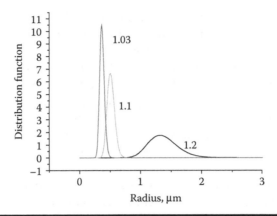

Figure 5.22 Examples of lognormal distribution functions for mists with different parameters *g*.

$$\phi(r, r_m, g) = \frac{N}{\sqrt{2\pi} r \ln g} \exp\left[\frac{1}{2}\left(\frac{\ln r - \ln r_m}{\ln g}\right)^2\right]. \tag{5.70}$$

The parameter determining the particle number density N is taken into account in the scaling coefficient α_1, and therefore it is taken equal to unity.

We propose the following approach: A sample of pairs of radii r_m and parameters of width of the lognormal function g is created (see Equation 5.70). For every pair, the mist scattering coefficients are calculated for different wavelengths. Then the calculated coefficients are substituted into Equation 5.71

$$\min_{\alpha_1, \alpha_m > 0, \beta_{174.5}(\lambda, r_{mj}, g_j)} \sum_{i=1}^{n} \left(\frac{I_e(\lambda_i)}{I_s(\lambda_i) T(\lambda_i)} - \alpha_1 \beta_{174.5}(\lambda_i, r_{mj}, g_j) - \alpha_2 \beta_{174.5}^{a+m}(\lambda_i)\right)^2. \tag{5.71}$$

Then, search for the coefficients α_1 and α_2 is performed. It is obvious that they can be found through solution of the overdetermined system of linear Equations 5.69 with two unknowns.

Among the obtained coefficients, positive pair α_1 and α_2 are selected at solution of system (5.69). Then among them, we select the pair and the parameters of the lognormal function r_m and g, at which functional (5.71) is minimal. It is obvious that the accuracy of solution depends on previously calculated $\beta_{174.5}(\lambda_i, r_{mj}, g_j)$ and noise present in the spectral signal, as well as errors in determination of the spectral profile of model scattering coefficients.

In the case of consideration, we have carried out the study for modes of lognormal distributions from 0.2 to 2 µm in the range of radii from 0.0001 to 4 µm and for the parameter g from 1.01 to 1.3. For matching of half-widths as functions of the modal radius, the parameter g is specified as $1 + a/r_m$, where a varies from 0.01 to 0.3. Thus, 5000 readouts for radii from 0.001 to 4 μm are created along with 50 ones for the parameter a. Enumeration of all versions takes about 15 min in a 2.5 GHz processor. A disadvantage associated with the computation time can be readily eliminated through parallelizing of algorithms and narrowing of the number of previously calculated $\beta_{174.5}(\lambda_i, r_{mj}, g_j)$. Although, as the experience shows, the decrease of the number of steps in radii more often can lead to other results, because the problem to be solved is ill-posed and allows multiple solutions. A possible way is to take a certain set of separated solutions providing a minimum as optimal solutions and then to search for the more accurate solutions in their vicinity.

Then, we have analyzed the possibilities of reconstruction of the distribution functions with allowance for the different level of noise in a signal.

Figure 5.23a shows the relative contributions to scattering, at which the guaranteed reconstruction of the particle size distribution is possible at a noise level of 10% for the parameter a equal to 0.03, 0.1, and 0.2. As the width of the distribution function increases, it is necessary to increase the contribution from mist to scattering. For the parameter a equal to 0.2, the scattering by mist should exceed the scattering by clear atmosphere two and more times. For narrow widths of the distribution functions, the situation is better, when the order of scattering can be the same or even several times smaller. This is caused by the fact that for narrow widths the scattering spectrum has high-frequency singularities, whereas at an increase of the width the scattering at a wavelength is a slightly varying function.

Figure 5.23b shows possible errors of reconstruction of the distribution mode for the noise level of 10%–20% at different widths of the lognormal function. Thus, for example, for the 0.2-μm mode the reconstruction 0.34 μm, while for 0.5 μm, it gives 0.6 and 0.49 μm. In the previous case, the reconstruction from experimental data by the genetic algorithm yielded the 0.6-μm mode for particles with a diameter of 1 μm, which is in agreement with the obtained results. For

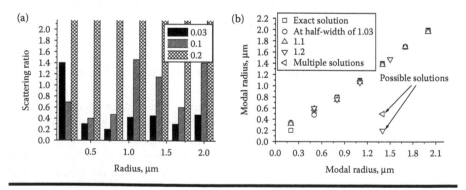

Figure 5.23 Ratio of the mist contribution to scattering to the scattering by clear atmosphere (a) and deviations of the reconstructed modal radii from preset model radii (b).

larger radii, more accurate results were obtained, but near the mode of 1.4 μm multiple solutions are obtained.

References

1. Measures, R. M., *Laser Remote Sensing. Fundamentals and Applications*. New York: John Wiley & Sons, 1984.
2. Luther, G. G., Newel, A. C., Moloney, J. V., and Wright, E. M., Short-pulse conical emission and spectral broadening in normally dispersive media, *Opt. Lett.*, vol. 19, pp. 789–791, 1994.
3. Woste, L., Wedekind, C., Wille, H., Rairoux, P., Stein, B., Nikolov, S., Werner, Ch., Niedermeier, S., Schillinger, H., and Sauerbrey, R., Femtosecond atmospheric lamp, *Laser und Optoelectronics*, vol. 29, pp. 51–53, 1997.
4. Rairoux, P. et al., Remote sensing of the atmosphere using ultrashort laser pulses, *Appl. Phys.*, vol. 71, pp. 573–580, 2000.
5. Wolf, J.-P. et al., Teramobile a nonlinear femtosecond terawatt lidar, *Proc. ILRC 21*, Quebec City, Canada, Part 1, 2002, pp. 47–50.
6. Weitcamp, C. Ed., *LIDAR: Rang-Resolved Optical Remote Sensing of the Atmosphere* (foreword by Herbert Walther). Springer Science & Business Media, 2005, 455 p. ISBN 0387400753, 9780387400754.
7. Zuev, V. E., Zemlyanov, A. A., Kopytin, Yu. D., *Modern Problems of Atmospheric Optics, Vol. 6, Nonlinear Atmospheric Optics*. Leningrad: Gidrometeoizdat, 1989, 256 p.
8. Zuev, V. E., and Zuev, V.V., *Modern Problems of Atmospheric Optics. Vol. 8, Remote Optical Sensing of the Atmosphere*. St. Petersburg: Gidrometeoizdat, 1992, 232 p.
9. Zuev, V. E., and Naats, I. E., *Modern Problems of Atmospheric Optics. Vol. 7, Inverse Problems of Atmospheric Optics*. St. Petersburg: Gidrometeoizdat, 1990, p. 287.
10. Kryukov, P. G., *Femtosecond Pulses*. Moscow: Fizmatlit, 2008, p. 208.
11. Kandidov, V. P., Kosareva, O. G., Mozhaev, E. I., and Tamarov, M. P., Femtosecond nonlinear optics of the atmosphere, *Atmos. Ocean. Optics*, vol. 13, No. 5, pp. 394–401, 2000.
12. Kandidov, V. P., Golubtsov, I. S., and Kosareva, O. G., Supercontinuum sources in a high-power femtosecond laser pulse propagating in liquids and gases, *Quant. Electron.*, vol. 34, No. 4, pp. 348–354, 2004.
13. Kandidov, V. P., Shlenov, S. A., and Kosareva, O. G., Filamentation of high-power femtosecond laser radiation, *Quant. Electron.*, vol. 39, No. 3, pp. 205–227, 2009.
14. Militsin, V. O., Kouzminsky, L. S., and Kandidov, V. P., Stratified-medium model in studying propagation of high-power femtosecond laser radiation through atmospheric aerosol, *Atmos. Ocean. Optics*, vol. 18, No. 10, pp. 789–795, 2005.
15. Zemlyanov, A. A., and Geints, Yu. E., Spectral, energy, and angular characteristics of supercontinuum formed by a femtosecond-duration pulsed laser radiation in air, *Atmos. Ocean. Optics*, vol. 20, No. 1, pp. 32–39, 2007.
16. Geints, Yu. E., Zemlyanov, A. A., Kabanov, A. M., Matvienko, G. G., and Stepanov, A. N., Self-action of sharply focused femtosecond laser radiation in air in filamentation regime. Laboratory and numerical experiments, *Optika Atmosfery i Okeana*, vol. 22, No. 2, pp. 119–125, 2009 [in Russian].

17. Geints, Yu. E., Zemlyanov, A. A., Krekov, G. M., and Matvienko, G. G., Femtosecond laser pulse scattering on spherical polydispersions: Monte-Carlo simulation, *Optika Atmosfery i Okeana*, vol. 23, No. 5, pp. 325–332, 2010 [in Russian].

18. Apeksimov, D. V. et al., Interaction of GW laser pulses with liquid media. Part 1. Explosive boiling up of large isolated water droplets, *Optika Atmosfery i Okeana*, vol. 23, No. 7, pp. 536–542, 2010 [in Russian].

19. Krekov, G. M., Krekova, M. M., and Sukhanov, A. Ya., Broadband lidar potentiality estimate for remote sensing of the molecular atmosphere, *Optika Atmosfery i Okeana*, vol. 22, No. 5, pp. 482–493, 2009 [in Russian].

20. Krekov, G. M., Krekova, M. M., and Sukhanov, A. Ya., Estimate of the promising white-light lidar efficiency for sensing of the stratus cloud microphysical parameters, *Optika Atmosfery i Okeana*, vol. 22, No. 7, pp. 661–670, 2009 [in Russian].

21. Krekov, G. M., Krekova, M. M., and Sukhanov, A. Ya., Estimate of perspective white-light lidar efficiency for sensing of the stratus clouds microphysical parameters: 2. Parametric modification of the iteration method lidar equation solution, *Optika Atmosfery i Okeana.*, vol. 22, No. 8, pp. 795–802, 2009 [in Russian].

22. Krekov, G. M., Krekova, M. M., and Sukhanov, A. Ya., Estimate of perspective white-light lidar efficiency for sensing the stratus cloud microphysical parameters: 3. Inverse problem solution., *Optika Atmosfery i Okeana*, vol. 22, No. 9, pp. 862–872, 2009 [in Russian].

23. Geints, Yu. E., Zemlyanov, A. A., Kabanov, A. M., Matvienko, G. G., and Pogodaev, V. A., Propagation of high power laser radiation in the atmosphere, *Optika Atmosfery i Okeana.*, vol. 22, No. 10, pp. 931–936, 2009 [in Russian].

24. Bourayou, R. et al., White-light filaments for multiparameter analysis of cloud microphysics, *Opt. Soc. Am. B.*, vol. 22, No. 2, pp. 211–215, 2005.

25. Rairoux, P. et al., Remote sensing of atmosphere using ultrashort laser pulses, *Appl. Phys. B.*, vol. 71, pp. 573–580, 2000.

26. Kasparian J., and Wolf, J.-P., A new transient SRS analysis method of aerosols and application to a nonlinear femtosecond lidar, *Opt. Com.*, vol. 152, pp. 355–360, 1998.

27. Faye, G., Kasparian, J., and Sauerbrey, R., Modifications to the lidar equation due to nonlinear propagation in air, *Appl. Phys. B.*, vol. 73, pp. 157–163, 2001.

28. Luo, Q., Yu, J., Abbas Hoseini, S., Liu, W., Ferland, B., Roy, G., and Chin, S. L., Long-range detection and length estimation of light filaments using extra-attenuation of terawatt femtosecond laser pulses propagation in air, *Appl. Optics*, vol. 44, No. 3, pp. 391–397, 2005.

29. Samokhvalov, I. V. et al., Remote determination of atmospheric aerosol parameters, Chapter 5, in *Laser Sensing of the Troposphere and the Underlying Surface*, Zuev, V. E. Ed. AS USSR, Sib. Branch, Novosibirsk: Nauka (Russian), 1987, p. 258.

30. Rozhdestvin, V. N. Ed. *Opto-Electronic Systems for Ecological Monitoring of the Environment.* by, Moscow: Bauman MSTU Press, 2002, p. 528.

31. Karasik, V.E., and Orlov, V.M., *Laser Vision Systems: Training Aid.* Moscow: Bauman MSTU Press, 2001, p. 352.

32. Orlov, V. M., Samokhvalov, I. V., Krekov, G. M., Mironov, V. L., Balin, Yu. S., Banakh, V. A., Belov, M. L., Kopytin, Yu. D., and Lukin, V. P., Operative control of atmospheric channel parameters by laser sensing methods, in *Signals and Noise in Lidar Detection and Ranging.* Zuev, V. E. Ed. Moscow: Radio i Svyaz', 1985, p. 264.

33. Zuev, V. E., Kaul', B. V., Samokhvalov, I. V., Kirkov, K. I., and Tsanev, V. I., *Laser Sensing of Industrial Aerosols.* Novosibirsk: Nauka, 1986, p. 188.

34. Naats, I. E., *Theory of Multifrequency Laser Sensing of the Atmosphere*, Novosibirsk: Nauka, Sib. Branch, 1980, p. 157.
35. Klett, J. D., Stable analytical inversion for processing lidar returns, *Appl. Optics*, vol. 20, pp. 211–220, 1981.
36. Britton Chance, B., Alfano, R. R., Tromberg, B. J., Tamura, M., Sevick-Muraca, E. M., Optical tomography and spectroscopy of tissue VI, *Proc. SPIE. Int. Soc. Opt. Eng.*, vol. 5693, p. 301, 2005.
37. Weitkamp, C. Ed. *LIDAR: Range-Resolved Optical Remote Sensing of the Atmosphere*. Singapore: Springer, 2005, p. 451.
38. Moreno, N., Bougourd, S., and Haseloff, J., Imaging plant cells, in *Handbook of Biological Confocal Microscopy*, Pawley, J. B. Ed. New York: Springer Science, 2006, p. 790.
39. Zheltikov, A. M., *Ultrashort Pulses and Methods of Nonlinear Optics*. Moscow: Nauka, 2006, p. 261.
40. Yoo, K. M., and Alfano, R. R., Time-resolved coherent and incoherent components of forward light scattering in random media, *Opt. Lett.*, vol. 15, pp. 320–323, 1990.
41. Liu, F., Yoo, K. M., and Alfano, R. R., Ultrafast laser-pulse transmission and imaging through biological tissue, *Appl. Optics*, vol. 32, pp. 554–558, 1993.
42. Wang, L., Ho, P. P., Zhang, G., and Alfano, R. R., Ballistic 2-D imaging through scattering walls using an ultrafast optical Kerr gate, *Science*, vol. 253, pp. 769–771, 1991.
43. Andreoni, A., Bondani, M., Brega, A., Paleari, F., and Spinelli, A. S., Detection of nondelayed photons in the forward-scattering of picosecond pulses, *Appl. Phys. Lett.*, vol. 84, pp. 2457–2460, 2004.
44. Calba, C., Mees, L., Rose, C., and Girasole, T., Ultrashort pulse propagation through a strongly scattering medium: Simulation and experiments, *J. Opt. Soc. Am.*, vol. A25, pp. 1541–1550, 2008.
45. Podgaetsky, V. M., Tereshchenko, S. A., Smirnov, A. V., and Vorob'ev, N. S., Bimodal temporal distribution of photons in ultrashort laser pulse passed through a turbid medium, *Opt. Commun.*, vol. 180, pp. 217–223, 2000.
46. Tereshchenko, S. A., Podgaetskii, V. M., Vorob'ev, N. S., and Smirnov, A. V., Conditions during passage of short optical pulses across a strongly scattering medium, *Quant. Electron.*, vol. 23, No. 3, pp. 265–268, 1996.
47. Tereshchenko, S. A., Podgaetskii, V. M., Vorob'ev, N. S., and Smirnov, A. V., Axial and diffusion models of laser pulse propagation in a highly scattering medium, *Quant. Electron.*, vol. 34, No. 6, pp. 541–544, 2004.
48. Ishimary, A., *Wave Propagation and Scattering in Random Media*. New York: Willey, 1991, p. 600.
49. Kopeika, N. S., *A System Engineering Approach to Imaging*. Bellingham, WA: SPIE Press, 1998.
50. Patterson, M. S., Chance, B., and Wilson, B. C., Time resolved reflectance and transmittance for the invasive measurements of tissue optical properties, *Appl. Optics*, vol. 28, pp. 2331–2336, 1989.
51. Krekov, G. M., Krekova, M. M., and Samokhvalov, I. V., On issue of deformation of short optical pulses in model scattering media, *Izv. Vyssh. Uchebn. Zaved. Fizika*, vol. 12, No. 5, pp. 150–153, 1969.
52. Bucher, E. A., Computer simulation of light pulse propagation for communication through thick clouds, *Appl. Optics*, vol. 12, pp. 2391–2400, 1973.

53. Zaccanti, G., Monte Carlo study of light propagation in optically thick media: point source case, *Appl. Optics*, vol. 30, pp. 2031–2037, 1991.
54. Jacques, S. L., Time resolved propagation of ultrashort laser pulses within turbid tissues, *Appl. Optics*, vol. 28, pp. 2223–2229, 1989.
55. Sergeeva, E. A., Kirillin, M. Iu., and Priezzhev, A. V., Propagation of femtosecond pulse in a scattering medium: theoretical analysis and numerical simulation, *Quant. Electron.*, vol. 36, No. 11, pp. 1023–1031, 2006.
56. Krekov, G. M., Orlov, V. M., and Belov, V. V., *Imitation Modeling in Problems of Optical Remote Sensing.* Novosibirsk: Nauka SO, 1988, p. 164.
57. Mihailov, G. A., *Optimization of Weighted Monte Carlo Methods.* Moscow: Nauka, 1987, p. 187.
58. Vorob'ev, N. S., Podgaetskii, V. M., Smirnov, A. V., and Tereshchenko, S. A., Observation of the separation of photons in ultrashort laser pulse, *Quant. Electron.*, vol. 28, No. 2, pp. 181–182, 1999.
59. Kirkby, D. R., Design and construction of a tissue-like optical phantom, Chapter 7, in *A Picosecond Optoelectronic Cross Correlator using a Gain Modulated Avalanche Photodiode for Measuring the Impulse Response of Tissue*, PhD thesis, 1999, pp. 177–184. Department of Medical Physics and Bioengineering, University College London. http://www.medphys.ucl.ac.uk/research/borg/homepages/davek/phd/chapter7
60. Kutis, I. S., Sapozhnikova, V. V., Kuranov, R. V., and Kamenskii, V. A., Study of the morphological and functional state of higher plant tissues by optical coherence microscopy and optical coherence tomography, *Russ. J. Plant Physiol.*, vol. 52, No. 4, pp. 559–564, 2005.
61. Reeves, A., Eeves, A., Parson, R. L., and Hettinger, J. W., In vivo three-dimensional imaging of plants with optical coherence microscopy, *J. Micros.*, vol. 208, pp. 177–189, 2002.
62. Johanson, J., Berg, R., Pifferi, A., Svanberg, A., and Bjorn, L. O., Time-resolved studies of light propagation in Crassula and Phaseolus leaves, *Photochem. Photobiol.*, vol. 69, pp. 242–247, 1999.
63. Krekov, G. M., Krekova, M. M., Lisenko, A. A., and Sukhanov, A. Ya., Radiation characteristics of a plant leaf, *Optika Atmosfery i Okeana*, vol. 22, No. 4, pp. 397–415, 2009 [in Russian].
64. Zuev, V. E., and Naats, I. E., *Inverse Problems of Laser Sensing.* Novosibirsk: Nauka, 1982, p. 242.
65. Zuev, V. E., and Naats, I. E., *Inverse Problems of Lidar Sensing of the Atmosphere.* Berlin: Shpriger-Verlag, 1983, p. 260.
66. Bockmann, C., Mironova, I., Muller, D., Schneidenbach, L., and Nessler, R., Microphysical aerosol parameters from multiwavelength lidar, *J. Opt. Soc. Am.*, vol. 22, pp. 518–528, 2005.
67. Matvienko, G. G., Veretennikov, V. V., Krekov, G. M., and Krekova, M. M., Remote sensing of atmospheric aerosols with a white-light femtosecond lidar. Part 1. Numerical simulation, *Atmospheric and Oceanic Optics*, vol. 16, No. 12, pp. 1013–1019, 2003.
68. Zuev, V. E., Krekov, G. M., Krekova, M. M., Makienko, E. V., and Naats, I. E., Theory and numerical experiment on remote sensing of cloud aerosol, in *Radiophysical Investigations of the Atmosphere*, Zuev, V. E. Ed. Leningrad: Gidrometeoizdat, 1977, pp. 6–15.
69. Deirmendjan, D., *Electromagnetic Scattering of Spherical Polydispersions.* New York: Elsevier, 1969, p. 166.

70. Veretennikov, V. V., Kostin, B. S., and Naats, I. E., On selection of the number of measurements at optical sensing of atmospheric aerosol, in *Issues of Laser Sensing of the Atmosphere*, Zuev, V. E. Ed. Novosibirsk: Nauka SO, 1976, pp. 92–104.

71. Tikhonov, A. N., and Arsenin, V. Ya., *Methods for Solution of Ill-Posed Problems*. Moscow: Nauka, 1979, p. 286.

72. Veselovski, I., Kolgotin, A., Muller, D., and Whiteman, D. N., Information content of multiwavelength lidar data with respect to microphysical particle properties derived from eigenvalue analysis, *Appl. Optics*, vol. 44, pp. 5292–5303, 2005.

73. Pornsavad, P., Bockmann, C., Ritter, C., and Rafler, M., Ill-posed retrieval of aerosol coefficient profiles from Raman lidar data by regularization, *Appl. Optics*, vol. 47, pp. 1649–1661, 2008.

74. Krekov, G. M., Krekova, M. M., Makienko, E. V., and Naats, I. E., Optical location of the microphysical characteristics of scattering media, *Radiophys. Quant. Electr.*, vol. 20, pp. 358–365, 1980.

75. Zuev, V. E., and Krekov, G. M., *Optical Models of the Atmosphere*. Leningrad: Gidrometeoizdat, 1986, p. 256.

76. Voutilainen, A., Statistical inversion methods for the reconstruction of aerosol size distributions, *Rep. Ser. Aerosol Sci.*, vol. 52, p. 137, 2001. ISSN 0784-3496.

77. Gillespie, J. B., Ligon, D. A., Pellegrino, P. M., Fel, N. N., and Wood, N. J., Development of a broadband lidar system for remote determination of aerosol size distributions, *Meas. Sci. Technol.*, vol. 13, pp. 383–390, 2002.

78. Ligon, D., Chen, T. W., and Gillespie, J. B., Determination of aerosol parameters from light-scattering data using an inverse Monte Carlo technique, *Appl. Optics*, vol. 35, pp. 4297–4304, 1996.

79. Blaunstein, N., Arnon, Sh., Zilberman, A., and Kopeika, N., *Applied Aspects of Optical Communication and LIDAR*. New York: CRC Press, Taylor & Francis Group, 2010, p. 262.

80. Ramachandran, G., and Leith, D., Extraction of aerosol-size distribution from multispectral light extinction data, *Aerosol Sci. Tech.*, vol. 17, pp. 303–325, 1992.

81. Voutilainen, A., and Kaipio J. P., Statistical inversion of aerosol size distribution data, *J. Aerosol. Sci.*, vol. 31, No. Suppl. 1, pp. 767–768, 2000.

82. Trahan, M. W., Wagner, J. S., Shokair, I. R., Tisone, G. C., and Gray, P. C., The use of intelligent algorithms in multispectral UV analysis, *CALIOPE Tech. Rev. Proc.*, vol. 1, pp. 358–375, 1998.

83. Madkour, A. A., Hossain, M. A., Dahal, K. P., and Yu, H., Real-time system identification using intelligent algorithm, *IEEE SMC UK_RI Confer. Proc.*, Londonderry, 2004, pp. 236–241.

84. Krekov, G. M., and Sukhanov, A. Ya., Application of artificial intelligence methods in remote sensing problems, *Proc. of XV International Symposium Atmospheric and Ocean Optics, Atmospheric Physics*, June 22–29, 2008, Krasnoyarsk, Ser. 100. CO-05.

85. Ressom, H., Miller, R. L., Natarajan, P., and Slade, W. H., Computation intelligence and its application in remote sensing, in *Remote Sensing of Coastal Aquatic Environments*, Miller, R. L. Ed. Berlin: Springer, 2005, pp. 205–227.

86. Galushkin, A. I., *Neural Network Theory*. Berlin: Springer, 2007, p. 402.

87. Goldberg, D., and Sastry, K., *Genetic Algorithms*. Berlin: Springer, 2007, p. 350.

88. Lienert, B. R., Porter, J. N., and Sharma, S. K., Repetitive genetic inversion of optical extinction data, *Appl. Optics*, vol. 40, No. 21, pp. 3476–3482, 2001.

89. Lienert, B. R., Porter, J. N., and Sharma, S. K., Aerosol size distributions from genetic inversion of polar nephelometer data, *J. Atmosph. Sci.*, vol. 20, pp. 1403–1410, 2003.

90. Mera, N. S., Elliott, L., and Ingham, D. B., A multi-population genetic algorithm approach for solving ill-posed problems, *Comput. Mechanics*, vol. 33, pp. 254–262, 2004.
91. Dolenko, S. A., Gerdova, I. V., Dolenko, T. A., and Fadeev, V. V., Laser fluorimetry of mixtures of polyatomic organic compounds using artificial neural networks, *Quant. Electron.*, vol. 31, pp. 834–838, 2001.
92. Ye, M., Wang, S., Lu, Y., Hu, T., Zhu, Z., and Xu, Y., Inversion of particle-size distribution from angular light-scattering data with genetic algorithms, *Appl. Optics*, vol. 38, pp. 2667–2685, 1999.
93. Kataev, M. Yu., and Sukhanov, A. Ya., Capabilities of the neural network method for retrieval of the ozone profile from lidar data, *Atmosp Ocean. Optics*, vol. 16, No. 12, pp. 1020–1023, 2003.
94. Grishin, A. I., Krekov, G. M., Krekova, M. M., Matvienko, G. G., Sukhanov, A. Ya., Timofeev, V. I., Fateyeva, N. L., and Lisenko, A. A., Study of organic aerosol of phytogenic origin with fluorescent lidar, *Atmos. Ocean. Optics*, vol. 20, No. 4, pp. 294–302, 2007.
95. Krekov, G. M., and Sukhanov, A. Ya., Improved genetic algorithm of multiwavelength lidar sounding of the atmospheric aerosol, *Atmos. Ocean. Optics*, vol. 24, No. 9, pp. 754–758, 2011.
96. Apeksimov, D. V. et al., Generation of femtosecond laser pulses, in *Femtosecond Atmospheric Optics*, Bagaev, S. N. and Matvienko, G. G. Eds. Novosibirsk: SB RAS Press, 2010, p. 238.
97. Golubitskii, B. M., and Tantashev, M. V., On local estimates in the Monte Carlo method, *Zh. Vychisl. Mat. Mat. Fiz.*, vol. 17, No. 6, pp. 1374–1379, 1977 [in Russian].
98. Holland, J. H., *Adaptation in Natural and Artificial Systems*. Ann Arbor: University of Michigan Press, 1975.

Chapter 6

Isoplanarity Problem in Vision Theory

Vladimir V. Belov and Mikhail V. Tarasenkov

Contents

6.1 Introduction to the Problem

The problem of the isoplanarity or shift-invariant problem in the vision theory begun to be discussed with the process of imaging of stationary objects through scattering and absorbing media taken as an example [1–10]. The considered approach to estimation of the size of isoplanarity zones is also applicable, from our point of view, to the observation of objects through turbulent media [11–28]. The main equation used in the vision theory for the construction of key functions and characteristics allowing analysis of the influence of a scattering medium on the imaging process is the stationary radiation transfer equation

$$(\boldsymbol{\omega}, grad\ I) = -\beta_{ext}I + \beta_{sc}\int_{\Omega} I(\mathbf{r},\boldsymbol{\omega}')g(\mathbf{r},\boldsymbol{\omega},\boldsymbol{\omega}')d\boldsymbol{\omega}' + \Phi_0(\mathbf{r},\boldsymbol{\omega}). \qquad (6.1)$$

Here, ω is the radiation propagation direction, I is the radiation flux intensity, β_{ext} and β_{sc} are, respectively, the extinction and scattering coefficients, $\mathbf{r}$ is the vector of point coordinates in space, ω' is the direction of radiation propagation after scattering, g is the scattering phase function, and Φ_0 is the source function.

The particular formulation of the problem of vision theory is determined, as known, by the boundary conditions. The general solution of Equation 6.1 is the Green's function $G(\mathbf{r}_0, \omega_0; \mathbf{r}, \omega)$ with the boundary conditions specified in the form of a omnidirectional point source $\delta(\mathbf{r} - \mathbf{r}_0)\delta(\omega - \omega_0)$ placed at the point $\mathbf{r}_0$ and emitting in the direction ω_0 (see definitions also in Chapter 2). Let they correspond to the night (no external sources of illumination) observation of objects on the surface $z = 0$ with a projection-type optical system being on the surface $z = z_1$ and oriented, for example, perpendicularly to the object plane (nadir observation). Then they can be written in the form

$$\begin{cases} I(r,\omega; z = 0) = \delta(\mathbf{r} - \mathbf{r}_0)\delta(\omega - \omega_0), & (\omega, n) < 0, \\ I(r,\omega; z = z_1) = 0, & (\omega, n_1) < 0, \end{cases} \qquad (6.2)$$

where n, n_1 are external normals to the surfaces $z = 0$ and $z = z_1$.

For the construction of the projective or scanner image of an object observed through the scattering medium under these observation conditions, it is sufficient to know (at least, for the central isoplanarity zone of the image) the function $G(\mathbf{r}_0, \omega_0; \mathbf{r}^*, \omega^*)$ only at the center of the entrance pupil of the optical system, that is, at the point $\mathbf{r}^*$ and for one given direction ω^*. If ω_0 is fixed, then the function $G = G(\mathbf{r}_0, \omega_0; \mathbf{r}^*, \omega^*) = G(\mathbf{r}_0)$ is identical to the point spread function $h(\mathbf{r})$ from the viewpoint of the theory of linear systems (it is applicable in this case because Equation 6.1 is linear with respect to the intensity I), and the two-dimensional Fourier transform of this function $F^2[G(\mathbf{r})] \equiv F^2[h(\mathbf{r})] = \overset{\circ}{H}(\omega, \gamma)$ is the optical transfer (in general case, complex) function, where ω, γ are the spatial frequencies. The subscript 0 of $\mathbf{r}$ is usually omitted from here on in this chapter to emphasize that it is an argument of the Green's function or the point spread function (PSF) rather than a fixed point at the line of sight of the optical system.

Are the functions $G(\mathbf{r}) \equiv h(\mathbf{r})$ and $\overset{\circ}{H}(\omega, \gamma)$ individual characteristics of this particular system independent of the properties of objects and (even ideal) optical systems? Is it possible to construct images of some objects with the aid of the function $G(\mathbf{r})$ (or $\overset{\circ}{H}(\omega, \gamma)$)? Obviously, yes, but only for the central point of the image plane of the ideal optical system oriented in the direction ω^* and *only* for objects, whose every point emits radiation energy strictly in the direction ω_0. Evidently, it is hard to find objects with such properties in reality. The only exclusion is the trivial case of observation of a collimated laser beam propagating along the optical axis of imaging system through the scattering medium.

As a rule, elementary parts of natural objects are characterized by the angular directional pattern markedly different from $\delta(\omega - \omega_0)$. For simplicity, we consider homogeneous (in the optical sense) object surfaces, that is, objects, in which radiation from every point is described by the same angular directional pattern $Q(\omega,\mathbf{r}) = Q(\omega)$.

Thus, to the describe the process of imaging through the scattering medium with the aid of the Green's functions, it is necessary to solve Equation 6.1 for every of N $\delta(\omega - \omega_i,\mathbf{r})$ sources. This set of the radiators should allow reconstruction of the real angular directional pattern $Q(\omega)$ of radiation from object surface elements, that is

$$Q(\omega) \approx p_i \delta(\omega - \omega_i), \quad i = 1,2,...,N. \tag{6.3}$$

This, in its turn, allows the total PSF to be constructed, for example, as follows:

$$h(\mathbf{r}) = h_{\sum}(\mathbf{r}) \approx \sum_{i=1}^{N} p_i G_i(\mathbf{r}) \Delta\Omega_i, \tag{6.4}$$

where p_i is the "weight" of the i-th direction of radiation from a monodirectional point source in Equation 6.3, and $G_i(\mathbf{r})$ is the Green's function corresponding to this direction, $\Delta\Omega_i$ is the solid angle, within which it is assumed that $G_i(\mathbf{r}) = \text{const}$. The accuracy of $h(\mathbf{r})$ reconstruction obviously depends both on the accuracy of the $G_i(\mathbf{r})$ estimation method and on the choice of N and $\Delta\Omega_i$ in Equation 6.4.

In the cases when the imaging requires consideration of the sizes of scattering spots arising because projective systems are used to construct plane images of volume objects [22], it is necessary to find the Green's functions already for the entire plane of the entrance pupil and all directions within a hemisphere rather than for the single point $\mathbf{r}^*$ and the selected direction ω^*. That is, it is necessary to find either $G = G(\mathbf{r},\omega; \mathbf{r}^*,\omega^*)$, where $\mathbf{r}^* \in \{R\}$ (a point in the coordinate space), $\omega^* \in \{\Omega^+\}$ (unit vector in the directional half-space) or $G = G(\mathbf{r},\omega;\xi)$, where ξ is the current coordinate along the section of the optical axis of the receiver between the center of the entrance pupil and the intersection with the object plane.

Remember that in the linear-system approach, as defined, for example, in Reference 23, the point spread function is understood as a solution of Equation 6.1 with the following boundary condition:

$$\begin{cases} I(\mathbf{r},\omega; z = 0) = \delta(\mathbf{r} - \mathbf{r}_0)Q(\omega), & (\omega, n) < 0, \\ I(\mathbf{r},\omega; z = z_1) = 0, & (\omega, n_1) < 0, \end{cases} \tag{6.5}$$

That is, in this case, Equation 6.1 is solved only once, and this solution is the point spread function, allowing us to construct, using the convolution integral, the image of any optically homogeneous objects (at least, in the central isoplanar zone), whose every pony emits by the law $Q(\omega)$.

Thus, the method of Green's functions allows the solution of problems of the vision theory in the more general form than within the framework of the linear-system approach. However, it is not always possible to implement this approach in practice, because usually there is no exact solution of Equation 6.1 in the analytical form (for example, when solving particular problems of vision in the atmosphere or through the atmosphere). Therefore, in the cases when the complexity of solution of Equation 6.1 with boundary conditions (6.2) for the only one given direction ω_i is comparable with solution of this equation with boundary conditions (6.5), the linear-system approach becomes more efficient. It can be easily shown, as in Reference 22 that within the framework of this approach, the finite focal depth of the space imaged by an optical system can be taken into account rather readily. To the full extent, the advantages of the linear-system approach show themselves in solution of Equation 6.1 by the Monte Carlo technique, the principle of which is described in Chapter 5.

It is obvious that the above statements are also true for the case of observation under daytime conditions. The similar reasoning can also be given for transfer properties of the channel of additional illumination of the object plane [11] with the only difference that for this channel there is no need to take into account any characteristics of the optical imaging system. This influence function (or the corresponding optical transfer function) allows us to take into account the processes of radiation reflection (rereflection) by the surface of objects with the following scattering in the medium toward this surface.

6.2 Algorithm for Statistical Simulation of Pulsed Reactions in Spherical Geometry

The PSF simulation algorithm can be easily constructed with the use of the algorithm for the calculation of the intensity of radiation reflected from the surface for the case of an optically homogeneous surface in the spherical geometry proposed in Reference 24. The statistical simulation of PSF reduces to the following (Figure 6.1). The trajectory of photon motion starts from the point A $(0,0,R_e)$ by the law of reflection, (e.g., Lambertian) with the unit weight, and their propagation in the medium is simulated by the standard algorithm from Reference 25. At every photon collision in the medium, the photon trajectory is turned so that the point of the last collision to be at the line of sight DN (Figure 6.1). In this case, the point A moves to some point on the surface with the coordinates (x_1, y_1). Prior to this, the Earth's surface is divided into cells by the polar coordinate system (r, φ) centered at

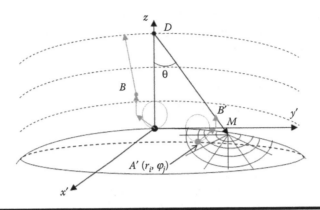

Figure 6.1 Geometry of statistical simulation of PSF.

the sensed point N. Then the cell, the point A falls in after the turn, is determined, and the local estimate of the radiation intensity is calculated as

$$I_k = \omega_k \frac{\beta_{sc,a} g_a(\cos(\mathbf{B'D},\mathbf{a'})) + \beta_{sc,m} g_m(\cos(\mathbf{B'D},\mathbf{a'}))}{(\beta_{ext,a} + \beta_{ext,m})\cos\theta} \exp(-\tau_{B'D}), \qquad (6.6)$$

where ω_k is the photon weight at the collision, $\beta_{sc,a}$, $\beta_{sc,m}$, $\beta_{ext,a}$, $\beta_{ext,m}$ are respectively, the coefficients of aerosol scattering, molecular scattering, aerosol extinction, and molecular extinction at the point of collision, θ is the angle between the nadir direction and the direction to the sensed point M, $\mathbf{a'}$ is the vector of photon motion direction after the trajectory turn, g_a, g_m are, respectively the aerosol and molecular scattering phase functions at the photon scattering from the direction $\mathbf{a'}$ to the direction $\mathbf{B'D}$. The photon weight $\omega_k = \omega_{k-1} \times \chi$, where k is the serial number of collision, $\chi = \beta_{sc}/\beta_{ext}$, and $\omega_k = 1$ if $k = 0$.

The statistical estimate of the radiation intensity at the point of detection as a function of (r_i, φ_j) is

$$I_{i,j} = \frac{1}{S_{i,j}} \left(\frac{1}{N} \sum_k I_{i,j,k} \right), \qquad (6.7)$$

where $S_{i,j}$ is the area of the corresponding cell, N is the total number of trajectories, k is the subscript of summation over all collisions, $I_{i,j,k}$ is the intensity of radiation coming to the point D from the cell (i,j) at the k-th collision.

It is obvious that, in the chosen coordinate system at the nadir observation and the stratified model of the atmosphere, PSF is characterized by the axial symmetry.

To check this algorithm, a series of numerical experiments was carried out in Reference 26, which confirmed the reliability of the results obtained.

It should be noted that in the single approximation in the axisymmetric case the solution is presented, for example, in Reference 27 and has the following form for the plane geometry of the problem:

$$h_1(r) = \frac{1}{2\pi^2} \int\limits_0^H \frac{\beta_{sc}(z)}{s^2(z)} \exp(-(\tau_1(z) + \tau_2(z))) g(\mu(r,z)) dz, \qquad (6.8)$$

where τ_1, and τ_2 are the optical distances from the collision point to the source on the surface and to the optical system, respectively; H is the height of position of the optical system, g is the scattering phase function, μ is the cosine of the angle between the vertical direction and the direction from the source to the collision point, β_{sc} is the scattering coefficient, s is the distance from the scattering point to the source.

It is shown in Reference 26 that the difference of statistical calculations of the PSF integral characteristics, that is

$$m_{00}(H,\theta) = \int\limits_{-\infty}^{+\infty} \int\limits_{-\infty}^{+\infty} h(x,y,H,\theta) dx dy,$$

obtained analytically in the single scattering approximation, from the statistical calculations (with allowance for only the first order of photon scattering) in the spherical geometry is no larger than 5%.

It is also shown there that the difference of PSF calculations by the Monte Carlo technique at the spherical geometry from calculations for the plain geometry begins to increase with an increase of the zenith angle θ and becomes significant starting from the angles $\theta > 75°$. It should be noted that PSF is not an axisymmetric function at $\theta \neq 0$. An example of calculation of this function is shown in Figure 6.2 for the case of space observation of the Earth's surface at $\theta = 45°$, which is characterized by an appearance of the second maximum (the first one is associated with the nonscattered radiation) caused by the effect of multiple scattering in the atmosphere.

6.3 Criteria for the Estimation of the Size of Image Isoplanarity Zones

Consider the isoplanarity of images formed in observations through scattering media. Remember that, for ideal optical systems, the pulsed reaction $h(\mathbf{r})$ is understood as an image of a point, and it is applicable to description of an image over the whole frame or the field of view. For vision systems, it is not usually true, even if the

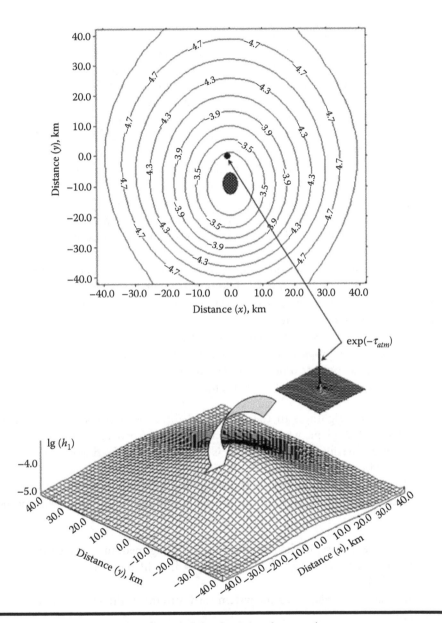

Figure 6.2 Example of simulation of the function $h_1(x,y;\omega_1^*)$.

optical source is assumed to be ideal. The point spread function of vision systems is characterized by the far wider half-width, which can exceed the frame size many times [11].

If we know the influence function determined for the central point of the frame, then this is by no means always possible to reconstruct (with the required accuracy) the image of an elementary volume—a radiant point against the background of an absorbing surface. This theoretically and experimentally established fact was used in Reference 27 to find criteria of the size of the central isoplanarity zone in vision systems. At the same time, this fact indicates the principal difference in the physical meaning of the point spread function used in the theory of analysis of optical systems and the characteristics used in the vision theory, in which the point spread function does not always correspond to the image of a point source.

Criterion 1 [11,27]. Let the observation be carried out through the scattering medium with an optical (or opto-electronic) system having the large field-of-view angle. In this case, it is worth asking how many isoplanarity zones can be separated in a given image or how many linear-system characteristics (pulsed reactions or optical transfer functions) should be used for the correct construction (with the required accuracy) of object images or for removal of a "trace" of scattering medium from an image.

We can answer this question using criteria [27] for estimates of the size of the *central* isoplanarity zone of images. It should be noted that even if we know the entire set of the Green's functions necessary for particular formulation of the problem, we cannot answer this question until the corresponding pulsed reactions of the imaging channels are determined with the use of these functions, that is, until real reflective (emissive) properties of the object plane are taken into account.

The image of some or other object can be obtained in different ways. The following procedures are the well known [28]:

1. Radiation brightness in the same fixed direction is measured at different points of the plane, where the detector is placed (spatial scanning)
2. Angular distribution of brightness is measured at one fixed spatial point (projective images)
3. The image is formed with a raster-type system when the angular scanning is carried out along one coordinate, while the detector moves along another axis

In the general case, in the presence of a scattering medium between the object and the receiving device, all the three images are different, even if the optics is assumed ideal.

Thus, for the isoplanar vision system (or invariant to the shift [21]), only one function

$$h(x, y; x', y') = h(x - x'; y - y') \tag{6.9}$$

can be used for image construction in place of the infinite set of the functions $h(x,y; x',y')$.

This simplifies significantly the image reconstruction process. However, the property of isoplanarity is inherent, strictly speaking, to only the imaging system employing the procedure (a) and under condition of horizontal homogeneity of the medium. Nevertheless, in the most applied problems of the vision theory for the procedures (b) and (c), it is possible to find the ranges of (x,y), within which Equation 6.9 is fulfilled with a certain degree of accuracy.

Two criteria for the estimation of the size of the central isoplanar zone (isozone) for the scheme of object observation through a scattering layer are proposed in Reference 27. Consider the image of the simplest object $Q(\omega;\xi,\eta) = Q(\omega)\delta(x - \xi)\delta(y - \eta)$—the point source of radiation placed at the point (ξ,η), formed by the method of spatial scanning. From here on, we assume that the coordinates in the object and image planes are reduced to the same scale. Then we can write

$$q(x - \xi, y - \eta) = Q(\omega) \int\int_{-\infty}^{\infty} \delta(x' - \xi)\delta(y' - \eta)h(x - x', y - y')dx'dy'$$

$$= h(x - \xi, y - \eta). \tag{6.10}$$

Here, ξ and η are coordinates in the image plane or in the object plane.

This means that in the case of imaging by the procedure (a), the image of point is, accurate to a constant, the point spread function and the image is isoplanar all over the frame.

The point spread function (or the optical transfer function) determined for the center of the frame in case of the schemes (b) and (c) allows reconstruction of the image of any optically homogeneous object at only one, namely, central point of the frame. Or, in other words, if condition (6.9) is not fulfilled, then the images formed by the algorithms (b) and (c) are not isoplanar in the strict sense.

Let us analyze the above mentioned geometrical explanation presented in Figure 6.3, which shows the scheme of formation of the image of a point through the scattering medium by the projective optical system L (Figure 6.3a) and the scheme of determination of PSF of the vision system (Figure 6.3b) for the central point of the frame.

We assume that the axis of the system L is directed along the vector ω_0^* (Figure 6.3b), $Q(\omega) = Q(\nu,\varphi) = Q(\nu)$ (here, ν is the zenith angle, and φ is the azimuth angle of the direction ω), and the scattering medium is homogeneous along the planes $z = const$. Then $h(\theta;\omega_0^*) = h(x, y;\omega_0^*) = h(r;\omega_0^*)$, where $r = \sqrt{(x^2 + y^2)}$, that is, the pulsed reaction has the axial symmetry and decreases monotonically with an increase of the argument. Here

$$0° < \theta_i = arctg(r_i/H) < 180°, \tag{6.11}$$

θ_i is the angle between the direction ω_0^* and the direction ω_i^* to the point A_i from the center of the entrance pupil of the optical system, H is the distance between the plane of the entrance pupil of the optical system and the object plane.

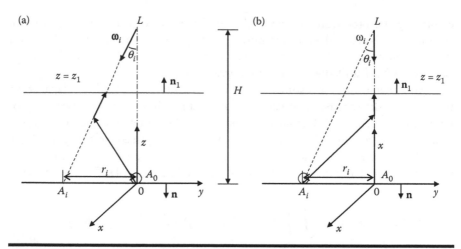

Figure 6.3 Scheme of formation of the image of point (a) and PSF (b) for the central isoplanar zone.

If PSF can be defined as dependence of brightness on the angle θ_i, then the measure of PSF $h(\theta)$ mismatch with the image of the point $q(\theta)$ can be introduced as follows:

$$\varepsilon(\theta) = \left| h\left(\theta; \omega_0^*\right) - q\left(\theta; \omega_0^*\right) \right| \Big/ q\left(\theta; \omega_0^*\right) \tag{6.12}$$

It should be noted that in this case $h(0) = q(0)$. Then the angular size θ_{iso} of the central isoplanarity zone can be found from the condition

$$\varepsilon(\theta_{iso}; \omega_0^*) = \varepsilon_0 \tag{6.13}$$

Then, using Equation 6.11, we can determine the radius $R_{iso}^0 = R_{iso}^0\left(\omega_0^*\right)$ of this zone around the point A_0 or (using the scaling coefficient) around the center of the image plane. An increase in the level of error ε_0 predetermines the obvious growth of R_{iso}^0.

It follows from the data presented in References 11,27 that criterion (6.12) is very sensitive to variations of optical-geometric parameters of the observation scheme. The integral criterion is somewhat more sensitive in this respect. Following to this criterion, the size of the isoplanarity zone can be determined from the condition

$$\delta\left(R_{iso}^0\right) = \left| \eta\left(R_{iso}^0\right) - \eta_1\left(R_{iso}^0\right) \right| \Big/ \eta_1\left(R_{iso}^0\right) = \delta_0 \tag{6.14}$$

where

$$\eta\left(R_{iso}^0\right) = \int\limits_0^{R_{iso}^0} rh(r)\mathrm{d}r, \quad \eta_1\left(R_{iso}^0\right) = \int\limits_0^{R_{iso}^0} rq(r)\mathrm{d}r.$$

Here, r and θ are related by Equation 6.11.

The results of laboratory (observation through a cell with the scattering medium) and numerical experiments (for the scheme of vertical observation through the cloudless atmosphere and under overcast conditions) reported in References 11,27 can be reduced to the following conclusions. The size of the central isozone is a complex, ambiguous function of optical-geometric conditions of observation. It depends on the optical thickness of the medium, the distribution of scattering and absorption coefficients at the line of sight, and the scattering phase function. The isoplanarity zone $D(x,y)$ converges to the center with a decrease of ε_0 or δ_0. In particular, in the scheme of numerical experiments considered in References 11,27 at $\varepsilon_0 \approx 20\%$, the angular dimensions of the radius of the $D(x,y)$ zone took values from few degrees to tens of degrees depending on the initial optical-geometric conditions. We assume that the radius of the central isozone is smaller than the frame size. This means that we cannot construct the image of even a point object with the required accuracy starting from $r > R_{iso}^0$.

Consequently, keeping within the framework of the theory of linear systems, we should construct the following pulsed reaction $h_1(x,y;\omega_1^*)$, that is, PSF for the direction ω_1^* (Figure 6.4b) oriented, for example, to the boundary of the central

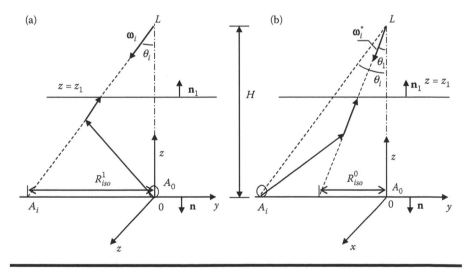

Figure 6.4 Scheme of formation of the image of a point (a) and PSF (b) for the peripheral isoplanar zone.

isoplanar zone. This allows us to construct the image of a point object at $r > R_{iso}^0$, and at $r = R_{iso}^0$ it coincides exactly with $h_1(0,0;\omega_1^*)$.

It should be noted that PSF for this isoplanar zone differs significantly from $h(r;\omega_0^*)$ determined for the image center. This difference consists, first of all, in the fact that the function $h_1(x,y;\omega_1^*)$ does not have axial symmetry, although it is independent of the azimuth angle between the projection of the direction ω_1^* onto the plane $z = 0$ and any point on the circle of the radius R_{iso}^0 on this plane (even if we assume axial symmetry of the function $Q(\omega)$ and homogeneity of optical properties of the medium in the directions perpendicular to the vector ω_0^*).

How can we determine the size of the following isoplanar zone of the image? Figure 6.4 shows the scheme of formation of the image of a point object for the second isoplanar zone (Figure 6.4a) and determination of the pulsed reaction $h_1(x,y;\omega_1^*)$ for this zone (Figure 6.4b). For determination of the angular or linear size of the second isoplanar zone, we can use criterion (6.12) upon the replacement of $h(\theta) = h(\theta;\omega_0^*) = h(x,y) = h(r)$ with $h_1(\theta;\omega_1^*)$. Then the criterion for estimation of the size of this zone has the following form:

$$\varepsilon\left(\theta_{iso}^1\right) = \min_{\varphi}\left|q\left(\theta_{iso}^1,\omega_1^*\right) - h_1\left(\theta_{iso}^1,\varphi,\omega_1^*\right)\right|\Big/q\left(\theta_{iso}^1,\omega_1^*\right) \tag{6.15}$$

where φ is the azimuth angle.

Then the angular size θ_{iso}^1 of the second isoplanar zone can be found from the condition

$$\varepsilon\left(\theta_{iso}^1,\omega_1^*\right) = \varepsilon_0.$$

The linear size of the radius R_{iso}^1 can be determined using either Equation 6.11 or

$$\varepsilon\left(R_{iso}^1\right) = \min_{x,y}\left|q\left(R_{iso}^1\right) - h_1\left(R_{iso}^1\right)\right|\Big/q\left(R_{iso}^1\right) = \varepsilon_0. \tag{6.16}$$

The integral criterion in this case takes the form

$$\delta\left(R_{iso}^1\right) = \left|\eta\left(R_{iso}^1,\omega_1^*\right) - \eta_1\left(R_{iso}^1,\omega_1^*\right)\right|\Big/\eta_1\left(R_{iso}^1,\omega_1^*\right) = \delta_0 \tag{6.17}$$

where

$$\eta\left(R_{iso}^1,\omega_1^*\right) = \int_{R_{iso}^0}^{R_{iso}^1}\int_{R_{iso}^0}^{R_{iso}^1} h_1(x,y)dxdy; \quad \eta_1\left(R_{iso}^1,\omega_1^*\right) = 2\pi\int_{R_{iso}^0}^{R_{iso}^1} rq(r)dr.$$

It should be noted that, using the same values of ε_0 (or δ_0) for estimation of the sizes of all isoplanarity zones, we reconstruct the image of a point with the same error.

Following this line of reasoning, we can estimate the size of the following zones of image isoplanarity.

Criterion 2 [26]. Thus, there are situations when it is necessary to separate the observed surface into zones of isoplanarity. If the spherically homogeneous atmosphere is considered, then the isoplanarity zones are rings around the center of the observed fragment of the surface.

Assume that the optical system is directed at the boundary of i-th isoplanarity, and the use of PSF of previous zone for construction of the image in the *i*-th zone leads to an intolerable error. This error can be determined by the equation

$$\frac{I_i'(x,y) - I_{i+1}'(x,y)}{I_{i+1}'(x,y)} = \frac{(E^{**}h_i)(x,y) - (E^{**}h_{i+1})(x,y)}{(E^{**}h_{i+1})(x,y)}, \qquad (6.18)$$

$$(E^{**}h_i)(x,y) = \int\limits_{-\infty}^{+\infty}\int\limits_{-\infty}^{+\infty} E_0(x-x', y-y')\rho(x-x', y-y')h_i(x',y')dx'dy' \qquad (6.19)$$

Here, I_i' is the radiation intensity reconstructed with the use of PSF of the previous zone, I_{i+1}' is the exact value of intensity reconstructed with the aid of PSF with the optical system oriented at the zone boundary, E is the surface brightness, E_0 is the surface illumination, and ρ is the distribution of the reflection coefficient over the surface.

If we expand $E(x - x', y - y')$ in Equation 6.18 into the Taylor series, then obtain

$$\frac{I_i'(x,y) - I_{i+1}'(x,y)}{I_{i+1}'(x,y)} = \frac{E\left(m_{00}^{(i)} - m_{00}^{i+1}\right) - \dfrac{\partial E}{\partial x}\left(m_{10}^{(i)} - m_{10}^{(i+1)}\right) - \dfrac{\partial E}{\partial y}\left(m_{01}^{(i)} - m_{01}^{(i+1)}\right) + \cdots}{Em_{00}^{(i+1)} - \dfrac{\partial E}{\partial x}m_{10}^{(i+1)} - \dfrac{\partial E}{\partial y}m_{01}^{(i+1)} + \cdots},$$

$$(6.20)$$

where

$$m_{00}^{(i)} = m_{00}(\theta_i) = \int\limits_{-\infty}^{+\infty}\int\limits_{-\infty}^{+\infty} h_i(x,y)dxdy = \int\limits_{-\infty}^{+\infty}\int\limits_{-\infty}^{+\infty} h(x,y,\theta_i)dxdy, \qquad (6.21)$$

$$m_{10}^{(i)} = m_{10}(\theta_i) = \int\limits_{-\infty}^{+\infty}\int\limits_{-\infty}^{+\infty} h_i(x,y)x\,dx\,dy = \int\limits_{-\infty}^{+\infty}\int\limits_{-\infty}^{+\infty} h(x,y,\theta_i)x\,dx\,dy, \qquad (6.22)$$

$$m_{01}^{(i)} = m_{01}(\theta_i) = \int\limits_{-\infty}^{+\infty}\int\limits_{-\infty}^{+\infty} h_i(x,y)y\,dx\,dy = \int\limits_{-\infty}^{+\infty}\int\limits_{-\infty}^{+\infty} h(x,y,\theta_i)y\,dx\,dy. \qquad (6.23)$$

The similar expansion of the convolution can be found, for example, in Reference 21, p. 36.

If we neglect all the expansion terms but the first one, we obtain approximately

$$\frac{I_i'(x,y) - I_{i+1}'(x,y)}{I_{i+1}'(x,y)} \approx \frac{m_{00}^{(i)} - m_{00}^{(i+1)}}{m_{00}^{(i+1)}} \equiv \delta. \qquad (6.24)$$

Therefore, the relative difference of the intensities reconstructed with different PSFs can be estimated approximately by Equation 6.24.

The integral characteristics $m_{00}(\theta)$, in their turn, can be approximated by the angles of orientation of the receiving system θ. In solution of particular problems, we recommend the approximations of the integral characteristics m_{00} to be constructed. In Reference 26, for the following ranges of optical-geometric parameters of numerical experiments: wavelength $\lambda = 0.35$–0.8 µm, meteorological visibility range $S_M = 1$–50 km, height of position of the optical systems above the Earth's surface $H \geq 10$ km, and zenith angle (determining its orientation) $\theta = 0$–$60°$, it is proposed to use the following approximation equation:

$$m_{00}(\lambda_0, S_{M,0}, H_0, \theta) = m_{00}(\lambda_0, S_{M,0}, H_0, \theta = 0) - \exp(A)\cdot(1 - \cos\theta)^N, \qquad (6.25)$$

Then, upon transformation of Equation 6.24 and expression of θ_{i+1}, we obtain

$$\theta_{i+1} = m_{00}^{-1}\left[\frac{m_{00}(\theta_i)}{\delta + 1}\right], \qquad (6.26)$$

where m_{00}^{-1} is the function inverse to $m_{00}(\theta_i)$.

If we substitute Equation 6.25 into Equation 6.26, then we get

$$\theta_{i+1} = \arccos\left(1 - \left(\left(m_{00}(\theta_0) - \frac{m_{00}(\theta_i)}{1+\delta}\right)\bigg/\exp(A)\right)^{1/N}\right). \qquad (6.27)$$

Thus, we obtain the following criterion for determination of boundaries of iso-planarity zones:

$$\begin{cases} m_{00}(\theta_i) = m_{00}(\theta_0) - \exp(A) \cdot (1 - \cos\theta_i)^N \\ \theta_{i+1} = \arccos\left(1 - \left[\left(m_{00}(\theta_0) - \dfrac{m_{00}(\theta_i)}{1+\delta}\right)\Big/\exp(A)\right]^{1/N}\right) \end{cases}. \qquad (6.28)$$

Using Equation 6.28, we obtain the values of the orientation angles of the optical system θ_i, which determine the boundary of the isoplanarity zones. Upon the calculation of PSF for given θ_i and using Equation 6.29, we can find the image of the observed surface

$$I'(x, y, h_d) = \int\limits_{-\infty}^{+\infty}\int\limits_{-\infty}^{+\infty} E_0(x', y') r(x', y') h(x - x', y - y') dx' dy', \qquad (6.29)$$

where $r(x',y')$ is the distribution of the reflection coefficient on the surface, $E_0(x,y)$ is the distribution of Earth illumination by the sun.

6.4 Example of Reconstruction of the Image of Test Object

To illustrate the procedure of application of criterion 2 in the problem of separation of isoplanar zones in an image, the following example is considered. Let $\lambda = 0.35\ \mu m$, $S_M = 10$ km, $H = 100$ km, and the Earth's surface be illuminated homogeneously with the illumination $E_0 = 1$ W/(m²μm). Let the distribution of the reflection coefficient on the surface have the following form (Figure 6.5) (triangle test object):

$$\rho(x, y) = \begin{cases} 1, & \left(|y| \le x/2\right) \cap (x \ge 0) \cap (x \le 200) \\ 0, & else \end{cases}. \qquad (6.30)$$

The task is to reconstruct the image of the object along the axis x (in this case, the image is understood as the angular distribution of radiation intensity at the center of the entrance pupil of the optical system).

The sought intensity distribution can be obtained directly (without invoking the theory of linear systems) with the use of the corresponding algorithm from Reference 24. This was made in Reference 26 by the Monte Carlo method. The

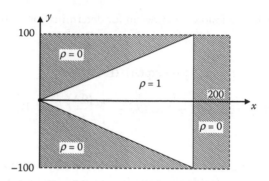

Figure 6.5 Distribution of the reflection coefficient on the surface (test object).

results of Monte Carlo calculation of m_{00} for the set of the node points $\theta = 0$, 15, ... ,75° are shown in Figure 6.6 (dots). The approximation constants A and N in Equation 6.25 take the following values in this case: $A = -1.785$, B $= 1.055$. The dependence of m_{00} on θ is demonstrated in Figure 6.6 as well.

Using criterion (6.28), we can easily determine the boundaries of isoplanarity zones. At $\delta = 0.05$, they are tabulated below for the case under consideration.

The Monte Carlo method, described in detail in Chapter 5, was used for each zone to calculate PSF at the inner boundary of the zone 10 functions in this example. In the calculation of each PSF, the following boundaries of the surface cells were chosen: $r = 0.01, 0.1, 0.5, 1, ... , 10, 11, ... , 20, 25, ... ,50$ km. But the values of φ were taken with in steps of 5°.

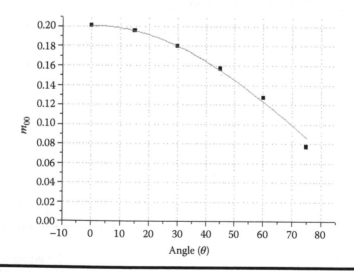

Figure 6.6 Calculated values of m_{00} (dots) and those obtained by Equation 6.25 (curve) as functions of the angle θ at $\lambda = 0.35$ μm, $S_M = 10$ km, $H = 100$ km.

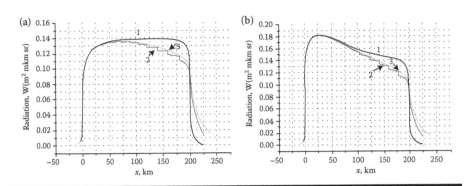

Figure 6.7 Images of the test object along the axis x at $\lambda = 0.35$ μm, $S_M = 10$ km, $H = 100$ km. (a) Scattered radiation. (b) Total radiation (scattered and nonscattered). Curve 1—image reconstructed with the first PSF ($\theta = 0°$); Curve 2—image reconstructed with 10 PSFs (from Table 6.1); Curve 3—image calculated by the program from Reference 24 for the inhomogeneous surface with the surface distribution of the reflection coefficient (6.24).

Figure 6.7a and b show the results of image reconstruction along the axis x of object (6.30) with the application of PSF and through direct calculation with the use of the program based on the algorithm from Reference 24. In addition, Figure 6.7 shows the image resulting from the use of only one axisymmetric PSF for the case of nadir sensing. It follows from the figure that 10 isoplanarity zones chosen by the criterion are sufficient for reconstruction of the image of test object with an error within 6% at $0 \le x \le 200$ km. One can also see that the use of only one axisymmetric PSF leads to large relative errors (within 25%) at the reconstruction of the image of test object.

Table 6.1 **Boundaries of Isoplanarity Zones at** $\delta = 0.05$ **and** $\lambda = 0.35$ μm, $S_M = 10$ km, $H = 100$ km

Number of Isoplanarity Zone	Outer Boundary of the Zone		Number of Isoplanarity Zones	Outer Boundary of the Zone	
	θ (deg)	r (km)		θ (deg)	r (km)
1	20.6	37.6	6	50.6	121.7
2	29.3	56.1	7	54.4	139.7
3	36.0	72.7	8	57.9	159.4
4	41.5	88.5	9	61.1	181.1
5	46.3	104.6	10	>61.1	>181.1

Thus, the image $P(x,y)$ of an extended object $p(x,y)$ observed through the scattering medium can be reconstructed as follows:

$$P(x,y) = \begin{cases} \displaystyle\int\limits_{-\infty}^{+\infty}\int p(x',y')h_0(x-x',y-y')dx'dy', & \sqrt{x^2+y^2} \leq R_{iso}^0 \\[2ex] \displaystyle\int\limits_{-\infty}^{+\infty}\int p(x',y')h_1(x-x',y-y')dx'dy', & R_{iso}^0 \leq \sqrt{x^2+y^2} \leq R_{iso}^o \\[1ex] \cdots\cdots \\[1ex] \displaystyle\int\limits_{-\infty}^{+\infty}\int p(x',y')h_n(x-x',y-y')dx'dy', & R_{iso}^{n-1} \leq \sqrt{x^2+y^2} \leq R_{iso}^n \end{cases} \qquad (6.31)$$

or, using the concepts of optical transfer functions and spatial spectra of objects:

$$P(x,y) = \begin{cases} F^{-2}[H_0(\omega,\gamma)k(\omega,\gamma)], & \sqrt{x^2+y^2} \leq R_{iso}^0 \\[1ex] F^{-2}[H_1(\omega,\gamma)k(\omega,\gamma)], & R_{iso}^0 \leq \sqrt{x^2+y^2} \leq R_{iso}^1 \\[1ex] \cdots\cdots \\[1ex] F^{-2}[H_n(\omega,\gamma)k(\omega,\gamma)], & R_{iso}^{n-1} \leq \sqrt{x^2+y^2} \leq R_{iso}^n \end{cases} \qquad (6.32)$$

where $H_i(\omega,\gamma)$ is the optical transfer function of the i-th isoplanar zone ($i = 0, 1, 2, \ldots, n$), $k(\omega,\gamma)$ is the spatial spectrum of the object, F^{-2} is the two-dimensional inverse Fourier transform.

The same reasoning can also be repeated for the imaging of optically inhomogeneous objects. In this case, the object plane is divided into zones of optical homogeneity, for every zone the necessary number of optically transfer functions (point spread functions) is determined, and the image of objects is formed with the use of equations of type (6.31) or (6.32).

Thus, we can state the following:

■ The concept of the optical transfer function of pulsed reaction, point spread function, or the influence function of external channel of image formation and transfer is not defined, if the reflective (emissive) properties of the object surface are not defined.

■ There is no characteristic of the scattering medium determining its transfer properties and independent of optical properties of the object surface.

■ External channels of formation and transfer of images of objects observed through turbid media can be described by a series of optical transfer functions (point spread functions), rather than by one function depending on the field of view (or angular amplitude of the scan) of the optical (optical-electronic) system and on the number of classes of optical homogeneity of objects forming the observed scene.

References

1. Bravo-Zhivotovskii, D. M., Dolin, L. S., Luchinin, A. G., and Savelèv, V. A., Some problems of the theory of vision in turbid media, *Izv. AN SSSR. Ser. FAO*, vol. 5, No. 7, pp. 672–684, 1969.
2. Levin, I. M., On observation of objects illuminated by narrow radiation beam in a scattering medium, *Izv. AN SSSR. Ser. FAO*, vol. 5, No. 1, pp. 62–76, 1969.
3. Tanre, D., Herman, M., and Deschamps P. Y., Influence of the background contribution upon space measurements of ground reflectance, *Appl. Opt.*, vol. 20, No. 20, pp. 3676–3684, 1981.
4. Tanre, D., Herman, M., Deschamps, P. Y., and de Leffe, A., Atmospheric modeling for space measurements of ground reflectances, including bidirectional properties, *Appl. Opt.*, vol. 18, No. 21, pp. 3587–3594, 1979.
5. Kopeika, N. S., Solomon, S., and Gencay, Y., Wavelength variation of visible and near-infrared resolution through the atmosphere: Dependence on aerosol and meteorological conditions, *JOSA*, vol. 71, No. 7, pp. 892–901, 1981.
6. Kopeika, N. S., Effects of aerosols on imaging through the atmosphere, *JOSA*, vol. 75, No. 4, pp. 707–712, 1985.
7. Zege, E. P., Ivanov, A. P., and Katsev, I. L., *Image Transfer in Scattered Medium*. Minsk: Nauka i Tekhnika, 1995, 328 p.
8. Katsev, I. L., Estimation of vision characteristics in warm clouds from data about their microstructure, *Izv. AN BSSR. Ser. Fiz.-Mat.*, vol. 2, pp. 93–98, 1984.
9. Ishimaru, A., Limitation on image resolution imposed by a random medium, *Appl. Optics*, vol. 17, No. 3, pp. 348–352, 1978.
10. Frazer, R. S., and Kaufman, Y., The relative Importance of aerosol scattering absorption in remote sensing, *IEEE. Trans. Geoscience Remote Sensing*, vol. GE-23, No. 5, pp. 625–633, 1985.
11. Zuev, V. E., Belov, V. V., and Veretennikov, V. V., *Systems Theory in Optics of Disperse Media*. Tomsk: Spektr, Publishing House of the Institute of Atmospheric Optics SB RAS, 1997, 402 p.
12. Belov, V. V., Statistical modeling of imaging process in active night vision systems with gate-light detection, *Appl. Phys.*, vol. 75, No. 4–5, pp. 571–576, 2002.
13. Sushkevich, T. A., Strelkov, S. A., and Ioltuhovskii, A. A., *Method of Characteristic in Problems of Atmospheric Optics*. Moscow: Nauka, 1990, 296 p.
14. Fraser, R. S., Ferrare, R. A., Kaufman, Y. J., and Mattoo, S., Algorithm for atmospheric corrections of aircraft and satellite imagery, *Int. J. Remote Sensing*, vol. 13, pp. 541–557, 1992.
15. Belov, V. V., and Afonin, S. V. *From Physical Grounds, Theory, and Modeling to Thematic Processing of Satellite Images*. Tomsk: Spectr, Publishing House of the Institute of Atmospheric Optics SB RAS, 2005, 266 p.
16. Thome, K., Palluconi, F., Takashima, T., and Masuda, K., Atmospheric correction of ASTER, *IEEE Trans. Geosci. Remote Sens.*, vol. 36, No. 4, pp. 1199–1211, 1998.

17. Sobrino, J. A., Jiménez-Muñoz, J. C., and Paolini, L., Land surface temperature retrieval from LANDSAT TM 5, *Remote Sens. Environ.*, vol. 90, No. 4, pp. 434–440, 2004.

18. Afonin, S. V., and Solomatov, D. V., Solution of problems of atmospheric correction of satellite IR measurements accounting for optical-meteorological state of the atmosphere, *J. Atmos. Ocean. Optics*, vol. 21, No. 2, pp. 125–131, 2008.

19. Golovko, V. A., Current technologies of removal of the atmospheric effect from multispectral measurements with high spatial resolution from space, *Issl. Zemli iz Kosmosa, Moscow*, vol. 2, pp. 11–23, 2006.

20. Belov, V. V., Optical transfer functions of external channels and image isoplanarity in vision systems, *J. Optika Atmosfery i Okeana, Moscow*, vol. 22, No. 12, pp. 1101–1107, 2009.

21. Papoulis, A., *Systems and Transforms with Applications in Optics*. New York: McGraw-Hill, 1968, 474 p.

22. Belov, V. V., Statistical modeling of image of 3D objects in problems of the vision theory, *Izv. AN SSSR. Ser. Fiz. Atmos. Okeana, Moscow*, vol. 18, No. 4. pp. 435–437, 1982.

23. Belov, V. V., Theory of linear vision systems. Modeling of the linear-systems characteristics, *J. Atmos. Ocean. Optics, Moscow*, vol. 2, No. 8, pp. 649–660, 1989.

24. Belov, V. V., and Tarasenkov, M. V., Statistical modeling of the light fluxes reflected by the spherical Earth surface, *J. Optika Atmosfery i Okeana, Moscow*, vol. 23, No. 1, pp. 14–20, 2010.

25. Marchuk, G. I., Mihailov, G. A., Nazaraliev, M. A., Darbinian, R. A., Kargin, B. A., and Elepov, B. S., *Monte Carlo Method in Atmospheric Optics*. Novosibirsk: Nauka, 1976, 284 p.

26. Belov, V. V., and Tarasenkov, M. V., Statistical simulation of the point spread function in the spherical atmosphere and criterion for detecting the image isoplanarity zones, *J. Optika Atmosfery i Okeana, Moscow*, vol. 23, No. 5, pp. 371–377, 2010.

27. Belov, V. V., Krekov, G. M., and Makushkina, I. Yu., Isoplanarity in vision systems, *J. Atmos. Oceanic Optics*, vol. 2, No. 10, pp. 1011–1018, 1989.

28. Dolin, L. S., and Savelèv, V. A., Equation of optical image transfer in a scattering medium, *Izv. AN SSSR. Ser. FAO*, vol. 15, No. 7, pp. 717–723, 1979.

Index